MARINE CONSERVATION

PEOPLE, IDEAS AND ACTION

BOB EARLL

PELAGIC PUBLISHING

Published by Pelagic Publishing
PO Box 874
Exeter
EX1 9YH, UK
www.pelagicpublishing.com

Marine Conservation: People, Ideas and Action

ISBN 978-1-78427-176-3 Paperback
ISBN 978-1-78427-177-0 ePub
ISBN 978-1-78427-178-7 PDF

A catalogue record for this book is available from the British Library

Cover photographs from top left: *Tompot blenny*, Keith Hiscock; *Earth from space*, NASA; *Diver & marine life*, Shutterstock; *Sharks fins drying in Hong Kong*, Stan Shea/ BLOOM; *Greenpeace blocking pipe*, Greenpeace; *Bleached coral*, Richard Vevers, The Ocean Agency, Catlin Seaview Survey; *Whale being rescued*; Megan Whittaker, Elding Whalewatching; *Planting a mangrove sapling*, Shutterstock; and *Brent Spar*, Greenpeace.

Quotation from Linda Lear © 1998 Linda Lear; reprinted by permission of Beacon Press, Boston, Massachusetts.

This book is dedicated to my parents Elizabeth and Walter, without whose commitment and encouragement my career would not have been possible. And to Louise, with love, who from the outset of our marriage and from the early years subsidising the Marine Conservation Society to the present day has contributed so much to my life and work.

Foreword

It was John Quincy Adams who said, 'if your actions inspire others to dream more, learn more, do more and become more, you are a leader.' This book is about what is being done to save the ocean, but it is also about leaders – individuals who have risen to meet the challenges of ocean decline and are inspiring others to learn more, do more and become better stewards of the blue heart of the planet.

While taking journeys to the depths of the sea, readers will also probe the depths of the minds of leaders whose discoveries and insights have appeared in textbooks and have become the basis for enlightened policies. Their narratives reflect the realities of what it takes to achieve positive change in hugely diverse sets of circumstances. In their own words they share what has driven them to commit their lives to safeguarding the ocean. Some are known professionally as conservationists, while others are achieving conservation outcomes through various disciplines. By whatever name and whatever pathways, all have made a significant difference for the planet through their dedication, perseverance, perception, actions and sheer determination.

Like many of the contributors to this book, I have witnessed decades of changes in the sea as a scientist, with special opportunities to observe the ocean during years at sea and thousands of hours of diving, including using scuba, more than 30 submarines and numerous remotely operated vehicles, and living in underwater laboratories ten times. From the surface, the ocean looks much as it has for many millennia, but those of us who have been face to face with whales, fish, octopus and squid on their own terms and explored their homes in reefs, canyons and broad deep-sea plains are particularly motivated to convey the reality of what is under the waves.

The evidence is clear. Natural ocean systems hundreds of millions of years in the making began unravelling in the twentieth century owing to human actions, and this process is continuing now at an accelerating pace. Ocean chemistry is changing, with increasing warming, acidification and pollution and decreasing levels of oxygen. About half of the tropical coral reefs, mangroves and seagrass meadows are gone, along with about 90% of many commercially taken species of wildlife, from tunas and swordfish to groupers and cod. Even notoriously abundant creatures such as herring, menhaden and krill have been fished to historically low levels. The good news is that many marine mammals, sea turtles, and sea birds, once in serious decline, are now – with protection – showing signs of recovery. Entire ecosystems that have been given full protection are also being restored to significantly improved health.

It is easy to observe, document and lament the problems now affecting the health of the ocean, and in this volume those issues are spelled out with authority. But, most importantly, solutions are offered that provide hope for a reversal of today's alarming trends. A new ocean ethic is emerging, largely powered by the response to evidence acted upon and voiced to the world by the contributors to this very timely volume.

Sylvia Earle
National Geographic Explorer in Residence; Founder, Mission Blue; and Founder, Deep Ocean Exploration and Research

Acknowledgements

It has been a great privilege to work with all the interviewees, who have contributed so much to the wide range of material in this book, including text, images, timelines and text boxes and I would like to thank them for all their efforts. In particular I would like to thank Keith Hiscock and Sue Gubbay, who went through the interview process twice. I am grateful to Sylvia Earle, who in 1984 came to the UK to help us launch the Marine Conservation Society, and who graciously agreed to write the foreword to this book – and to Dan Laffoley for setting this up. I am also very grateful to Sue Wells for contributing the *marine protected areas* timeline and John Hartley for the *life-cycle approach to environmental management techniques* diagram and text. Glenn Hall, Derek Pagett and Justin Taberham from the outset of this project steered me in the right direction and provided great support. I would also like to thank Sue Gubbay for her considerable advice and help with reviewing the manuscript, Alice Earll for preparing and editing some of the transcripts and Louise Earll for reading the drafts. For their help with the diagrams and timelines I would also like to thank Kirstie and Rod Dive. I would also like to thank my publisher Nigel Massen for taking the risk with this rather different approach to marine conservation books, and Hugh Brazier, whose exacting editing has transformed the final version.

I owe considerable thanks to a large number of photographers and other people who have provided the images for the book, including all the interviewees – and in particular Keith Hiscock, who came to the rescue with many of the images used in other chapters. The contributors include ABP, Nikoloa Bodova, British Divers Marine Life Rescue, Commonwealth of Australia (GBRMPA), the Cousteau Society, Alan Davis, Nick Davison (Scottish Marine Animal Stranding Scheme, SRUC Veterinary), Jon Day and K. Blomvik, *Diver* magazine, Jamie Dyer, Joan Edwards, Tony and Jenny Enderby, the European Union, Bud Ehler and Fanny Douvere (UNESCO), Chris Gomersall (RSPB Images), John Hartley, Neil Hope, Paul Horsman, Morag Hunter, IMCC, IWC, Gordon James, Sam Jones, Alan Knight, Heather Koldewey, Dan Laffoley, Piers Larcombe, Lundy Management Forum and the Landmark Trust, Marine Conservation Philippines, the Marine Conservation Society, Christine McGuinness, NASA, NOAA, the Ocean Agency, Octavio Aburto-Oropeza, F. Pharand-Deschênes/Globaïa Stockholm Resilience Centre, Fabiano Peppes (Projeto Albatroz), Tom Probert, Project Seahorse, Callum Roberts, Sue Sayer, Stan Shea/Bloom, Tony Sheehan, Jeremy Stafford-Deitsch, Alec Taylor, Amanda Vincent, Peter Welsh, Megan Whittaker (Elding Whalewatching), XL Catlin Seaview Survey, ZSL.

Contents

Acronyms and abbreviations

ABP	Associated British Ports
ACFA	Advisory Committee on Fisheries and Aquaculture
AES	American Elasmobranch Society
AIS	Automatic Identification System
AMO	Atlantic Multidecadal Oscillation
ASW	aboriginal subsistence whaling
ATF	Albatross Task Force
AUV	autonomous underwater vehicle
BANC	British Association of Nature Conservationists
BAS	British Antarctic Survey
BBC	British Broadcasting Corporation
BDMLR	British Divers Marine Life Rescue
BMAPA	British Marine Aggregate Producers Association
BPEO	Best Practicable Environmental Option
CaBA	catchment-based approach
CCAMLR	Commission for the Conservation of Antarctic Marine Living Resources
CCS	Centre for Coastal Studies, Provincetown, USA
CEE	Center for Environmental Education (USA)
CEQ	Council on Environmental Quality (USA)
CFC	chlorofluorocarbon
CFP	Common Fisheries Policy
CITES	Convention on International Trade in Endangered Species
CMS	Convention on Migratory Species
COFI	Committee on Fisheries (FAO)
COP	Conference of Parties
CPOA	Community Plan of Action
CPR	Continuous Plankton Recorder
CPRE	Campaign to Protect Rural England
CRoW	Countryside and Rights of Way Act
CSGRT	Cornwall Seal Group Research Trust
CSR	corporate social responsibility
CWT	Cornwall Wildlife Trust
CZM	Coastal Zone Management
DDT	dichloro-diphenyl-trichloroethane
DG-MARE	Directorate-General of Maritime Affairs and Fisheries of the European Commission
Defra	Department for Environment, Food and Rural Affairs
DoE	Department of the Environment

DSFC	Devon Sea Fisheries Committee
EA	Environment Agency
EEA	European Elasmobranch Association
EEZ	exclusive economic zone
EIA	environmental impact assessment
EMFF	European Marine Fisheries Fund
EMS	environmental management system
ENG	Ecological Network Guidance
EPA	Environmental Protection Agency (USA)
ESA	Ecological Society of America
EUNIS	European Union Nature Information System
FAO	UN Food and Agriculture Organization
FKNMS	Florida Keys National Marine Sanctuary
FoE	Friends of the Earth
GBR	Great Barrier Reef
GBRMP	Great Barrier Reef Marine Park
GBRMPA	Great Barrier Reef Marine Park Authority
GDP	gross domestic product
GFCM	General Fisheries Commission for the Mediterranean
GIS	geographic information system
GLORES	Global Ocean Refuge System
GPS	Global Positioning System
GSRI	Global Sharks and Rays Initiative
HCFC	hydro-chlorofluorocarbon
IAR	International Animal Rescue
ICCAT	International Commission for the Conservation of Atlantic Tunas
ICES	International Council for the Exploration of the Sea
ICZM	integrated coastal zone management
IFCA	Inshore Fisheries and Conservation Authority
IFAW	International Fund for Animal Welfare
IFOMA	International Fishmeal and Oil Manufacturers Association
IMCC	International Marine Conservation Congress
IMM	Intermediate Ministerial Meeting
IMO	International Maritime Organisation
IOC	Intergovernmental Oceanographic Commission
IOI	International Ocean Institute
IPCC	Intergovernmental Panel on Climate Change
IPOA	International Plan of Action (FAO)
IROPI	Imperative Reasons of Overriding Public Interest
IUCN	International Union for Conservation of Nature
IWC	International Whaling Commission
JMP	Joint Marine Programme
JNCC	Joint Nature Conservation Committee
JR	judicial review
LOIS	Land Ocean Interface Studies
MAFF	Ministry of Agriculture, Fisheries and Food
MARC	Marine Animal Rescue Coalition
MarLIN	Marine Life Information Network

MARPOL	Marine Pollution Convention
MBA	Marine Biological Association
MBA	Master of Business Administration
MCBI	Marine Conservation Biology Institute
MCCIP	Marine Climate Change Impacts Partnership
MCS	Marine Conservation Society
MCZ	Marine Conservation Zone
MMO	Marine Management Organisation
MNCR	Marine Nature Conservation Review of Great Britain
MNR	Marine Nature Reserve
MoU	Memorandum of Understanding
MPA	marine protected area
MSC	Marine Stewardship Council
MSFD	Marine Strategy Framework Directive
MSP	marine spatial planning
NBN	National Biodiversity Network
NCC	Nature Conservancy Council
NCEAS	National Center for Ecological Analysis and Synthesis (USA)
NCME	Nature Conservation in the Marine Environment
NCR	A Nature Conservation Review
NDF	non-detriment finding
NE	Natural England
NEAFC	North East Atlantic Fisheries Commission
NERC	Natural Environment Research Council
NGO	non-governmental organisation
NOAA	National Oceanic and Atmospheric Administration (USA)
NOF	Norwegian Ornithological Society
NRA	National Rivers Authority
NRSMPA	National Representative System of Marine Protected Areas
NSAC	North Sea Advisory Council
NSM	North Sea Ministerial meetings
NTZ	no-take zone
NZCPS	New Zealand Coastal Policy Statement
OPRU	Oil Pollution Research Unit
ORCA	Office of Ocean Resources, Conservation and Assessment
OSPAR	Oslo/Paris convention (Convention for the Protection of the Marine Environment of the North-East Atlantic)
PCB	polychlorinated biphenyl
PCR	polymerase chain reaction
PDV	phocine distemper virus
PIB	poly-isobutylene
PISCO	Partnership for Interdisciplinary Studies of Coastal Oceans
RAC	Regional Advisory Council
RCP	Representative Concentration Pathway
RFMO	regional fisheries management organisation
RMA	Resource Management Act (New Zealand)
RSNC	Royal Society of Nature Conservation
RSPB	Royal Society for the Protection of Birds

RSWT	Royal Society of Wildlife Trusts
SAC	Special Area of Conservation
SAHFOS	Sir Alister Hardy Foundation for Ocean Science
SANA	Scientists Against Nuclear Arms
SCB	Society for Conservation Biology
SMBA	Scottish Marine Biological Association
SMRU	Sea Mammal Research Unit
SNH	Scottish Natural Heritage
SPA	Special Protection Areas
SSSI	Site of Special Scientific Interest
SUDG	Seabed User and Developer Group
TAC	total allowable catch
TBT	tributyl tin
TCM	traditional Chinese medicine
TED	Technology, Entertainment and Design
UCL	University College London
UCLA	University of California, Los Angeles
UCY	Underwater Conservation Year
UNCBD	United Nations Convention on Biological Diversity
UNCCD	United Nations Convention to Combat Desertification
UNCED	United Nations Conference on Environment and Development
UNCLOS	United Nations Convention on the Law of the Sea
UNESCO	United Nations Educational, Scientific and Cultural Organisation
UNEP	United Nations Environment Programme
UNFCCC	United Nations Framework Convention on Climate Change
VMNR	voluntary marine nature reserve
WCPA	World Commission on Protected Areas
WRT	Welland Rivers Trust
WTO	World Trade Organization
WVP	Welland Valley Partnership
WWF	World Wide Fund for Nature (formerly World Wildlife Fund)
ZSL	Zoological Society of London

CHAPTER 1

Introduction

Today, marine conservation is a widely recognised human endeavour, but when I was at university in the late 1960s, it barely existed. When I started work in 1978 for what became the Marine Conservation Society you could count on the fingers of one hand the number of people employed full time in the UK with marine conservation in their job titles. For many it was a very small part their main jobs. Since then, there has been a massive transformation, and marine conservation is now recognised as a mainstream activity by governments worldwide, advocated by hundreds of non-governmental organisations (NGOs) and engaging many thousands of people.

The development of marine conservation is a story about people whose ideas and actions have challenged the status quo and have been translated into tangible protection for the marine environment. I have had the privilege to work with many who have played a part in this transformation, and as the idea for this book grew it seemed worthwhile to explore the development of marine conservation through their eyes. Accordingly, I interviewed nineteen marine conservation practitioners, and their chapters form the main content of the book. All of the interviewees have been directly involved in activities that have made a difference, often in difficult circumstances, and they have pioneered developments in many areas.

The central aim of this book is to describe and scope the development of marine conservation over the last fifty years through the very different perspectives of the interviewees. The book also explores some of the common themes to emerge from their chapters in a series of crosscutting chapters.

There are four main themes to this book:

- *Marine conservation* – its scope and development

- *People* – marine conservationists, their rationale, motivation, diversity of approach and skills

- *Ideas* that have influenced the way we protect our seas and undertake marine conservation

- *Actions* that have made a difference

MARINE CONSERVATION

The term *marine conservation* is used throughout the book in a deliberately open way, often coupled with the phrase *protecting the marine environment*. *Environmentalists* in its widest sense is another word that could usefully be applied to the interviewees. It is also clear that a large body of work surrounds managing human activities in the marine environment. You do not need to be a conservationist to work on protecting the marine environment, and many disciplines have been brought to bear to achieve significant gains for conservation as well as protecting the wider marine environment from the most damaging of human activities.

Chapter 2 describes what marine conservation involves, its scope and development. It explores how people frame and define their approach to marine conservation to guide their work, and reveals a wide range of viewpoints far richer than are found in textbook definitions. The chapter also includes a systematic structure for the content of marine conservation, revealing the richness of the subject (Earll 2016). Similarities and differences between terrestrial and marine conservation are explored, and the chapter finishes by outlining the broad challenges of marine conservation as it has developed.

PEOPLE

The key idea of the book was to involve a wide range of people. This is an approach which is entirely consistent with the way marine conservation is undertaken, because a major difference between marine and terrestrial conservation is that the former involves far more work with a wide range of stakeholders to achieve change.

People often come to marine conservation with an interest in a particular topic – maybe corals, cetaceans, birds or fish – but this book is more about marine *conservationists* and the way they work, than what they work on. Another reality is that many of the people who have made a huge contribution to protecting the oceans, including the interviewees, do not have conservation in their job description, let alone their job title, because managing and protecting the marine environment can be done in many ways.

For the nineteen interviewees, this book describes, in their own words, how their interest in protecting the environment and marine conservation started, and how it developed. Their work relates to different interests, from seahorses to whales, and from habitat protection and marine protected areas to management of large areas of sea and the mitigation of damage from pollution, fishing and many other human uses of the sea.

As important and interesting is the wide variety of disciplines and styles they have adopted in their conservation work in order to achieve change. Chapter 3 describes a number of elements of this, including:

- The personal development and inspiration of the interviewees and their mentors

- Their personal qualities and skills, such as passion, commitment, ambition and innovation

- The nature of marine conservation as a career or a vocation

- Their experience of building organisations and capacity building

Chapter 3 also explores how people have worked together in different ways, including multi-sectoral partnerships, active collaborations of organisations working together to find solutions to problems, as well as more focused work with particular sectors, such as the fishing industry, to find solutions to conservation problems.

IDEAS

The ideas that drive marine conservation have been heavily influenced by the wider context of thinking on conservation, the environment and sustainability. There are at least six major drivers, as described in Chapter 4:

1. Terrestrial and marine biodiversity conservation

2. Science

3. Environmental management

4. Sustainability and its principles

5. Other cultural inputs, including welfare, non-violent protest and social sciences

6. Events, planned and unplanned

ACTION

Something that sets conservation apart from many other disciplines is the desire to act and respond to the status quo in order to try and achieve change in activities that harm the environment, its species and ecosystems. Every chapter reflects the reality of delivering such change, covering an enormous range of case studies from UK, European and international perspectives. The book deliberately covers an enormous range of styles, including the science-policy approach, direct action, welfare, advocacy, innovation, capacity building, campaigning and working with people – for this is the reality of working in conservation.

After the nineteen interviews, Chapter 24 draws out a number of common themes on marine conservation actions which have emerged in the narratives, including:

- The threats to the marine environment and priorities for future action

- The barriers to action

- The different ways people have achieved change

- The importance of innovation

Looking forward, the insights from this chapter underline some of the main difficulties in making progress and the lessons we need to learn to progress more effectively in the future.

HOW THE BOOK WAS PREPARED

How were the interviewees selected?

My marine conservation career started with organising citizen science project for sports divers, and gradually this developed to include capacity building by developing an organisation, the Marine Conservation Society, by building a membership, developing a constitution, communicating with newsletters, prospectuses and conferences, developing information sources and identification guides. The conservation programme themes arose from working to support the developing Wildlife and Countryside Act (1981) and working on issues from basking shark conservation to sewage and tributyl tin (TBT) pollution, from protected areas to bathing beaches, and working with people on local and whole-sea scales. It exposed me to people with completely different ways of thinking about the environment and the diversity of what marine conservation involves. Consequently, during my career I have come across many people who have made a difference. Their thinking has come from very divergent roots, ranging from the mainstream of government terrestrial nature conservation through to Greenpeace – but this barely describes the scale of the differences.

As the process of the interviews developed I used a number of criteria to select the interviewees, the four main ones being as follows:

1. *History.* The time span of the development of marine conservation, as covered in this book, is from the late 1960s to 2017. Many of the interviewees have careers which cover this time frame, and the majority bring at least 20 years of perspective to their narratives, with more than 600 years' experience in total.

2. *Specialist expertise.* There is a very wide spread of expertise covered in the book, from the protection of species and habitats, through to seas and oceans and the mitigation of human impacts.

3. *Style and approach.* Whilst the majority of the interviewees trained as natural or social scientists, I deliberately chose people who had come from very different backgrounds and used very different styles and approaches, well 'beyond science'. I have included people who use direct action, people with a strong welfare background, and people who proposed the spatial management of the seas in which conservation can be set. The contrast in working environment – from large government organisations and NGOs, through creating new organisations to acting individually – was also something that I wanted to reflect.

4. *A world view.* It is clear that marine conservation is now truly international in its scope. Whilst virtually all of the interviewees have worked on projects all over the world, the contributions from Jon Day (Australia), Keith Probert (New Zealand), Bud Ehler and Elliott Norse (USA) bring perspectives from beyond Europe.

The interview process

There are rigid research protocols for interview-based work, but that was not my purpose. Rather, this has been a personal interest – my aim being to look back and try and make sense of the development of marine conservation during my working life.

The methodology developed as I carried out the initial interviews, and the questions for the most part focused on four themes:

- An insight into how the interviewees began their careers
- An initial set of questions on marine conservation and its development
- A middle section that focused on the major themes of their work
- A final set of questions about the future

For the interviews with Jon Day, Keith Probert and Elliott Norse, a simpler style was adopted, asking about the key steps in the development of marine conservation in their countries and what lessons could be drawn from their experience.

As well as telling their individual stories, I was interested in looking at common themes arising from the interviews. These themes were certainly not clear at the outset, and the process has produced some surprising and fascinating insights into the way marine conservationists think and work. The interviews were videoed and transcribed, and chapters were then sent to the interviewees to be signed off. The videos of the interviews made the task of transcription much more enjoyable, and the archive of these videos may be of long-term interest. A few comments made by the interviewees appear in the common themes chapters although they are not included in the interview chapters themselves.

Five timelines have been included in the book as a way of recording the key developments. These are linked to the narratives of the interviewees as well as the wider context of marine conservation. Some of the events may seem remote in time or context, but their influence has been profound in various ways, and this is well illustrated by the debate over Brexit, which risks undermining many of the environmental gains that have been achieved over the last thirty years through our close connections with Europe.

LIMITS TO COMPLETENESS – WHAT THE BOOK DOES AND DOESN'T COVER

This book cannot be comprehensive, and this reality struck me many times. Nevertheless it does cover a number of important points that arise when considering marine conservation in a more holistic way.

The activities surrounding marine conservation are so extensive and far-ranging that it is absurd to imagine that they could be covered by a single book. Four points illustrate this:

- I did not select any lawyers or economists to interview, or a host of other specialists, and so the book is lacking in those and many other perspectives.
- I have not tried to cover marine conservation systematically, although Figure 2.2 shows how a book on the subject could be structured. Such a structure is more likely to be used in the future for a website that covers this task.
- Individual topics now have so many strands, that even if you were working on them full time it would still be very difficult to keep up with developments.
- Given its sheer scale and richness, the totality of the subject matter of marine conservation is well beyond any single volume.

WHAT YOU WILL LEARN FROM THIS BOOK

This book outlines the context of the development of marine conservation and scopes the diversity of the subject, as well as pointing to the issues that will continue to challenge us. It includes:

- An introduction to how marine conservation has developed

- A systematic structure to describe marine conservation

- An insight into the people, ideas and actions that have led to significant developments in marine conservation

- An understanding of the varied scale and focus of marine conservation – from individual species to oceans, and from detailed studies of ecosystems to the mitigation of damaging human activities

- Multiple voices – insights into the motivation of practitioners, their views and values, and the wide range of beliefs and approaches that have helped make significant gains

- An insight into the realities and difficulties of what marine conservation involves in practice, through the perspectives of successful practitioners

- The diversity of approaches to delivering marine conservation, from science and policy work through to direct action and welfare thinking

- A realistic picture of the skills and motivation required to work in marine conservation

- An understanding of the ongoing need for innovation, capacity building and developing programmes of action at all scales

- The importance of working with people in a variety of ways, from partnership to confrontation, to make progress in marine conservation

Put simply, the book describes the work of some remarkable people who, often in difficult circumstances, have made a difference.

CHAPTER 2

Marine conservation

WHAT IS MARINE CONSERVATION?

This looks like a simple question, but even after a short period of reflection it becomes rather more complicated and interesting. At a personal level it has many answers. At the start of this project I gave this a good deal of thought, not least because if you look at the breadth and coverage of marine conservation in a systematic way its scope is enormous – and this led to the development of the structural approach described later in this chapter.

Initially the question put to interviewees was 'How do you define marine conservation?' It quickly became apparent that their responses were not so much about definitions, but more to do with how they 'framed their views' in terms of their work. Usually, after some hesitation, interviewees commented on the difficulty of answering: 'I see you've started with the most difficult question' (Joan Edwards). The question of course encourages people to think through their own attitudes and beliefs, and prompts them to give a personal explanation of what they do. There was often a recognition that their views had changed over time. Over the course of the interviews, eight recurring themes emerged, and these are described below.

HOW DO WE FRAME OUR VIEWS? THE KEY THEMES

1. Definitions

I used the words 'define' and 'frame' when asking people what they believed marine conservation was all about, but it was surprising how few people responded with formal definitions. More common responses were 'defining things is always difficult' (Callum Roberts) and 'I don't use any formal definition' (Sue Gubbay). Definitions, by definition, are a clever use of a small number of words to describe really complicated things. Even so, one of the fascinating points to emerge from the responses was how many of the elements described in people's understanding simply do not arise in the formal definitions. Words like *inspiration, passion, magical* and *fairness* – which are entirely missing from the definitions and yet commonplace in the rationale of the interviewees.

Keith Hiscock referred to a definition: 'I use a definition that is from a time before "biodiversity" had entered our vocabulary and become fashionable: *the regulation of human use of the global ecosystem to sustain its diversity of content indefinitely* (Nature Conservancy Council 1984).'

Elliott Norse was part of a group that sowed the seed for the use of the term *biodiversity*. This was linked with conservation in 1992 at the United Nations Earth Summit, where, using earlier work, biological diversity was defined as '*the variability* among living organisms from all sources, including, inter alia, terrestrial, marine and other aquatic ecosystems and the ecological complexes of which they are part: this includes diversity within species, between species and of ecosystems.' This definition was used in the United Nations Convention on Biological Diversity.

Dan Laffoley responded that 'the classical view would be that marine conservation is about protection and preservation of wildlife and biodiversity in the ocean and its ecosystems, and about lessening and removing impacts.' He then went on to qualify this statement, noting that the danger with the classical view is that it focuses on describing the things we want to conserve, whereas for him marine conservation is 'much more about people, places, priorities and action, and in particular the reality and need to manage human behaviour.'

Such is the interconnectivity of the marine environment and its species that the protection of the environment has to extend to the oceans as a whole and not just 'special places'. With growing awareness that humanity is challenging the main planetary boundaries outlined by the Stockholm Resilience Centre (Rockström *et al.* 2009; see Figure 25.1), our work should also be directed toward protecting the ecosystem services that the healthy oceans provide.

Simon Brockington also describes the need to go beyond standard definitions. For example, 'IWC uses a standard fisheries definition very much akin with the language that was in use in the late 1940s when the organisation was established. It defines conservation as *the numbers of animals that can be removed from a stock or a population before there is some consequence for future viability*.' This sits alongside statements of values trying to achieve an equitable balance – for example, areas for whaling, or for marine conservation, areas which support whale watching or which support indigenous communities in the high Arctic. He considers that 'the basic discourse is one of fairness, which is probably not in any dictionary definition of conservation.'

2. Inspiration
In the autumn of 2017 the BBC released the second series of *Blue Planet*. This series encapsulates the wonder of the marine world, showing spectacular images and encounters with wildlife that are diverse, interesting and inspiring. Inspiration was a shared theme across the interviewees, both for what started their interest and also, as importantly, for what continues to drive their work in the marine environment. Both of these aspects are developed further in Chapter 3.

3. Passion and commitment
The number of times the word *passion* arose in the responses to how people framed their approach to marine conservation surprised me. In many ways marine conservation is a vocation and requires a commitment well beyond what is expected in normal jobs. This theme is developed in Chapter 3.

4. The need to act to change the status quo
Implicit in the rationale of every interviewee was the recognition that in their work they have come across situations where the status quo has been unacceptable, and that this motivated them to try and change things to better protect species or the wider environment. There are many examples in the interviews, summarised in Chapter 24.

5. The desire to be objective, scientific, with decision making based on evidence
All the interviewees had either a science or a technical background to degree level, and so it was hardly surprising that the belief in science and evidence was strongly articulated.

Indeed the preparation of evidence in various forms is a strong common theme emerging in the narratives in each chapter. But there are other aspects of this as well – for example, the frustration that the scientific approach can lead to. Dan Laffoley summarises as follows: 'A lot of people want to debate and describe what we should do, in my mind seemingly endlessly, but I think marine conservation must be, especially now, about setting priorities, really getting on with it and delivering effective action and outcomes.' There was also the recognition that science on its own is not enough but rather just one of the points that decision makers consider when arriving at their conclusions (Larcombe 2006; see Figure 24.5).

6. The importance of people

Unlike on land, there is no ownership of sea, and physical management is largely impossible in the marine environment. Recognising this was routinely covered in the responses to this question. Sue Gubbay described it as follows: 'It is about people and having a positive interaction with wildlife and the environment.' And Heather Koldewey responded: 'When people ask me to make sense of my work there are two elements which, simply put, are nature for nature and nature for people.' Working with people is a key element of marine conservation, because influencing their actions and decisions is the key to securing change. Working with key stakeholders, especially fishermen, to devise solutions to conservation problems is a recurring theme throughout the book.

7. Recognising change and acting to enable recovery

Sue Gubbay and others made the point that marine conservation is not just about preserving things, but that it is important to recognise that natural systems change. Unlike on land, where management practices can be designed to attempt to hold a particular habitat in a particular state, this is simply not an option in the marine environment. Even a basic understanding of marine ecology and natural systems teaches us this. We also now have a very clear understanding of how human activities have changed marine ecosystems, from the insights into shifting baselines described by Daniel Pauly (1995) and by Callum Roberts in his book *The Unnatural History of the Sea* (2007). Far from being untouched or pristine, many of our marine environments have been changed beyond recognition, not least by fishing. The issue for marine conservation now is summarised well by Dan Laffoley: 'We need to allow breathing space in the ocean where recovery can take place' – and by Callum Roberts: 'Marine conservation is about giving nature relief from the adverse effects of human influence. It's all about revitalising the oceans, and for that you need to appreciate history.'

8. The importance of ideas

A wide range of ideas underpins how the interviewees describe their approach to marine conservation, and these are discussed in Chapter 4. A core belief in science and evidence was clear, but also an emphasis on the importance of high-level principles such as sustainability, precaution and integration, and the principles in sustainability packages like the ecosystem approach, integrated coastal zone management (ICZM) and marine spatial planning. Other ideas stemming from different traditions surrounding animal welfare and non-violent direct action also have a part to play in the organisational ethos and personal beliefs of what marine conservation means. In addition, there is the interesting question of whether the personal views you hold are similar to or different from those of the organisation you work for. Where the organisation has a strong philosophical ethos then it is crucial that personal and organisational beliefs coincide.

Figure 2.1 Lots of words, including inspiration, come to mind during encounters with sharks, and such memories live with you and inspire your actions. These are Caribbean reef sharks. Source: Jeremy Stafford-Deitsch

Conclusions

The richness of views expressed by the interviewees shows that whilst definitions are helpful, the reality of delivering change to protect the marine environment involves a much wider and stronger set of beliefs and commitment which every individual needs to work through for themselves. It also suggests that textbooks on marine conservation often place rather too much emphasis on the formal definitions rather than on understanding what actually drives and motivates people working to protect and manage the marine environment.

Chris Rose in his interview explains how Max Nicholson, who helped establish the Nature Conservancy Council in the UK and the WWF, had deliberately set out to frame conservation in terms of natural science in the 1950s, because there was a great belief in science at that time. By the end of the first decade of the twenty-first century, conservation in the UK has increasingly come to be framed in terms of enabling people to enjoy the health and wellbeing benefits of being in natural surroundings, an approach that should have a very high appeal for those interested in the marine environment.

These insights also reveal an inherent contradiction in our approach to the marine environment: being passionate but at the same time believing in objectivity. How can we, on the one hand, be effusive and inspired by a first encounter with sharks (Figure 2.1) or the sight of a great fish shoal, and on the other hand enjoy eating them? Roger Mitchell summed this up neatly: 'There is a tension between objectivity and subjectivity, so I'd describe what motivated me as a passion moderated by science and evidence and a desire to do something about things that weren't right about the marine environment.' Choosing not to eat fish, for example, is a logical extension of this thinking. Simon Brockington describes the contradiction in terms of the recognition of different views held by delegates at the International Whaling Commission. Sarah Fowler describes a very real dilemma for practical shark conservation, where those who propose management are directly countered by those asking for a total ban on removing shark fins (Shiffman & Hueter 2017). Asking for bans is a much easier message to communicate. This contradiction is a very real issue, with many practical ramifications.

MARINE CONSERVATION – A STRUCTURAL APPROACH

At the outset of my work on this book, the sheer number of issues covered by marine conservation was obvious, and it was clear that this would play an important part in the way I approached the book. Two issues in particular were important. The first, covered in this section, was whether there was a systematic description of the scope of marine conservation. The second was how the diversity of subject matter and disciplines would affect the selection of interviewees. This challenge was partly resolved by my attendance at the International Marine Conservation Congress (IMCC) held in Glasgow in the summer of 2014. With an audience of more than 700 delegates, and over 500 presentations from around the world, this meeting was stimulating in many ways. Whilst some of the sessions were structured around themes, a large proportion of the topics seemed to be distributed through the programme almost at random. This then raised the question of what that structure of marine conservation might look like, which was exactly the same question that had struck me as I started to plan the book.

After the IMCC I spent some time thinking about this, and commissioned work to see whether there were websites or diagrams that summarised the scope and structure of marine conservation. When I didn't find such a structure, I decided to develop a website and a diagram myself (Figure 2.2), and this was published in 2016 (Earll 2016). My background has been influenced by diagrams like the periodic table and the evolutionary tree, and it was this tradition that I drew upon for the marine conservation diagram. Such diagrams have developed over a long time and in many versions, and if this is the first for marine conservation it is bound to be refined and improved.

The diagram illustrates three main segments, representing the subjects that marine conservationists work on:

- *Marine life, places and seas* – the range of natural resources, from species and habitats to ecosystems, whole seas and oceans

- *Threats* – the threats and the mitigation required

- *Actions* – the fast-expanding range of disciplines that are being applied to undertake marine conservation

There are 93 categories in this diagram, going some way towards reflecting the diversity of the subject. During this process – now three years – I have not found anything similar. The reason for this is probably that most marine conservationists and organisations are too busy working on a limited number of these content categories and do not look at the whole picture.

The diagram has many potential applications, for example:

- The diagram is a two-dimensional representation of marine conservation. The power of computing and its multidimensional links is still accessed through a two-dimensional screen, and so this diagram provides a way of accessing, ordering and understanding its complexity.

- It describes the work marine conservationists undertake, such as reducing impacts on a particular resource, whether that be whales or the North Sea. Almost all of them will be using a variety of methods or disciplines to address the issues – for example, marine protected areas to protect fish, or legal provisions to study high seas conservation measures.

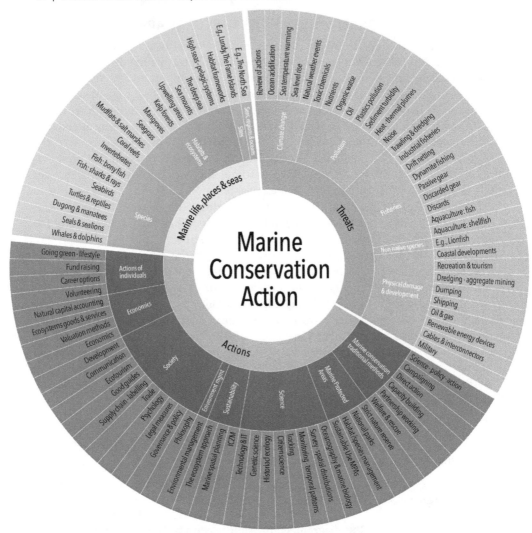

Figure 2.2 Structure diagram of marine conservation. Source: Bob Earll (2016)

- It reveals why this and other books on marine conservation are unlikely to be comprehensive. No single volume could cover the detail of all the efforts that are currently directed to this subject; most books cover perhaps ten of the topics which it incorporates.

- The diagram might also provide food for thought for anyone who organises large-scale meetings on marine conservation, as it provides a clear structure by means of which topics could be organised into themed and productive sessions.

I have drawn on this structure and diagram throughout the book to reflect upon the diversity and richness of marine conservation.

TERRESTRIAL AND MARINE BIODIVERSITY CONSERVATION – SIMILARITIES AND DIFFERENCES

A question that arose early in planning the book concerned the similarities and differences between terrestrial and marine conservation. This was because of the influence terrestrial conservation has had on its marine counterpart – usually leading the way as a mainstream activity recognised by governments long before marine conservation gained traction. Interviewees were asked to describe what they saw as the main differences between marine and terrestrial conservation, but they also frequently added their views on the similarities. There are closely shared beliefs and values and many generic skills and approaches across conservation work, which was much in evidence in their responses. Many of the interviewees have worked on terrestrial and coastal conservation for parts of their careers, and their chapters reflect rich and complex relationships between the two.

Common ground – the similarities

1. Shared ideas

Often the principles and overarching ideas of conservation apply equally on land and in the marine environment. Sustainability and welfare are two such examples. Sue Gubbay put it like this: 'I believe there is a lot of common ground with terrestrial conservation – similar difficulties (politics, funding, priorities), similar values to promote (the importance of wildlife and the environment), and conservation in both environments embodies the same ideas (taking a long-term view and safeguarding for the future). You could debate whether it is harder or easier than terrestrial conservation, but in reality I believe this depends on the level of motivation and support, not on any fundamental differences in the ideas.' Alan Knight expressed this as well: 'I work on marine and terrestrial animal rescue. The same welfare ethic applies, and there is no difference in our approach to animals in either environment.'

Dan Laffoley described an interesting case stressing the commonality of approach: 'In my global work, for example, in relation to the marine–terrestrial question, I have been very careful not to automatically resort to special pleading for marine. In the protected area sphere, whilst there are clear physical and locational differences, the principles are similar on land and in the sea. With the whole conservation community behind one, an attack on MPAs can be seen as an attack on the idea of protected areas and the system in general, and that can be a great strength. In other situations ... the differences are so great that special pleading has been helpful.'

2. From terrestrial to marine: common ground but a time lag

In the story of the development of marine conservation there has often been a significant lag between what has been done in the terrestrial and what has been done in the marine environments, and many examples were provided. Terrestrial conservation has often led the way with its major initiatives and legal frameworks. For example, Dan Laffoley points out that in relation to the sequestration powers of marine and coastal habitats, taking the idea of 'blue carbon' from a terrestrial context can be incredibly useful in helping advocacy in relation to marine protection (Laffoley & Grimsditch 2009).

Other examples include the JNCC *Marine Nature Conservation Review* which was published in 1998, 21 years after its terrestrial counterpart the *Nature Conservation Review* (Ratcliffe 1977), and the first time CITES was used cover the trade in marine fish was in 2004, 41 years after CITES came into force. In the case of MPAs in the UK it was as recently as 2008 that the legal machinery became available to create a network of MPAs, 27 years after the Wildlife and Countryside Act (1981) had thoroughly revised the terrestrial protected area network; even now there is no adequate recognition of no-take zones. The UK has also been years behind on MPAs when viewed from a global perspective. New Zealand's Marine Reserves Act (1971), which included provisions for no-take zones, was in place four years before the UK even had a marine official in its conservation agency, the NCC (see Timeline 4, page 3).

It has often been a great battle to get 'marine' recognised. In the early days (the 1970s and 1980s) in the NCC it was simply a battle over money; for other government departments, especially fisheries, it was a turf war about protecting their 'patch'. As most battle-hardened UK marine conservationists recognise, even after legislation is passed there is another time lag before implementation. This was highlighted by Peter Barham and Joan Edwards, who observed that the effective implementation of laws takes a disproportionately long time, with interpretation and implementation of those laws creating all sorts of challenges and unforeseen consequences.

The differences

Responses to the question of differences between marine and terrestrial conservation were very rich in detail. Keith Hiscock in his book *Marine Biodiversity Conservation* (2014) lists seven differences, but whilst many of these were shared by the interviewees, this list is different in a number of respects, and each of the points is more fully developed.

1. Size and connectivity

The world's oceans are vast, occupying 70% of the planet's surface, and the science of oceanography has always been undertaken at a world scale. This has probably led to the relative ease with which people working on marine environmental protection have sought to tackle issues at the largest geographic scales. Through the development of the science of oceanography, we also consider the total body of water as well as the seabed as part of an integrated three-dimensional environment which has very high levels of connectivity. Our knowledge of the global ocean is developing fast, illustrating how the main bodies of water circulating around the globe – the ocean conveyor belt – are highly connected (Figure 2.3). Whilst species and their reproductive stages often rely on this connectivity, unfortunately a great deal of pollution, most recently highlighted by plastics, is also being transported by these very same currents all around the world. As Keith Hiscock points out, the concept of wildlife corridors is irrelevant in the marine environment.

Dan Laffoley summarises as follows: 'The processes in the sea that regulate ecology are much less predictable. On land you can simply see the effects of particular actions. The fluid dynamics and 3D structure of the oceans means that connectivity is much more important than on land. Many marine species move very significant distances and across many administrative boundaries in their routine behaviour. Things can affect places from a much greater distance. Toxic chemicals exemplify this, with the polychlorinated biphenyls (PCBs) turning up in the very extremities of the Arctic and Antarctic many thousands of miles away from points of origin, and years after some may have thought we had "solved" this problem.'

Landscape conservation became a buzzword among terrestrial conservationists in the 2000s, but marine has always been far ahead of terrestrial conservation in this regard. This was because of the widespread recognition that many of the issues were trans-

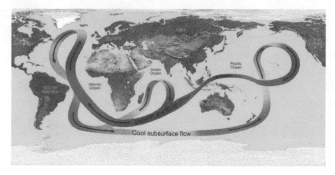

Figure 2.3 The thermohaline circulation, the global ocean conveyor belt. Source: NASA

boundary in nature and applied to mobile marine species that move vast distances with no regard for our administrative boundaries. Euan Dunn summarises this succinctly: 'Marine conservationists have been working on this scale for a very long time and it sets them apart from their terrestrial colleagues. I see marine conservation working at a series of spatial scales – a bit like Russian dolls – so whilst there are areas of special importance that require protection it is also essential to work at a wider spatial scale.'

Given this understanding of the sheer size, scale and connectivity of the marine environment, from the earliest days people concerned with marine protection have sought to protect the totality of the seas from pollution in the first instance. Some examples are the UNEP Regional Seas Programme (1975), the Great Barrier Reef Marine Park Act (1975) and in the Antarctic the CCAMLR convention (1980).

2. Describing the marine environment – not enough information

Two issues, gaining access and visualising the environment, are largely taken for granted by terrestrial conservationists but pose colossal problems for marine conservationists. 'Most terrestrial ecologists would only get a sense of the difference by working in a dense fog where you can't see much around you' (Dan Laffoley). There is no doubt that it is much harder and more expensive to collect information in the marine environment. As a consequence it is no coincidence that the rise of interest in marine conservation has come alongside the development and accessibility of scuba diving. Diving has served to enable us to describe and understand the shallow seas and make the wider public much more aware of marine life and habitats and the issues of concern.

Keith Hiscock summarises it as follows: 'Our understanding of what is where – the basic distribution of marine species and habitats – is very poor and it is very expensive to fill in the information gaps. Likewise, our knowledge of change brought about by human activities and our understanding of long-term natural fluctuations in abundance and distribution of marine species and of natural change in the character of habitats is very poor compared to the land.

The scale of movement of mobile species and the consequently enormous practical difficulties in studying many marine species are well described by the interviewees – Euan Dunn for birds, Sarah Fowler for sharks, Sue Sayer for seals, Simon Brockington for cetaceans. Technology is helping rewrite our understanding of the ecology of many of our well-known species. Euan Dunn describes how satellite tags show that Irish nesting puffins feed off the eastern seaboard of America, and Simon Brockington described in the interview how tagged grey whales migrated across the Pacific, completely contrary to a century of common understanding.

A great difficulty is that techniques routinely applied on land, such as the IUCN Red Listing process, cannot easily be applied to marine species, simply because so many marine species are 'data deficient'. Sarah Fowler highlights the issues with species conservation: 'When you are trying to assess a population, certain things are taken for granted in the terrestrial world. You generally know how long a mammal lives, how quickly it breeds, and what area of ground it needs to live in. Marine animals are so much harder to count and to track. You don't know how many there are, where they are, how far they travel. We don't even know how long some species live, or when they become mature.'

3. Managing marine areas for conservation – wildlife gardening

The concept of protected areas is the most significant individual idea that has been transferred from terrestrial to marine environments, and it has come to dominate much marine conservation work.

Joan Edwards summed it up thus: 'On land there is more certainty. You can see what you're doing. Lots of people can get physically involved and you can easily see what measures such as restoration look like. It's much easier to convey messages of a healthy environment to the public and give them a feeling of ownership about it.' It is also possible to purchase, own and therefore control what happens to land.

Physically managing the land – sometimes referred to as ecological gardening – holding vegetation at a particular point to enable particular species to flourish, has developed as the central core of the UK's approach to terrestrial nature conservation. But there is a fundamental conflict in this approach when applied to the sea, which Callum Roberts succinctly summarises by pointing out that there is no marine equivalent of releasing a herd of sheep to maintain the vegetation. The current application of terrestrial thinking to UK Marine Conservation Zones in what some call the 'features-based' approach of notionally favouring particular habitats is impossible in the sea, because once human pressure is removed change is inevitable.

4. Working with people – the greater need for advocacy and policy change

Generally speaking there is no private ownership of the sea or seabed. As a result, in marine conservation, achieving change requires much greater emphasis on working with people, whether it be in regard to advocacy, developing policies and networks, or working with user sectors to mitigate damaging activities. This is described in more detail in Chapter 3.

5. Public awareness – it is harder to visualise marine species and habitats

There is no doubt that general public awareness of the habitats and species in the marine environment is very different from their understanding of the terrestrial environment. A number of the interviewees highlighted this major difference and the importance of overcoming it as a first step to gaining public support. Chris Rose pointed out that in the clearer, warmer and more tropical seas, the public have an awareness of the sea and its marine life which is much more akin to terrestrial ecology.

In any event, if awareness is the basis of public support, then before persuading people of the need to conserve and protect the marine environment, much more work is needed in the colder seas. Heather Koldewey put it this way: 'There is a big difference in connection and visibility, so whilst many people have an intrinsic love for the ocean it is a lot harder to see what's going on below the surface … It is a challenge to engage people in marine conservation both conceptually and practically. First they need to know what is there, the wildlife and the habitats, but getting them to care about it is much harder. The Thames Estuary is a perfect example, teeming with wildlife – but people think it's brown and dead.'

6. Differences in legal frameworks
Legal frameworks used in the marine environment are very different from those used on land and have had to recognise shared interests with other countries from the outset – as in the Regional Seas concept. The sheer geographic scale of their coverage, which also recognises the connectivity and three-dimensional nature of the seas, is very different to terrestrial legislation. On another occasion, Sue Gubbay suggested to me that the main differences the main differences are working in a system of no ownership (the 'tragedy of the commons'), complicated jurisdictions, international rights and obligations, and the need for international cooperation at least at Regional Seas level. UNCLOS (the United Nations Convention on the Law of the Sea) has been in operation since 1982, and the UNEP Regional Seas Programme has been developed since 1975 (see Timeline 2, page 45). Conservation has come late to many of the agreements with sector-specific interests first (fishing, mineral exploitation etc.) and we are still trying to address this in relation to conservation under UNCLOS (Sue Gubbay). It is only relatively recently that marine conservation objectives have been recognised in the work of a range of conventions.

In Europe, although the Oslo and Paris Conventions (OSPAR) and the International Council for the Exploration of the Sea (ICES) were in place, when the new ideas of the environmental movement struck the decision makers in the 1980s they were found wanting, and it took twenty years of North Sea Ministerial (NSM) meetings to enable them to recognise and enact fundamental changes in environmental thinking and transparency. This coincided with the European Union developing a number of measures including the Urban Waste Water Treatment Directive, the Water Framework Directive, the Habitats and Birds Directives, the Marine Strategy Framework Directive and most recently the Marine Spatial Planning Directive (see Timeline 5, page 93).

In the UK, the major rationalisation of the management of the marine environment, recognising the need to overhaul much long-standing legislation on sectoral use and licensing and bringing in measures for marine spatial planning and a network of marine protected areas, only took place with the introduction of the Marine and Coastal Access Act (2009).

Conclusions
There are clearly many shared core values between terrestrial and marine conservationists, although the differences in approach still cast long shadows. One major problem is thinking that terrestrial conservation thinking can be simply applied to marine – it cannot. Much misplaced time, money and effort has been wasted learning this lesson. The pioneering and novel efforts highlighted by many of the interviewees demonstrate how innovative thinking can pay dividends in marine conservation.

THE DEVELOPMENT OF MARINE CONSERVATION
The question of how marine conservation developed reflects my personal interests over the past forty years and the exponential growth of the subject during this period. Rather than look at the historical development of marine conservation in any one country, I have drawn on a number of common developmental stages and themes that recur in the

narratives of the interviewees. The main emphasis of this book is the UK and European context, although the majority of interviewees have worked elsewhere in the world and have experience of practice from South America to the Southern Ocean and from Asia to the Caribbean. The inputs of Keith Probert from New Zealand, Jon Day from Australia and Bud Ehler and Elliott Norse from the USA also reflect on the development of marine conservation thinking in those countries. Whilst the development of marine conservation is often thought of as mainly a product of Western cultural thinking, the interests of indigenous peoples in coastal and marine resource management – and their long-standing expertise – is now increasingly recognised (Day, Probert), and recognised routinely by the International Whaling Commission (Brockington). Some of the major themes involved with marine protection and conservation which have involved considerable bodies of work are summarised in Box 1, and discussed below at greater length, under seven themes.

BOX 1. THE CHALLENGES OF FIFTY YEARS OF PROTECTING, CONSERVING AND MANAGING THE MARINE ENVIRONMENT

This list provides a summary of the main areas of work undertaken by the marine conservation sector. It supports a number of chapters in this book, covering several of the major elements – universal challenges – that the interviewees have worked on and are still engaged with. The development of protecting the marine environment and marine conservation over the last fifty years is described mainly for the UK and Europe. Clearly events, actors and key steps have varied in every country and region.

- Challenging the paradigm that the sea could be used to dilute and disperse pollution and as a dump for every type of human waste.

- Challenging the paradigm of uncontrolled exploitation of natural living resources driven by the primacy of the economic model to determine rates of use.

- Recognising marine environmental protection and conservation – 'marine' – as a legitimate political and scientific activity and mainstreaming this into law, policy and action.

- Raising awareness of the marine environment, ecosystems, habitats and their wildlife, and the need for action to protect them.

- Recognising that protecting marine species includes vertebrates, invertebrates and plants. Conservation measures have progressed from the larger charismatic species such as the great whales to include a much wider array of species.

- Recognising the need to protect habitats, ecosystem processes and functions using the marine protected area approach.

- Building the capacity of marine conservation in terms of ocean governance, including civil society, the environmental NGOs, the private sector and government – by making conservation a mainstream activity.

- Recognising the role of case work in mitigating the effects of development and human activities by other sectoral users – this links directly to environmental management practice over the life cycle of projects.

- Recognising marine protection and conservation in the context of other sectoral activities and management frameworks such as marine spatial planning, integrated coastal zone management, watershed management and land-use planning, in the context of sustainability and its related principles.

1. When did marine conservation start?

In his book *The Unnatural History of the Sea*, Callum Roberts describes how, from as early as the seventeenth century, some people thought that marine resources should be managed, not just pillaged and exploited. Heather Koldewey describes how when London Zoo opened its doors in the nineteenth century there were people with a strong conservation ethos, and indeed Philip Henry Gosse, the Victorian naturalist who popularised marine life, commented on the damage to seashores caused by overzealous collecting. Timeline 4 (page 45, based on Wells *et al.* 2016) documents a series of steps to create what we would now call marine protected areas from the late nineteenth century.

It is clear, therefore, that some of the key ideas have been around for a long time, but it was probably only in the 1960s that the momentum for what we currently understand as marine conservation started to develop. It is also clear that the UK was not in the lead in this development. By the time the UK had appointed its first professional to a post that included 'marine conservation' in the job title – Roger Mitchell in 1975 – New Zealand already had its Marine Reserves Act (1971), the USA had the Coastal Zone Act (1972) and the Marine Sanctuaries Act (1973), and in Australia the Great Barrier Reef Marine Park Act (1975) had been passed.

2. Prevailing paradigms

In the 1950s and 1960s – but dating from the pre-war era – there were two prevailing paradigms of how we used and managed the seas, concerning pollution and the exploitation of natural resources.

The pollution paradigm

It was a commonly held view into the 1990s that the sea could be used to dispose of any type of human waste, a view embodied in the phrases 'dilute and disperse' and 'out of sight out of mind'. Governments, big industry and the water companies saw the sea and coastal waters as a legitimate dumping ground for a very wide range of wastes. There were many examples:

- Industrial toxic waste, including radioactive materials, either into the near shore via pipelines or into deeper water from dump ships
- Infrastructure, including military waste
- Sewage and much else that was mixed with it
- Persistent organic pesticides
- Oil pollution incidents
- Runoff of nutrients from agriculture
- From the 1970s, plastic pollution

Up until the 1990s, European scientists from the marine establishment would frequently rehearse the idea of the sea's carrying capacity to deal with this waste. This was seen as the 'Best Practicable Environmental Option' (BPEO), and much time was spent monitoring dump sites. Not surprisingly, therefore, many of the interviewees describe major campaigns and the response to associated events. One outcome was that in in the early phases of work on marine conservation, anti-pollution campaigns had a much higher profile than they do now in the developed countries. However, the same old bad practices of dumping are still occurring in South America (Paul Horsman).

Exploiting natural resources

The second paradigm concerns natural resources. There was an explicit assumption that the seas were a boundless source of fish and marine resources. This had already been widely debunked, not least by two world wars, which showed how commercial fish stocks recovered over the war periods. Then in the 1960s and 1970s the decline of the great whales caused by commercial whaling was the *cause célèbre* of early NGO efforts across the world, eventually leading to the IWC moratorium in 1982. Many fishing interests carried on as if stocks were boundless, however, and Sarah Fowler relates the surreal experience of a fisheries civil servant from MAFF in the early 1980s explaining that fisheries were 'self-regulating; the fishermen simply exploit the stock until it is uneconomic to carry on'. There has been a long-standing and ongoing tension between the fisheries and marine conservation interests (Callum Roberts, Sarah Fowler, Euan Dunn), and this is described in detail by Serge Garcia and colleagues (2014). Since the mid-1990s the developing awareness of the ecosystem approach has played a significant part in helping bring these two separate strands of thinking together.

In Europe, and especially in the UK, both of these paradigms were fully supported by the attitudes of government, business and the marine scientific establishment right up until the 1990s, with huge implications. The time period covered by the interviewees has seen both of these paradigms successfully challenged on many fronts in the UK, elsewhere in Europe and around the world, leading to a fundamental shift in attitudes.

However, Dan Laffoley refers to 'three eras of arrogance' in human attitudes to the ocean over the years, pointing out that the problem is far from over. We have moved on from believing that we cannot deplete the renewable resources or alter the fundamental processes and chemistry, but we are still under the illusion that we can never exhaust the non-renewable resources of the ocean. 'We still don't realise it is a finite resource that should be cherished,' he says. 'In taking these short-sighted approaches, humans have unfortunately significantly damaged the blue heart of the planet that keeps us alive.'

3. The marine environment not on the agenda

There were many significant steps towards conservation on land, but in the early twentieth century the marine environment seldom figured in this. Above, I describe how terrestrial conservation thinking often led the way and how there has been a significant time lag in relation to the implementation of ideas from terrestrial to marine conservation. This has been true throughout the development of marine conservation, when 'marine' simply was not on the agenda of the statutory conservation agencies, government or industry. 'Out of sight out of mind', 'the poor relation' and the 'Cinderella of conservation' are phrases that characterise this period. Considerable time was spent trying to ensure that 'marine' was on the agenda, and many of the interviewee narratives describe this. Sue Gubbay vividly remembers going to conservation meetings in the 1980s when the organisers had simply not included marine conservation at all – and so her input was 'what about marine?' An example of this was the late but completely inadequate inclusion of marine habitats into the EU Habitats Directive (1992) and the endless and continuing problems which this has raised. One of the recurring themes of the narratives is that it is necessary to implement legislative change in order to get formal recognition of marine species and habitat conservation (usually in relation to marine protected areas). This step was key to 'mainstreaming' marine conservation from being just the preserve of NGOs – who were usually in the lead in pressing for legislative change.

Another manifestation of this has been the fight to get marine species recognised by the conservation conventions, which had essentially been set up with terrestrial conservation in mind. This met with considerable resistance, because the people driving these conventions thought that other mechanisms should be doing the job; this made for very slow process. Sarah Fowler describes the efforts of the shark conservation community to get sharks included in CITES and the Convention on Migratory Species, and Heather Koldewey comments on the same issue in relation to seahorses. Similar problems arose when people tried to get birds or sharks included in fisheries agreements and conventions, with procrastination and obstruction being the order of the day. Euan Dunn describes attempts to get the bycatch of birds included in EU fisheries management policies. Sarah Fowler tells the same story concerning sharks and the International Commission for the Conservation of Atlantic Tuna (ICCAT), explaining that in the early days there was often great tension between conservation and the management of fish species. They were not viewed as being the same thing and there was a tendency for conservationists to be viewed as the enemy of fisheries management. Over this period battles had to be fought simply to get effective NGO representation at many forums. Now NGO delegates are present in many different influential forums, including OSPAR, ICES, IMO and the European Commission, and this has been a big step forward.

Whilst the major conventions covered whole seas from coast to offshore, it has taken the marine conservation community some time to fully extend its ambitions to the whole-sea scale. The progression is clear in terms of the conservation agenda moving from coastal to offshore waters, and covering the deep sea as well as species and habitats on the high seas. Incorporating the ecosystem approach to management of natural resources and the use of MPAs, high-level conventions have also been relatively recent. The transition from coastal zone management to concepts like marine spatial planning for the territorial seas also reflects this progression of thinking.

4. Public awareness of the marine environment
Blue Planet 2, the BBC's flagship wildlife project, was launched as this book was being written. The audience in the UK for the first programme was over 10 million, twice as big as that for *The X Factor*, a successful talent show on the other main television channel. There is no doubt that public awareness about the environment in general – and the marine environment – is very high, and this was reinforced by many interviewees. This awareness is fundamental for the support of marine conservation initiatives, and yet there are very clear challenges in helping the public make the links to the marine habitats and wildlife in the bleaker northern seas, a challenge described in some detail by Chris Rose.

Over time, conservationists have become much more aware and adept at using the media, especially the digital channels, to get their messages across. This change forms a fundamental part of modern conservation programmes.

At the same time various events and issues have pushed marine conservation onto the public and government agenda. The effect of the major oil spills in this process cannot be underestimated, and this features in the narrative of a number of the interviewees. The Save the Whales campaigns of 1960s and 1970s had a huge profile. In the UK, various local issues such as seal conservation and the over-collection of curios like urchins and sea fans by divers have risen to prominence. Issues such as the effects of mercury poisoning or tributyl tin pollution and direct action campaigns by Greenpeace in the 1980s and 1990s raised the profile of the need to curb marine pollution. These are described by Paul Horsman. From the 1980s onward the impact of commercial fishing, whether it was the giant floating 'walls of death' drift nets, or the bottom trawling gears going ever deeper in the ocean, became increasingly apparent as NGOs were able to observe and raise awareness of these issues.

5. Capacity building, the development of NGOs and the mainstreaming into government of marine conservation

The development of organisations, the generation of new NGOs and the development of departments within government to mainstream marine conservation objectives has characterised the explosive growth of marine conservation during the last forty years: this is covered in greater detail in Chapter 3. The narratives reflect capacity building for marine conservation, and management has involved the following elements:

- Creating new NGOs and capacity building within existing NGOs

- Mainstreaming marine conservation and management into government departments and agencies

- Developing marine conservation meetings at national and international scale

- Building multiple stakeholder groups at different spatial scales, from local to national to regional

- Developing international collaborations (NGOs and governments) to tackle specific issues such as plastic pollution

6. Recognising the multi-sectoral use of the marine and coastal environment and the need for environmental and conservation activities

The traditional approach to the management and regulation of the dozen or so large sectors operating independently in the marine environment, including oil and gas, aggregates and shipping, was that they should be managed by separate government departments. The first challenge to this came with new ideas on coastal zone management, initially in California but then in the USA as a whole with the Coastal Zone Management Act (1972), as outlined by Bud Ehler. It was clear that although there were many sectoral stakeholders, their interests were increasingly at odds, and the thinking started to evolve towards a more integrated approach.

Sue Gubbay describes her work for MCS and WWF, which effectively started the ICZM debate in the UK. A product of this process was a movement that brought stakeholders together routinely at both a national and a local scale to discuss issues before they became problematic. Peter Barham describes the work of these partnerships, which are still going strong today since their ability to effectively communicate and operate with many stakeholders is seen as a major strength. The idea of sea-use planning was bubbling away in academic circles in the 1980s and 1990s (Smith & Vallega 1991), and it really took hold in the 2000s with the developments in marine spatial planning. This systems-based view, which was led by Bud Ehler and Fanny Douvere, is now recognised around the world.

Another strand of systems thinking developed in the 1980s when environmental impact assessments (EIA) became commonplace. This included environmental management applied across the life cycle of developments (see Box 3, page 38), and across a range of sectors. The premise was that the impact on the environment had to be taken into consideration at each stage. Increasingly, this led to the need for government to employ marine environmental and conservation staff to assess the effects of developments on the marine environment. 'Case work', as this is known, became an increasingly important task for governments in the UK and elsewhere in Europe from the 1980s, and this required evidence. This led to the development of information systems such as MarLIN, described by Keith Hiscock, which gather together the evidence of the likely impact of particular actions on marine species and habitats.

7. The development of science, technologies and evidence

Developing the evidence base upon which to act is commonplace and described by all the interviewees, and it has been the subject of major and landmark projects in the development of marine conservation. The narratives reflect a range of approaches that have been used as the basis for evidence gathering, including the development of a number of important tools and methodologies:

- Identification and description guides for species and habitats (biotopes) (Hiscock, Gubbay, Koldewey)

- Citizen science projects using divers (Mitchell, Hiscock, Sayer)

- Species inventories at different spatial scales (Probert)

- Marine natural resource and marine site inventories and maps (Gubbay, Ehler, Hiscock)

- Status and sensitivity data for a wide range of species and habitats (Hiscock, Gubbay, Fowler)

- Practical protocols for working with species, and advice packages for case work (Knight, Sayer, Hiscock)

The recognition that there was no basic description of the distribution of species and habitats has driven many projects. Keith Hiscock describes the work of the Marine Nature Conservation Review to describe the species and habitats around the coast of the UK, which was published in 1998. Bud Ehler describes a major mapping and resource atlas programme for the USA conducted by NOAA. Recently Sue Gubbay led a team of over 350 European scientists describing the biotopes and their Red List status for European seas, which was published in 2017. Identifying marine species was a huge challenge, and it must be difficult for the current generation to imagine what it was like having to cope with species guides that were more than 100 years old, with line drawings, now that there are web-based guides full of colour images. Major computer-based inventories of marine life have been completed as our knowledge has expanded. The task of describing the status of mobile species for conservation purposes has been just as much of a challenge, and has led to many major breakthroughs of understanding.

These methodologies, whilst devised in the era before computing became widespread, have been transformed by the power of digital media. Other technological developments, such as digital cameras like the GoPro, will transform the way we think about data, information and its communication, not least because of the use of such technology in citizen science projects.

Conclusions

It would be good to think that some of these challenges, such as changing attitudes, are battles that have been won – but unfortunately many are still part and parcel of the work of marine conservationists. In particular, economic exploitation of the marine environment with little regard for environmental consequences is still alive and flourishing.

CHAPTER 3

People

One of the main themes of the book is people and the diversity of approach that different personal perspectives bring to marine conservation. In addition, one of the key messages to emerge from the interviews was how, much more than on land, marine conservation is all about working with people to achieve change. The process of starting work in marine conservation is seldom straightforward, and this chapter explores how the interviewees developed their interest, how they started their careers, and who and what inspired them. The personal qualities and skills that are required are also discussed. A common theme was just how often the interviewees had been involved in either developing new organisations or building capacity within existing organisations.

GETTING STARTED – EARLIEST INSPIRATIONS

The interviewees were asked what inspired them and how they got started. Whilst it was often a combination of things, there are many similarities between these influences. Some of the interviewees lived by the sea (Callum Roberts, Keith Hiscock, Dan Laffoley), but for the majority, visiting the seaside on holiday, rock-pooling and exploring the seashore provided lots of memories. A wider interest in nature – working as a volunteer conservationist, birdwatching and the first pair of binoculars (Chris Rose, Euan Dunn, Sue Sayer), pond dipping or collecting butterflies – and being outside 'in the environment', surfing and sailing are also commonly cited. Early on, a range of family members, relatives, family friends and inspirational schoolteachers frequently provided the encouragement to do natural sciences at university.

Books were an important source of inspiration too, and buying the Collins guide to the seashore, or discovering Gosse's works on seashore life from the nineteenth century or the New Naturalist series are mentioned. There is no doubting, however, that the major stimulus that raised awareness of the marine environment was television. From the 1950s to the 1970s, programmes with David Attenborough, Peter Scott, Hans and

Figure 3.1 Jacques Cousteau had a huge influence on marine conservation, not only through his TV series but also because he and his colleagues designed the aqualung. Source: the Cousteau Society

Lotte Hass, Lloyd Bridges and Jacques Cousteau played an important role. The Cousteau TV series was mentioned by virtually everyone and generated an interest in seeing the marine environment first-hand. Cousteau's influence was doubly important because of the work he and his colleagues did in developing the aqualung, which enabled sports diving as we know it today (Figure 3.1). By the mid-1960s diving equipment became cheap enough for sports diving to take off and for university diving clubs to thrive, and, in the UK, this was when scientific diving grew spectacularly.

For many of the interviewees, it was experiencing the environment first-hand, often through diving, that led to their careers, and that has continued to motivate and inspire them:

- 'I got involved in marine conservation because I find the ocean amazing with its magical and extraordinary species' (Heather Koldewey)

- 'From the outset, growing up in the suburbs of London, I had a love of looking for wildness and was excited by seeing golden plovers flying over in the winter. That was magical – for me they were emissaries from real, wild nature' (Chris Rose)

- 'I really enjoy being outside *in the environment*, and at university diving opened up a new horizon. I can recall feeling a tremendous sense of discovery and excitement after seeing fields of brittle stars, vibrant cuttlefish and huge conger eels' (Simon Brockington)

CAREER DEVELOPMENT

Decisions and chance

Career paths in marine conservation are seldom linear, and starting out is particularly challenging, with many twists and turns. Chris Rose was the only person interviewed who knew before taking sciences in the sixth form that he wanted to do conservation and, in particular, to attend the University College London MSc course in conservation before going on to work in this field. The realities of deliberate decision making on a career path versus the chance events abound. Simon Brockington was dragged to a lecture on the Antarctic by a friend instead of going to the pub. He was so inspired by the talk that he applied that night to work for the British Antarctic Survey and went there after his degree. A career in medicine thwarted because of fainting at the sight of blood, or as a vet by an 'over-enjoyable' sixth form, or as a priest by a desire to work in biology are some of the events which led interviewees to their career in marine conservation.

Mentors

An almost universal feature of the narratives is the role of what we have come to call 'mentors' who fundamentally influenced the direction interviewees took early in their careers. Every interviewee mentions people who nudged, led and prompted, who went out of their way to help, and who made a huge difference to their work. Many of the interviewees acknowledge the importance of mentors in helping them develop their ideas and programmes: for Chris Rose it was Max Nicholson; for Sue Sayer, Stephen Westcott; and Paul Horsman spent five years travelling the world's oceans on merchant ships on a contract set up by Frank Evans, a lecturer from his university. The interviewees' texts also describe the input of a host of people later in their careers who influenced their thinking. These include people like Daniel Pauly and Bill Ballantine.

The next steps

There are few common threads to the next step of working in marine conservation, but what is clear is that it is not easy. Rather it was often circuitous, taking different opportunities and absorbing different influences. There is a saying that the jobs people at school today will be doing have not yet been invented. This was particularly true for the interviewees. In many cases, neither the jobs nor the organisations they ended up working for existed when they graduated. It was 1975 when the UK got its first full-time marine conservation official in Roger Mitchell, who joined the Nature Conservancy Council (NCC). There are many more jobs today in marine conservation and marine management, in a huge array of organisations. There is also much more mobility between organisations with different styles. Chris Rose was warned off working for an NGO because it would 'ruin his career'. This mobility is important, as Chris Rose and Simon Brockington point out, because with every new job and organisation there is much to be learnt. Alan Knight, from CEO of a scientific instrument maker, and Sue Sayer, from teaching, came to their full-time work on welfare and conservation respectively from professional backgrounds with very different roots.

Particularly in the NGOs, at the heart of their different styles there are very different philosophical beliefs. The personal choice of buying into these beliefs is a major decision. Paul Horsman elaborates on the reality of being prepared to break the law and be arrested for doing what you think is right. Not many conventional job interviews expect candidates to have that degree of commitment.

The support, influence and commitment of partners is a recurring theme in many of the interviewees' narratives; without them their stories might have been very different. Alan Knight's partner supported him becoming a vegan and in their initial volunteering on welfare campaigns. Paul Horsman's partner was at the Greenham Common peace camp in the 1980s, and he worked though his ideas on non-violent protest with her. Sue Sayer decided mid-career to change tack and work on seals, and with the support of her partner left her high-level teaching post. Bud Ehler and Fanny Douvere have been very close partners in their work on the development of marine spatial planning.

Personal qualities

Joan Edwards points out that 'marine conservation is seldom dull and by its nature very varied', and Roger Mitchell found that the great attraction of working in marine conservation in the early days was its variety. This raises questions about the skills and personal qualities needed to work in this area. In one sense work in marine conservation is very much a vocation, and the frequency with which interviewees used the word 'passion' to frame their approach reflects this. Joan summed it up as follows: 'It seems to me that it's a vocation. The people that do it well – that are successful – are passionate about their subject; they want to see the marine environment managed in the right way, not only because it's the right thing to do, but for the benefit of humans as well … People need to be inspired, which is more than just a methodology, about being a marine biologist or knowing the regulations. Passion is important and marine conservation is certainly more than just being able to do an environmental impact assessment.'

High levels of skills are very important, but many other words such as *passion, resilience, leadership, ambition, optimism* and *commitment* also feature in the way one might describe the personal qualities associated with the interviewees. In a business context, commitment is key to success – and it is clear from the interviews that this is also a given in marine conservation work.

At the extremes, marine conservation can be very unpleasant, and a number of interviewees have been threatened with physical violence. Another aspect of the work is that there are plenty of setbacks. Commitment and resilience are therefore clearly very important qualities. Conservation is messy, and even when it goes well it does not necessarily work out how you might expect. Chris Rose describes how the work on the Brent Spar campaign continued for over three years and soon escalated well beyond Greenpeace's control, taking on a life of its own. One of the most interesting and instructive parts of preparing the book has been understanding the realities of how things *actually* happen. Not least of these is the need to show great determination, and if necessary to be cantankerous and bloody-minded, in order to get things done. This sort of approach stands in stark contrast to notions of objective or strategic decision making, or things happening in a logical and linear order.

Disciplines and skills

From my time at the Marine Conservation Society, watching a variety of organisations work and develop, it was very clear to me that there are many ways of protecting the marine environment and that it involves people with very different skill sets. One example was the work done to ban the use of the extremely toxic chemical tributyl tin (TBT), the active ingredient in antifouling paints used for shipping and fish farming. TBT was known to stunt oyster growth and cause sex changes in snails, but a later paper showed that it had decimated the biodiversity of a tidal creek by 50% (Waldock *et al.* 1999). Ministers, government scientists, the statutory nature conservancy agency, the shell fishermen and a number of NGOs worked together to get an effective ban. This illustrates an important point about marine conservation, which is that individuals and organisations working together with a shared goal but with different skill sets can make a significant difference.

All the interviewees are graduates, mainly with a natural science degree, often followed by an MSc or PhD. Three of the interviewees did the UCL conservation course, which was unique at one point although many universities now offer this subject. There are some social scientists too: Bud Ehler trained as an architect, Jon Day as a planner and Sue Sayer as a geographer. Qualifications are clearly very helpful – and given the range of approaches being applied to marine conservation these will increasingly come from many disciplines. But this should not be interpreted to mean that a degree is the only way in to marine conservation. Had it not been for Bernard Eaton – a journalist of the old school – who thought of Underwater Conservation Year (1977), which later led to what became the Marine Conservation Society, my own career would have been fundamentally different.

Figure 2.2 illustrates the ever-increasing number of disciplines that are being used to achieve change for marine conservation. Their output includes consumer guides (for fish or beaches), supply chain measures (Marine Stewardship Council) and ecotourism (for sharks and whales). Whilst science and information underpin these approaches, they all go well beyond the terrestrial tradition of species and habitat protection.

There were a very interesting assortment of views on the question of skills, and whether to specialise or be a generalist. Simon Brockington echoes a theme that runs throughout the narratives: 'Talking to friends and colleagues in other professions, it seems that it is possible to become more senior by becoming more specialised and more expert. In conservation almost the opposite is true. To progress you have to learn about more and more disciplines.' Although I have known some very reclusive people who have made a difference to marine conservation, all the interviewees have exceptionally good interpersonal skills. The ability to communicate, listen and share as well as energise people is very important, as is being able to work with people whose values or background you do not share.

As the timelines show, major new initiatives and campaigns have characterised the development of marine conservation over the last fifty years. Another reality shared by all the interviewees is their involvement in starting a host of initiatives or being ground breakers as the first people to take conservation into resource exploitation meetings. They have had to be self-starters or pioneers in their work. In other settings this would be called innovation, but this terminology is seldom used in the context of conservation. As Joan Edwards puts it, 'For my terrestrial colleagues, many of their activities are now well established, even traditional, whereas in marine conservation, techniques are often viewed as being slightly mad, and there is a constant need for new, innovative and inspirational approaches to take people on the journey.' This innovation can take many forms, and this is very evident, for example, in the response to unplanned events such as pollution incidents.

Sarah Fowler describes the tensions experienced by the first people to represent NGOs' interests at the international tuna meetings (ICCAT), and Heather Koldewey explains the challenges of getting the first fish into the CITES convention. Similarly, Dan Laffoley describes how the 'Blue Carbon' concept emerged from taking the idea of carbon sequestration and applying it to marine habitats.

Perhaps the best example in this book of all of these qualities and transferrable skills is described by Heather Koldewey. Networks of local people were developed to support work on seahorse conservation, marine protected areas and mangrove restoration in the Philippines, and then two massive natural disasters struck, a major earthquake and a typhoon. The project group transferred the skills that they had developed for the conservation work to a disaster relief programme, which proved very successful in this difficult situation (Figure 3.2). This pattern of transferrable skills covers the work of the interviewees, which has taken them all over the world, applying skills acquired at home to wide range of international policy and practical field projects.

Figure 3.2 Marine conservation field staff hand out relief packs to villagers on the island of Panay, Philippines. The packs contain food, clean water, soap and essential medicines. Source: ZSL, Heather Koldewey

CREATING AND BUILDING ORGANISATIONS – CAPACITY BUILDING

Few of the interviewees had any formal training in organisational development or business management, and they had to learn these skills quickly 'on the job'. The point that a degree in natural sciences wasn't exactly the best preparation for developing organisations was made by several interviewees. This is something of an understatement, but nevertheless many of the interviewees have been heavily involved in exactly this task, and the challenge seems unlikely to diminish. Developing the ethos of organisations and developing funding models that are sustainable is a considerable challenge. Joan Edwards was the first person employed in the Wildlife Trusts in a marine conservation capacity. She soon realised that the only way to do better was to get more people – and now there are about forty employed in marine conservation across the Trusts. In the early 1990s Sarah Fowler took on the challenge of starting two organisations simultaneously, the UK Shark Trust and the European Elasmobranch Association.

New organisations will continuously arise, filling particular niches and exploiting different opportunities. Alan Knight describes how British Divers Marine Life Rescue was set up in response to the seal distemper virus episode. Chris Rose describes how Media Natura evolved from understanding how NGOs could use the skills of the advertising sector. Its establishment was prompted by not having enough money to run a media campaign and created the capacity of getting PR professionals involved for a fraction of their ordinary rates. The campaign was such a success that Media Natura built and developed on the back of this idea. Sue Sayer describes how the Cornwall Seal Group Research Trust developed slowly from routine meetings to discuss seal observation work. Today it is a network of over 200 people with charitable trust status. Chris Rose explains how different organisations appeal to different audiences, even in what is apparently a crowded setting. Digital technologies are also leading to a range of new organisations capitalising on skills in this sector.

On other occasions a strategic gap needed to be filled, and organisations developed to meet that need. In the UK, funding bodies like NCC and WWF took very deliberate decisions to support capacity building in the NGOs, and agreed to support the development of the Marine Conservation Society (1980s), and then WWF to support the appointment of marine conservation officers in the individual county trusts (early 1990s). The creation of new NGOs is an ongoing reality, because there is an ever-increasing number of niches and issues to be pursued with innovative ideas and technologies. It has been especially challenging to do this from the start, when there are few precedents for focus on the topic of concern.

It is one thing to build organisations, but just as much of a challenge to keep them developing and on track over a long period. In this regard, Simon Brockington describes how undertaking a Master of Business Administration (MBA) course with the Open University provided him with many of the insights he needed to help manage the IWC. As a small organisation grows, it is likely to reach a critical point that is well described by the term 'founder's syndrome', where you move from needing an enthusiastic generalist to more specialist staff who can help develop the organisation. In this book I have not explored the issues of fundraising or funding models, but if there was one task most of the interviewees could write pages in blood about it would be the joys of preparing funding bids.

The growth in work has been described in many ways. Roger Mitchell points out, for example, that within three years of starting at NCC his work had led to the need to appoint coastal habitats and seabird experts to cover the developing aspects of his original job description. In the early days there were three NCC staff covering marine for the whole of the UK; today there are several hundred marine conservation staff. The success of NGO campaigns has often led to the mainstreaming of ideas and the development of government programmes to take forward their ideas.

WORKING TOGETHER – MEETINGS, PARTNERSHIP AND COLLABORATION

Bringing communities of interest together, whether it be specialist groups, groups of wider interest or international representatives, is an important step and a recurring theme of the narratives. The Underwater Conservation Society meeting in Manchester in 1979 was one of the first large meetings for people with an interest in marine conservation in the UK, and it continued to fulfil this role during the 1980s, a time which saw very considerable developments. Elliott Norse describes two similar meetings in the USA. He also points to the landmark international conference on marine conservation in 1996 and how such events have morphed into the International Marine Conservation Congress which takes place every two years and attracts hundreds of delegates from around the world. The development of international meetings on marine protected areas also shows their role in the development of MPAs (see Timeline 4, page 83). Similarly, Bud Ehler describes the 2006 UNESCO meeting on marine spatial planning as the launch pad for the development of this idea internationally. Similar events are a long tradition for the community of workers on coral reefs, who regularly meet at global meetings.

Collaboration between organisations has long characterised marine conserva- tion, but this is an increasing trend on many different governmental and NGO fronts. Such has been the development of marine conservation that it is now routine for very large groups (hundreds) of stakeholders to meet together to address specific subjects and projects. Simon Brockington describes the work of the International Whaling Commission (Figure 3.3), whose formal and scientific meetings involve more than 400 officials, observers and scientists. Sue Gubbay describes recent work on a European project to describe the biotopes of the North East Atlantic, and this involved over 350 scientists. For these projects and meetings to succeed the organisers need a wide range of skills, and personal qualities including patience, diplomacy and tact.

Figure 3.3 Delegates attending the International Whaling Commission meeting in Portoroz, Slovenia, 2014. Source: IWC

There has been long tradition for the UK's NGOs to work together on issues of a political nature. In 1990 MCS was able to join Wildlife and Countryside Link (Box 2). The Link model was at the front of my mind in my work with a variety of marine stakeholders leading to the formation of the Seabed User and Developer Group (SUDG) and its work on the Marine and Coastal Access Act (2009). The ongoing work of SUDG, which is very supportive of high environmental standards, is described by Peter Barham in Chapter 21. Another example of marine environmental NGOs working together was evident in the work on the North Sea, with European NGOs working in a group called Seas at Risk. Sarah Fowler describes another example at a European scale with the formation of the European Elasmobranch Society.

BOX 2. WILDLIFE AND COUNTRYSIDE LINK

Bob Earll and Joan Edwards
Wildlife and Countryside Link – 'Link' – is a remarkable UK organisation enabling effective and routine collaboration on political issues by more than forty NGOs coming from very different backgrounds, from nature conservation to access and welfare. Link covers a very wide agenda, from marine conservation to air quality.

Wildlife Link, later to become Wildlife and Countryside Link, was formed in 1979 from a previous loose alliance, the Council for Nature. This was timely, since it coincided with a major piece of wildlife legislation – the Wildlife and Countryside Bill – which became an Act in 1981. Link is an organisation that aims to influence the political processes of the day. Unlike conventional 'umbrella organisations', what enables Link to work is its *opt-in* (inverse umbrella) approach. This means that organisations cannot be linked to statements that they haven't explicitly agreed to. This doesn't make life all plain sailing, as there are still robustly held differences of opinion, but it does mean that when there is consensus a powerful array of organisations, representing millions of members, have a powerful voice. The Link Marine Group has played a very significant part in the development of marine conservation in the UK, especially when chaired by Joan Edwards in the 2000s, culminating in the passing of the Marine and Coastal Access Act (2009).

Marine and coastal stakeholder groups

There is another tradition in the UK which has developed since the earliest stakeholder groups formed to develop voluntary marine reserves. In the absence of any effective formal mechanisms for getting marine reserves off the ground these groups sprang up to advance the cause of particular sites. Sue Gubbay describes this in Chapter 7. The idea was subsequently scaled up with support from the Department for Environment, Food and Rural Affairs (Defra) into a multi-sectoral stakeholder forum which worked on the London North Sea Ministerial conference in 1987. This group, known as the Marine Forum, continued to meet until the mid-1990s. The idea of multi-sectoral stakeholder groups was also encouraged by work on the large estuaries and a variety of estuarine partnerships from 1990 onward supported by government agencies. Many of these still operate, although with different funding models, and they provide an effective forum for stakeholders to meet, work on issues and take forward projects. Peter Barham's work with the Solent Forum is an example. This multi-sectoral tradition supporting stronger environmental management and sustainability was also at the core of the idea for the Coastal Futures conference which I started running from 1993 and which will reach its 25th anniversary in 2018.

Large-scale collaborations

A more recent variant on this theme has been the involvement of the large charitable foundations like the Pew Charitable Trusts and the Gulbenkian Foundation to bring NGOs together to collaborate on focused campaigns to change frameworks or regulations. Euan Dunn explains how the Pew Trusts brought NGOs together to work on a more effective reform of the European Common Fisheries Policy in 2015; Sarah Fowler discusses a similar initiative by Pew to help reform the European shark finning regulations; and Heather Koldewey describes working with the Gulbenkian Foundation to investigate how collaboration could best be achieved. 'No egos or logos' – Heather's turn of phrase sums up the challenge of collaboration. One of the main difficulties is that each of the collaborating organisations has its own constituency and fundraising to do and, as Heather puts it, PR departments find it difficult to see beyond their own organisation. The central problem is that the scale of some of the challenges facing us is global. It is only through large-scale collaborations – between NGOs, and increasingly with governments as well – that there is any hope of devising specific actions to effect change.

CONCLUSIONS

This chapter – and the book as a whole – is a testament to the richness and diversity of inputs that people have brought to marine conservation. It also illustrates the importance of a range of personal qualities that have been brought to bear to effect change.

TIMELINE 1

Sustainability and environmental thinking: agreements, meetings and principles

1950s–60s   Systems thinking begins to be applied to concepts of ecology and ecosystems. Standing in contrast to reductionist thinking, systems thinking sets out to view the environment in a holistic manner.

1956   Minamata disease first discovered in Minamata city, Japan. It was caused by the release of industrial waste containing methyl mercury, a highly toxic chemical bioaccumulated in shellfish and fish, eventually killing hundreds of people.

1962   *Silent Spring*, Rachel Carson's book on the effects of DDT, was a landmark in environmental thinking. She also wrote about the marine environment in books like *Under the Sea Wind*, *The Sea Around Us* and *The Edge of the Sea*.

1969   Moon landing. The images of earth from space – the Blue Planet – start to be widely used.

1969   The Santa Barbara oil spill, the result of an oil-rig blowout, was the largest oil spill recorded at that time. It triggered environmental awareness in the USA.

1970s   The Gaia hypothesis. James Lovelock and Lynn Margulis propose that organisms interact with their inorganic surroundings on earth to form a synergistic, self-regulating, complex system that helps to maintain and perpetuate the conditions for life on the planet.

1970s   The Circular Economy. Economists and industrialists develop ideas around resource use and material flows in relation to reuse and recycling.

1970s   The 'polluter pays' principle emerges in response to environmental disasters like the oil spill from the *Amoco Cadiz* (1978).

1970   Earth Day. In response to the Santa Barbara oil spill, 20 million Americans take to the streets to demand a healthy environment.

1972   *The Limits to Growth*, a report on the results of computer simulations of growth and finite resources.

1972   Stockholm Conference: United Nations Conference on the Human Environment. The UN's first major conference on international environmental issues, and a turning point in the development of international environmental politics.

1972   UNEP, the United Nations Environment Programme, founded.

1980   *The World Conservation Strategy* prepared by the International Union for Conservation of Nature and Natural Resources (IUCN). The first international document on living resource conservation produced with inputs from governments, NGOs and other experts.

1985   Sinking of the Greenpeace ship *Rainbow Warrior* in Auckland harbour, New Zealand, results in worldwide coverage and awareness of environmentalism.

1987   *Our Common Future*, also known as the Brundtland Report, from the United Nations World Commission on Environment and Development (WCED) sets out the ideas for sustainable development.

1987   The precautionary principle is applied to pollution at the North Sea Ministerial Meeting.

1992   The Rio Earth Summit produced a number of significant conventions including the United Nations Framework Convention on Climate Change (UNFCCC) and the Convention on Biological Diversity (CBD).

1995   The ecosystem approach and its twelve principles developed by the CBD.

1998   Aarhus Convention on access to information, public participation in decision making.

2003   One Planet Living and its ten principles show that a simple way for us to plan, deliver and communicate sustainable development and a green, circular economy is possible.

2009   Planetary Boundaries. The Stockholm Resilience Centre proposes a framework of 'planetary boundaries' designed to define a 'safe operating space for humanity'.

2012   The Rio + 20 Earth Summit agrees the sustainable development goals.

2012   The first *World Happiness Report*, from the UN High Level Meeting: Well-being and Happiness: Defining a New Economic Paradigm.

2015   Paris Climate Change Conference produces a series of more constructive and optimistic ways to address climate change.

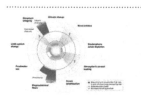

PARIS2015
UN CLIMATE CHANGE CONFERENCE
COP21·CMP11

CHAPTER 4

Ideas

I have been fortunate to live through a period when there has been a blooming of ideas about how we should manage our activities in a more environmentally friendly and sustainable way. If there was a moment that signified the point that this awareness took hold, it was the day images of the earth from space showed us our place (Figure 4.1). For those with an interest in protecting the seas it was also the time when we could visualise the blue planet and how the earth is dominated by the oceans.

The gestation from ideas to action is a recurring theme, and covered in Chapter 2, often with many years of work being put in before ideas make it onto a timeline as a published convention, law or academic publication. This is well illustrated by Callum Roberts' book *The Unnatural History of the Sea* (2007), and Sue Wells and her colleagues have demonstrated how the ideas around marine protected areas (MPAs) developed over the nineteenth and early twentieth centuries (Wells *et al.* 2016).

A higher level of concern for the environment was developing in the 1950s, but it was in the 1960s that this really gained momentum; this is illustrated by Timeline 1 (page 34), which looks at events surrounding the development of thinking on the environment and sustainability. It was Rachel Carson who led the way, with her books on the marine

Figure 4.1 The earth from space. Source NASA

environment in the 1950s and then her game-changing book *Silent Spring* (1962). For the marine environment it was from the 1960s that a whole host of initiatives started to be put forward, to be realised from the 1970s onwards.

The timelines developed for this book show how particular ideas have developed. In considering these, and the narratives of the interviewees, there seemed to be six reasonably coherent sets of ideas driving action; these included ideas taken from:

1. Marine and terrestrial biodiversity conservation

2. Science

3. Environmental management

4. Sustainability and its principles

5. Other cultural inputs, e.g. welfare

6. Events and issues, planned and unplanned

It is worth remembering here that I am deliberately using a very broad view of marine conservation. This is important, because protection of the marine environment and effective management of human activities there have always relied on much more than just a strict definition of biodiversity conservation. Many ideas have helped us frame and manage our activities in ways that benefit species and habitat conservation, not least by providing effective management frameworks for those who want use the marine environment responsibly.

MARINE AND TERRESTRIAL BIODIVERSITY CONSERVATION

The considerable shared thinking between the terrestrial and marine conservation communities is described in Chapter 2. Concepts like protected areas, controlling the trade in endangered species, protecting migrating species and Red Listing groups of species and habitats have permeated their way into marine conservation. The pattern of time lags and the difficulties of this transition process have also been described in Chapter 2.

Useful ideas taken from terrestrial conservation have been adapted for use in the marine environment, but there are often significant differences in translation and application. Within protected areas, for example, whilst the concept is equally applicable, the folly of trying to adopt the detailed terrestrial features-based approach is apparent when no-take zones and recovery strategies are being demonstrated to be much more appropriate to the marine environment (Callum Roberts). In the same vein, the poor state of knowledge of many mobile marine species of fish, birds, reptiles and mammals, and the difficulty of obtaining sufficient good-quality data to enable Red Listing for many species and habitats, point to the fundamental difference between working in terrestrial and marine environments. Not least of these problems is that deferring protection and management until the data issues are resolved is a recipe for doing nothing, and can be resolved by adopting the precautionary principle (see below). In some areas marine conservation has always been significantly ahead of terrestrial conservation, notably in the protection of large sea areas.

SCIENCE

Of all the disciplines used to deliver marine conservation in its first fifty years, science has probably played the greatest role in helping to frame and describe the issues. It is often said that we know more about the moon than our own oceans, but over this fifty-year period the observations of scientists have given us fascinating and growing insights into species, habitats and ecosystems in this extraordinary and inspiring environment. Scientific observations on the major human impacts of fisheries, pollution and more recently climate change have played key roles in the development of awareness of the need for action. One relatively recent example is the discovery of the huge levels of plastic debris in all the world's major ocean gyres (5 Gyres Institute: www.5gyres.org).

Science as a process has also been crucial to the way we try to describe and understand the marine environment, and emerging concepts such as historical ecology, shifting baselines and fishing down the food chain have changed the way we think about the effects of our activities on the marine environment. Whilst science as a process is crucial to the way we describe and understand the marine environment, many of the interviewees stress that science on its own does not necessarily lead to the changes that its results often call for.

ENVIRONMENTAL MANAGEMENT

Starting in the 1980s, systems and business management thinking inspired the development of environmental management and its application to every sector of human activity in the marine environment – with the notable exception of fishing. In Chapter 21, Peter Barham outlines many challenges that arose when incorporating 'the environment' into practical operations such as flood defences and port development. Environmental management has developed and matured into a mainstream activity, so that we can now look at the life cycle of developments and the environmental management techniques to be used at particular stages of the development of areas of sea for different activities such as wind farms or aggregate dredging. These stages include environmental impact assessments, and environmental management systems during the operational life through to decommissioning. Box 3 describes and illustrates these stages and management techniques for offshore oil, but they are equally applicable to many different types of development (Figure 4.2).

BOX 3. THE LIFE-CYCLE APPROACH TO ENVIRONMENTAL MANAGEMENT TECHNIQUES

John Hartley
Multiple and often independent strands led to the application of systems thinking to environmental management. The use of environmental impact assessment (EIA) by leading companies for the development of major North Sea oilfields in the mid-1970s was followed by the 1985 European EIA Directive (85/337/EEC, now superseded by Directive 2014/52/EU) requiring EIA for projects above set thresholds. The evolution of environmental management systems (EMS) has mirrored the increasingly international focus on environmental issues and the accompanying recognition of the benefits of a systematic consideration of the environment in routine business management. The innovative UK

Specification for Environmental Management Systems (BS 7750:1992 & 1994) was developed into and superseded by the international standard on *Environmental Management Systems* (ISO14001:1996, 2004 & 2015).

During this time the idea of applying environmental (and safety and health) management across the entire development process – from cradle to grave – gained momentum. In the late 1980s and early 1990s when I was working for BP and Amoco the idea for this graphic (Figure 4.2) emerged to illustrate the environmental management techniques applicable across the whole life cycle of an asset, from acquisition to decommissioning. These techniques allow the identification and effective mitigation of risks to the environmental and thus to a business.

All operators of UK offshore blocks licensed for hydrocarbon exploration and production are required to have an independently verified EMS in place in line with the UK offshore energy regulator guidance and OSPAR Recommendation 2003/5 to '*Promote the Use and Implementation of Environmental Management Systems by the Offshore Industry*'. The recommendation was focused on the oil and gas industry and preceded the recent major development of offshore renewable energy. It is, however, considered relevant to all operators of offshore installations (including wind turbines and other renewable energy devices) in the OSPAR area. The EMS frameworks outlined in the standards, enhanced through successive versions, provide valuable templates for companies to follow. The EMS and the constituent tools and processes apply variously throughout the business life cycle, as illustrated in the graphic. The example given relates to offshore energy developments, but the approach and techniques apply equally to any marine or terrestrial industry and also have relevance to conservation project management.

The mode of implementation of an EMS reflects the organisation's ethos. If their approach to the environment is internally driven they will be proactive and active participants in adaptive management initiatives and external discussions (leaders). They typically have a staged (gated) decision-making process for new projects, with each stage supported by documentation (including environmental) with detail appropriate to the level of project definition. Conversely, externally driven organisations are reactive, focused on legal compliance and what regulators require of them (followers). EMS implementation can be facilitated by tools such as a Permits, Licences, Authorisations, Notifications and Consents (PLANC) register to organise and track regulatory compliance requirements. Activity–pressures matrices (see also Jon Day) applied in a considered rather than a formulaic way can assist the identification of environmental aspects of activities which may give rise to significant environmental effects or opportunities.

Given the damaging effects that development has had over the years, this developing body of thinking is providing a way of mitigating the worst effects of development. More formally, many of the elements of environmental management are now incorporated into international ISO standards.

For business, environmental management thinking also reflects how industry approaches its activities. Does a company want to be on the front foot, being proactive

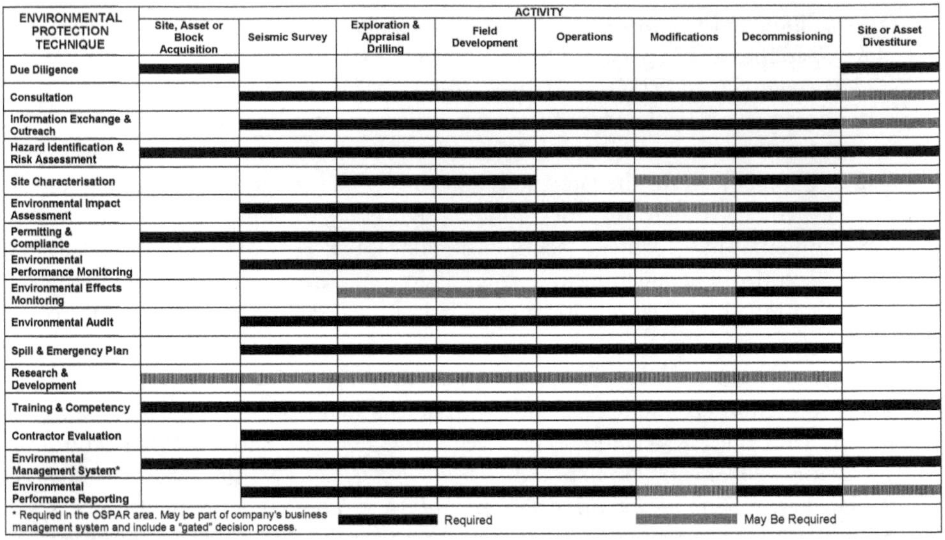

ENVIRONMENTAL PROTECTION TECHNIQUE	ACTIVITY							
	Site, Asset or Block Acquisition	Seismic Survey	Exploration & Appraisal Drilling	Field Development	Operations	Modifications	Decommissioning	Site or Asset Divestiture
Due Diligence								
Consultation								
Information Exchange & Outreach								
Hazard Identification & Risk Assessment								
Site Characterisation								
Environmental Impact Assessment								
Permitting & Compliance								
Environmental Performance Monitoring								
Environmental Effects Monitoring								
Environmental Audit								
Spill & Emergency Plan								
Research & Development								
Training & Competency								
Contractor Evaluation								
Environmental Management System*								
Environmental Performance Reporting								

* Required in the OSPAR area. May be part of company's business management system and include a "gated" decision process.

Required   May Be Required

Figure 4.2 The life cycle of environmental management techniques for the offshore energy industry. Source: John Hartley

and in control, or reacting to events? Serious and expensive disasters litter the history of marine developments, the most costly of which, at $40 billion and rising, was the failure of the Deepwater Horizon oil well in the Gulf of Mexico in 2010, which almost led to the demise of one of the world's leading oil companies. The message from these events is very simple. Doing the right thing might be expensive, but just look at the costs when things go wrong. Environmental systems and thinking are one of the tools that can be used to avoid these mistakes.

The concept of the circular economy also derives from systems thinking and is similar to the life-cycle or 'cradle to grave' idea. It is usually applied to manufacturing and product design so that end-of-life issues are incorporated into the product design at the outset. This can help solve many issues like enhancing resource use, facilitating recycling and avoiding pollution. The concept of the circular economy is now central to thinking about solving the issue of plastic pollution.

Adaptive management is another idea which has come to the fore in recent decades and which is normal practice for many sectors (Figure 4.3). It involves commitments, policies and plans set out in a programme of work. These are then carried out, reviewed, monitored and adapted, following which the cycle starts again to achieve more successful outcomes. The adaptive planning and management approach also enables a start to be made on projects. Many of the interviewees highlight the need to 'start' projects, because waiting for more information or better circumstances is not helpful. This iterative process is described by Jon Day in relation to the Great Barrier Reef Marine Park.

Another manifestation of this type of thinking has developed from the 'logical frameworks' (Logframes) approach that is now widely used in conservation project planning to systematically work through the developmental stages that projects should pass through so that progress and success can be audited. This approach dates back to the 1970s but a recent article highlights current thinking (Larsson 2015), and a report by Bill Jackson (1997) provides a useful reference from an IUCN perspective.

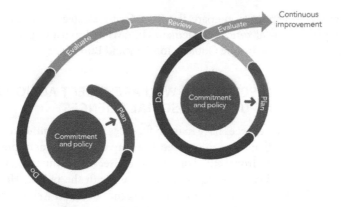

Figure 4.3 Adaptive management. Source: Bob Earll

SUSTAINABILITY – PRINCIPLES AND PACKAGES

The development of ideas about sustainable development crystallised around the Brundtland Report in 1987. Its main pillars or dimensions, including environmental, social and economic, continue to run through our thinking. The first Earth Summit in Rio (1992) saw these ideas take hold, and they are still manifest today in the sustainable development goals that emerged from the Rio +20 Earth Summit process in 2012 (see Timeline 1, page 34). The ideas around sustainability, including Jonathan Porritt's work on capitals (Porritt 2005), are embedded in a wide variety of assessment and accounting processes across an astonishing array of societal, business and environmental interests. They continue to have considerable resonance, not least because of the need to plan for the future and to take environmental and societal as well as economic considerations into account.

Alongside sustainability, a number of high-level *principles* have evolved over this period. These are powerful generic ideas that can be applied to all the sustainability dimensions and sectors. They include ideas such as integration, holism, taking a long-term view, equity, the 'polluter pays' principle, the precautionary principle (Earll 1992) and the concept of participation as expressed by the Aarhus Convention of 1998. They have had a profound effect on a wide range of policy and programmes including the way we approach protecting the marine environment. A good example is the way the precautionary principle was applied to the campaigns against pollution of the North Sea, as described by Paul Horsman, and subsequently it has been used in fisheries management and the application of the Habitats Directive to protect marine sites from fishing impacts. The precautionary principle is particularly helpful when there are problems of inadequate data (data-deficient situations), as it promotes a responsibility to act to prevent further loss of species or degradation of the environment.

These same principles have often been combined in *packages* such as integrated coastal management, marine spatial planning, the ecosystem approach and corporate social responsibility. The strength of these packages of principles is that whilst they set out broadly the same ideas they have resonance for different audiences or geographic settings. More recently, packages framed in terms of 'resilience' and 'health and wellbeing' routinely embody many of these same principles in a way that resonate with new audiences. Of these packages, the ecosystem approach and its application to North Sea fisheries is discussed by Euan Dunn, and Dan Laffoley describes how important it proved to be when helping to prepare the EU Marine Strategy Framework Directive (MSFD). Integrated coastal zone management (ICZM) and marine spatial planning are described later in the book by Sue Gubbay, Jon Day and Bud Ehler.

OTHER CULTURAL DISCIPLINES – WELFARE, DIRECT ACTION AND THE INPUT OF THE SOCIAL SCIENCES

The narratives of the interviewees have many references to other significant and well-reasoned lines of thought that have influenced the way different practitioners approach protecting the marine environment. For nature conservation purists, animal welfare has had little relevance despite being highly resonant with the public. However, as the interview with Alan Knight shows, there is convergence between welfare and conservation objectives. He describes them as follows: 'My beliefs are based absolutely around animal welfare, the idea that every animal counts and we should rescue individuals and stop their suffering.' Sue Sayer puts it like this: 'My views on welfare have developed in part because I routinely work with identifying individual seals using photo ID, and so for me the individual animal matters!' For Simon Brockington, 'a definition of conservation based on capacity for human enjoyment of environmental benefits necessarily includes concern for welfare.'

As populations of some marine mammals dwindle, for example the small porpoise, the vaquita, in the Gulf of California where there are only thirty animals left, the rescue and handling expertise of the marine animal welfare community comes increasingly into play when trying to implement rescue or relocation plans. Interestingly, the International Whaling Commission routinely considers whale welfare issues. Whilst the death of a whale by a harpoon or ship strike can be instantaneous, the issues of whale and dolphin entanglement has prompted IWC to build and support work on rescue from entanglement carried out by the welfare community. For many cetaceans 'every one does count (Alan Knight), 'which is why the US government has funded rescue programmes for many years.

In achieving change to the status quo, direct action has delivered some notable gains for organisations like Greenpeace and the marine conservation cause. Greenpeace has been influenced by other philosophical traditions, not least the Quaker movement and the approach espoused by Gandhi, which have shaped their views and the way they work. Paul Horsman describes his position on non-violent direct action as a legitimate way of registering protest, and how Greenpeace has used this approach. His description of his experience of direct action goes to the core of his beliefs and illustrates a much deeper connection to all the people involved with protests from the environmentalists to the police.

The influences of a host of other schools of thought abound throughout the interviews. The American space programme's use of systems thinking inspired Bud Ehler, and he describes how his work in environmental policy making was heavily influenced by this: 'Systems thinking is a holistic approach to analysis that focuses on the way that a system's constituent parts interrelate and how systems work over time and within the context of larger systems.' Jon Day trained and worked as a planner, which was of direct importance to his work on the Great Barrier Reef zoning, but another of his skills was experience of stakeholder engagement.

What is also remarkable in the interviewees' narratives is the input from social scientists which has brokered new ideas or ways of thinking about existing problems.

Heather Koldewey describes the development of the idea of ocean optimism: 'I went to a talk by Ellen Kelsey, who has worked extensively on the psychology of communicating science and the narrative of doom and gloom. This is a phenomenon well known in the behavioural sciences which recognises that a constantly depressing narrative is a turn-off and that we need to redress the balance.' Heather describes how this led to other meetings, and to the #OceanOptimism hashtag going viral on its launch in 2014.

Simon Brockington describes a similar critical intervention: 'In 2008 Calestous Juma, a social scientist and professor at Harvard University, came to the IWC's annual meeting. He identified four basic interests amongst the contracting governments and his insights led to important breakthroughs in the way IWC went about its work.' Chris Rose covers the importance of communication and campaigning, drawing on commercial and social thinkers in those fields whose methods have provided him with invaluable insights into the way to mount effective campaigns. These inputs are far removed from traditional nature conservation thinking.

EVENTS AND ISSUES

Events have also played a part in galvanising public awareness as to the need for action, and some stand out as landmarks in the development of environmental thinking (see Timeline 2, page 45). The moon landing programme and the resulting images of the earth from space were arguably one of the most significant spurs to environmental awareness. Then the work around the original Rio Earth Summit in 1992 led to a number of important conventions, including the Convention on Biological Diversity (CBD) of 1993. The impetus behind the North Sea conferences (see Box 6, page 170) provided the incentive for the more staid conventions (e.g. OSPAR) and institutions (e.g. ICES) and agreements to change the thinking behind their work and the processes they used to meet the needs of the twenty-first century. Unplanned events such as major oil spills (Box 4), Brent Spar or the phocine distemper outbreak in 1988 have also been landmarks for the development of ideas, and have led directly to measures to protect the marine environment (see Chapter 24). The massive public awareness such events lead to can provide major opportunities to change the way we think about the sea and how we use it.

BOX 4. OIL SPILLS – EVENTS THAT GALVANISED THE ENVIRONMENTAL MOVEMENT

Bob Earll, Bud Ehler and Roger Mitchell
Bad events can sometimes lead to good outcomes. In this respect major oil spills from tankers and offshore oil and gas installations have had a disproportionately huge effect, galvanising successive generations of the marine environmental movement and legislators in their work to create stronger protection for the marine environment. Their immediacy and very visual impacts, including images of dead wildlife, contribute to their impact.

In the UK the *Torrey Canyon*, wrecked off Cornwall in 1967, caused a massive spillage of oil and damage to wildlife and the local economy but at the same time stimulated a growing awareness of marine pollution. The forlorn attempts to try and control the spill played out on the TV news night after night. Significantly, it was discovered that the chemical dispersants used were more toxic than the oil itself.

The Santa Barbara oil spill off the California coast, arising from an offshore oil well in 1969, is cited by Bud Ehler and Elliott Norse as being key to mobilising the environmental movement in the USA and lent weight to the Earth Day programme in 1970 that drove millions of demonstrators onto the streets across the nation. Its effects on wildlife, and probably more importantly high-value Californian beach properties and tourism, gave rise to a raft of environmental legislation.

The Ekofisk Bravo platform blowout in the North Sea (1977) and then the grounding of the *Amoco Cadiz* off Brittany (1978) led the oil industry in the UK to get the message that something needed to be done about a whole range of oil pollution incidents, not least because of the reputational damage. Roger Mitchell describes the oil companies' response in his interview. Bud Ehler describes how a team from NOAA led work on estimating the environmental costs – and, for the first time, the social costs – of the *Amoco Cadiz* spill, which provided further grist to the mill of the polluter pays principle. People started to understand the full range of impacts, and the differences between predicted and actual impacts.

The *Exxon Valdez* tanker oil spill in Alaska (1989) was another massive event. Bud Ehler describes how it set the scene for a completely revised approach to damage assessment, escalating compensation requirements and renewed waves of tighter legislation. The processes used to clean the beaches were subsequently shown to be highly damaging.

It was the background research on a raft of smaller oil pollution incidents (such as the *Braer*, 1993) and contacts in the oil industry that enabled the leaked documents on the Brent Spar (1995) to be sent to Paul Horsman at Greenpeace. The subsequent Greenpeace campaign in 1995 fundamentally changed the rules on dumping in the sea through revisions to the London Dumping Convention (1996).

Fast forward to 2010. The Deepwater Horizon oil-well blowout in the Gulf of Mexico became the largest oil pollution event, killing eleven people and causing widespread pollution and damage to wildlife on an unprecedented spatial scale. Costs in excess of $40 billion almost brought the oil giant BP to the point of financial ruin. Once again a major oil spill had roused the environmental awareness of another generation and caused the oil industry to take stock.

CONCLUSION

These examples illustrate that the way we try and achieve conservation is informed by ideas from many influences and disciplines, including those which might not normally figure in a purely scientific consideration of biodiversity conservation. Perhaps the most remarkable thing, given the importance that marine conservationists have to place on working with people, is the relative lack of input from social scientists, who are trained in these very disciplines.

TIMELINE 2

International marine conservation: legal developments, ideas and events

1946	The International Convention for the Regulation of Whaling (ICRW) and the foundation of the International Whaling Commission (IWC).
1960s–1970s	'Save the Whales' campaigns. During this period awareness of the plight of the great whales was raised and then taken further by the anti-whaling campaigns and NGOs, including the newly formed Greenpeace, who undertook direct actions at sea.
1963	Convention on International Trade in Endangered Species (CITES). It is a multilateral treaty to protect endangered plants and animals.
1964	US Wilderness Act.
1969	Establishment of the International Commission for the Conservation of Atlantic Tunas (ICCAT), an intergovernmental organisation responsible for the management and conservation of tuna and tuna-like species in the Atlantic Ocean and adjacent seas.
1971	The Ramsar Convention on Wetlands of International Importance especially as Waterfowl Habitat. An international treaty for the conservation and sustainable use of wetlands including shallow waters.
1971	New Zealand Marine Reserves Act, 'to provide for the setting up and management of areas of the sea and foreshore as marine reserves for the purpose of preserving them in their natural state as the habitat of marine life for scientific study'.
1972	US Coastal Zone Management Act, to encourage coastal states to develop and implement coastal zone management plans.
1972	US Marine Mammal Protection Act. The first legislation to call specifically for an ecosystem approach to natural resource management and conservation.
1972	World Heritage Convention. The programme began with the Convention Concerning the Protection of the World's Cultural and Natural Heritage, which was adopted by the General Conference of UNESCO on 16 November 1972. By 2017 there were 49 marine World Heritage sites.
1974	UNEP Regional Seas programme launched the regional approach to controlling marine pollution. Now 143 countries participate in 18 Regional Seas Conventions and Action Plans.
1975	The Great Barrier Reef Marine Park Act.
1978	*Amoco Cadiz*: the largest major oil spill to date. For the first time, the social and environmental costs of a major spill are estimated at a time when the polluter pays principle is gaining traction.
1979	Bonn Convention on Migratory Species, commonly abbreviated to Convention on Migratory Species (CMS), aims to conserve terrestrial, marine and avian migratory species throughout their range.
1980s	Biodiversity. The term biodiversity comes into widespread use, later to be incorporated into the Earth Summit text and the Convention on Biological Diversity.
1980	Convention for the Conservation of Antarctic Marine Living Resources (CCAMLR), held in high regard for taking an ecosystem approach
1981–1990	CITES corals listed. Black corals listed on Appendix II in 1981, 17 genera of hard corals listed on Appendix II in 1985, remaining genera in 1990.

1982 United Nations Law of the Sea Conference (UNCLOS), and the Law of the Sea treaty, is the international agreement that defines the rights and responsibilities of nations with respect to their use of the world's oceans, establishing guidelines for businesses, the environment, and the management of marine natural resources.

1982 IWC introduces a 'moratorium' on commercial whaling from the 1985–1986 season, due to uncertainty over whale numbers.

1986 Moratorium on whaling

1989 *Exxon Valdez* oil spill in Alaska. Another major oil spill that had international ramifications on compensation and oil-spill recovery thinking.

1990s Fish Guides for Consumers. Many national and regional variations on this idea, and digital applications are now widely available.

1992 Convention on Biological Diversity – from the Rio Earth Summit.

1992 *Global Marine Biological Diversity: a Strategy for Building Conservation into Decision Making.* Elliott Norse's book, prepared for the Earth Summit.

1993 IPCC Coastal Zones and Small Islands Working Group second assessment.

1995 UNEP Global Action Plan for protection of the marine environment from land-based activities.

1995 Shifting baselines. A paper by Daniel Pauly highlights a key idea in the way we perceive information on the status of ecosystems and species.

1995 Marine Stewardship Council. Mike Sutton (WWF) and Simon Bryceson (PR consultant for Unilever) pull the key ideas together. This is now a global force influencing sustainable fisheries.

1995 Historical ecology: Jeremy Jackson's talk on 'Reefs since Columbus' – which later inspired Callum Roberts to write *The Unnatural History of the Sea* (2007).

1998 Fishing down the food web. Daniel Pauly and others describes the changes fishing is causing in the food web.

2000s Ocean acidification: awareness and publications begin in the early 2000s.

2004 CITES listing for seahorses: the first marine fish to be listed.

2006 First International Workshop meeting on Marine Spatial Planning – Visions for a Sea Change.

2009 Blue carbon, the carbon captured by the world's oceans and coastal ecosystems. Key report prepared for IUCN by Dan Laffoley and Gabriel Grimsditch (2009).

2010 Deepwater Horizon oil blowout, Gulf of Mexico. The sheer scale and cost ($40 billion and rising), almost resulting in the collapse of BP, sends a massive shockwave through the oil and gas sector.

2010 Aichi Biodiversity Targets: governments have committed to establishing national targets in support of the Aichi Goals for MPA coverage.

2012–2014 Ocean Optimism. Ellen Kelsey, Heather Koldewey and Nancy Knowlton launch the #OceanOptimism hashtag on World Oceans Day 2014.

2016 The Global Ocean Commission was an international initiative between 2013 and 2016 to raise awareness of the degradation of the ocean and promote action to help restore it to full health and productivity.

2016 IUCN ocean warming report. The report, *Explaining Ocean Warming: Causes, Scale, Effects and Consequences*, reviews the effects of ocean warming on species, ecosystems and the benefits oceans provide to humans.

2017 UN agrees a non-binding resolution to tackle plastics in the ocean, one of many agreements and collaborations in 2017.

CHAPTER 5

Roger Mitchell

Roger Mitchell (left) on diving fieldwork in the Outer Hebrides

'Portfolio career' is a popular phrase these days, but it doesn't do justice to the daunting diversity of conservation interests and work with which Roger has been involved, across terrestial and aquatic habitats. His commitment to nature conservation is deeply ingrained and now extends well beyond the marine environment. Since leaving English Nature, he has been International Chief Scientist of Earthwatch, worked for a philanthropist managing large estates and an environmental grants fund, worked to set up conservation projects in Estonia, Poland and Romania, chaired the Cambridge Conservation Forum (CCF), been a board member of the Cambridge Conservation Initiative (CCI), and co-chief editor of *Conservation Evidence*. Currently he's still involved with CCF and CCI, helps with the MPhil in conservation leadership at the University of Cambridge, is a board member of the Natural Cambridgeshire (the Local Nature Partnership), is Treasurer Trustee of the Amphibian and Reptile Conservation Trust, and carries out a great deal of pro bono and consultancy work with statutory bodies and conservation groups in Cambridgeshire and the East Anglian fens and river valleys.

His early influencers were his father, a financial advisor who was generally interested in nature and took him for long country walks, and his cousin John Mitchell, a professional musician who had an interest in butterflies and plants, eventually worked in conservation management and wrote the Collins New Naturalist volume on *Loch Lomondside*. Although living inland in Northamptonshire, many holiday hours were spent rockpooling under Hunstanton's magnificently coloured cliffs, snorkelling off the coast in south Devon and a first experience of scuba diving in Guernsey. It was in Brixham in 1958 that he saw and bought the Collins *Guide to the Seashore*, which he still has, and which introduced him to the animals and seaweeds, sparking his interest even more. In school and through A-levels his career path was clearly set on medicine. His father chaired the local hospital management board and took the medical journal *The Lancet*, and arranged temporary work for him in the hospital. However, the upsetting experience of watching operations and helping with post-mortems brought the realisation that he probably wasn't fitted to practising medicine. A gap year teaching, some travel and work helping a local National Nature Reserve warden – Ray Collier – and involvement in National Nature Week enhanced his interest in natural history and pointed him in the direction of London University and a degree combining zoology and botany with a specialisation in marine biology.

An MSc in nature conservation at University College London was the next step towards consolidating his interest in conservation. Next to Southampton University to do a PhD in the Oceanography Department. This was also where he trained to commercial diver level by an arrangement with the Royal Engineers Diving School at Marchwood, and he also founded the Southampton University Underwater Research Group, a forerunner of Seasearch, with four sets of diving gear he persuaded his very supportive and perceptive supervisor, Professor Raymont, to purchase from departmental funds. In these formative years, and alongside his PhD studies, he also pursued a business involvement in shell fishing and farming along the south coast, which introduced him the not always sustainable or legal practices of inshore fishermen. Then a very different tack, successfully applying for a Civil Service Commission and getting a post in the Department of the Environment which included investigating the potential environmental impacts of building freshwater reservoirs in estuaries, the seawater cooling of coastal nuclear power stations, desalination plant effluents, managing freshwater flows to estuaries, and coastal reclamation. Then in 1975 a move to the Nature Conservancy Council (NCC) to become the first marine officer in the Chief Scientist Team, a step that prompted many of the initiatives and projects described in this chapter and that have become the routine work of many of the people working in marine conservation in the UK to this day.

This chapter covers the very early days of marine conservation in the UK, when Roger played a big part in influencing its development. His work also involved mitigating the effects of oil pollution, wave-energy generation and tidal barrages, monitoring seabirds at sea, non-native species, marine fish farming, basking shark conservation and many other areas including citizen science. The concepts around marine protected areas in the UK began in the early 1970s, and Roger put the flesh on the bones of this thinking and the science required to support it. The family tree of marine conservation in the UK begins with many of the initiatives he started and the people he employed.

MARINE CONSERVATION

How do you describe or frame your beliefs in marine conservation?

These stemmed from my early love of the coast and sea and my passion about their conservation. Later, seeing clearly the increasing impacts of damaging policies and activities, I felt driven to play my part in influencing behaviour and promoting interventions to mitigate or avoid damage being done to the marine environment. Some say that, like religion, 'conservation' is a calling – well, it certainly called to me!

What are the differences between when you started and now?

When I started work at NCC in 1975, marine conservation was 'out of sight and out of mind'. In any case, it was overshadowed by the perceived more evident and urgent conservation need on land, which influenced the priorities for the allocation of effort and legislation. However, a growing number of marine issues, including the expansion of North Sea oil and gas and the collateral damage from oil pipelines, offshore structures and discharges, refining and transport, tributyl tin (TBT) in antifouling paint (which even MAFF declared was the most toxic compound ever introduced into the sea by man), the appearance of the non-native invasive seaweed *Sargassum muticum*, commercial bait digging, plans for massive tidal energy barrages and the impact of fisheries increasingly inserted 'marine' into government and NCC thinking. This was considerably aided by the NGOs who, ever the constructive irritant to government, did a very good job extending their interests into the sea and providing pressure on many issues.

How is marine conservation different from terrestrial conservation?
On land, conservation action to restore habitats and species is usually achieved by physical interventions, and management is also often aimed at maintaining that habitat in its seral successional state such as by the grazing or cutting of grassland, coppicing woodland, felling trees and clearing scrub on heathland or re-wetting of fens and bogs. Apart from some coastal habitats, like salt marshes, saline lagoons and oyster beds, the physical management of marine habitats is not so easy or so necessary; it's more a case of controlling any damaging activities. Now with landscape-scale restoration and connectivity rising up the terrestrial conservation action agenda, there is a recognition that management interventions are better directed, if possible, at enabling sustainable natural processes, which are often associated with 'rewilding'. In the sea, the natural processes of tides and currents are operating at a huge scale in the connected medium of the sea and the seabed substratum. Here, the successional states from past management practices are not really a consideration, and management and manipulations are not practical. I have often thought that the coastal nearshore marine environment with its rocky reefs and sediment is a 'last wilderness', because natural processes are to the fore and human impacts from fishing, etc., relatively slight. For marine conservation to succeed, relatively far more effort is required on advocacy, policy and mitigating harmful effects.

What was it like at the start? Early days at NCC – Chief Scientist Team
In NCC's Belgrave Square, London headquarters, my office neighbour was Deryk Frazer, who later wrote the Collins New Naturalist volume *Reptiles and Amphibians in Britain* and kept some natterjack toad pets in a pond in the garden behind the office. There were many others of a naturalist bent too in the NCC. I became the fifth member of the Chief Scientist Team (CST) led by Derek Ratcliffe, who gave me my induction brief. He ran through a great many areas on which work was required to set up projects and policies, including seabirds, shore birds, salt marshes, sand dunes and oil pollution, but I don't remember fisheries being mentioned – so I had a very broad remit from the outset.

As Derek was a hands-off manager and often out of the office and in the field, I had a pretty free hand in setting my own priorities in the job, which excited and empowered me. But I was on my own, as virtually everyone else in the organisation had a terrestrial focus. Nevertheless, in working alongside my terrestrially focused colleagues, I appreciated how they used survey and monitoring information in conservation site selection criteria and for habitat and species management. Since the Nature Conservancy had been recently split into the NCC and the Natural Environment Research Council (NERC), the NCC was required to operate a customer–contractor relationship with NERC for a few years – which meant, in the case of marine, that we had to contract some rather expensive survey work from the Marine Biological Association (MBA) and the Scottish Marine Biological Association (SMBA) to start work on a review of the intertidal habitats and species of Great Britain. Eventually we regained control over directing our research funds and were able to target funding on a wider range of pressing issues. I became increasingly overloaded as I commissioned and managed more and more projects, but as the CST expanded, others were recruited to take over the coastal–terrestrial and ornithological aspects of my role, and I was also able to start to recruit marine assistants.

One of my responsibilities within NCC was oil pollution, which was killing a lot of seabirds as well as causing damage to soft and hard shore habitats and species and the recreational amenity of shores. It consequently had a very high public profile, which helped to further marine conservation. When NCC first started to monitor the activities of the oil and gas industry, I joined a study tour arranged by British Petroleum for NCC senior staff and council members including David Attenborough around some North Sea offshore installations, the Shetland oil terminal and an oil refinery in the Firth of Forth. NCC continued to work closely with the oil industry, who were generally reasonably open and frank with us, and I helped to organise, and was the assessor for, the first International Conference on Wildlife and Oil Pollution in the North Sea. Then the largest blowout in the North Sea occurred from the Ekofisk Bravo platform (1977) and the very large oil tanker *Amoco Cadiz* lost steerage and drifted onto rocks in Brittany (1978) and spilled its entire cargo of crude oil as well as its fuel oil. These emergencies involved me in oil spill contingency arrangements with the industry and government officials in Britain, Norway and France. The source of other, seemingly frequent, oil incidents were various and included spills from refineries, shipping, rigs as well as the deliberate washing-out of tanks at sea.

These oil pollution incidents were causing both reputational and economic damage to the industry. Therefore we persuaded them to fund the establishment of an NCC Seabirds at Sea Team in Aberdeen – a sort of business and biodiversity initiative, though not then called as such – to determine where seabirds congregated in every month of the year, so we could produce meaningful, data-based oil-spill contingency planning charts to focus and prioritise clean-up efforts. We were also commissioned by the oil industry to investigate whether the flaring of gas at offshore rigs was killing significant numbers of migratory birds; it wasn't. During this time, I had several attractive offers of jobs with the oil industry. I was sorely tempted but stayed with the NCC.

Working with the marine community

As far as I know, I was the first person in the UK to have 'marine conservation' in my job title. From the start, it was clear that to do my job I would need the support and advice of a range of marine biological experts. I remember early visits to have discussions with, amongst others, Geoff Potts, Norman Holme and Gerald Boalch at the MBA in Plymouth, Harry Powell at the SMBA, Jenny Baker, Keith Hiscock, Robin Crump and Peter Hunnam at the Field Studies Council, Professor David Nichols at Exeter University, Dennis Crisp and Eiffion Jones at the university in Bangor, and the rocky shore expert Jack Lewis at the Wellcome Marine Laboratory. In those early years, the Underwater Association (from 1965) was a forum for people interested in marine science and held an annual conference for the diving community. Here was a constituency of interested marine experts with whom to discuss the first formative ideas of marine conservation – an important milestone (see Timeline 3, page 59).

Underwater Conservation Year and the development of the Marine Conservation Society

In 1975, Bernard Eaton, editor of *Diver* magazine, invited me to a meeting at the Wig and Pen Club to discuss the idea of having an Underwater Conservation Year (UCY) with David Bellamy, Paul Cragg (London University) and Geoff Potts (MBA). I was especially pleased that Bernard was keen to get divers involved in marine conservation, as this was something I had tried on a small scale with the Southampton University Underwater Research Group. Because this excellent initiative potentially increased the non-government involvement in marine conservation, it was accepted by NCC that helping to further this was part of my job. UCY was held in 1977, with Prince Charles as a very supportive patron, and then continued as the Underwater Conservation Programme, of which I was secretary. As there was no other UK NGO taking marine conservation forward in such a dedicated way, I then applied to the Charity Commission to form the Underwater Conservation Society (UCS) in 1978, of which Bernard Eaton and I became chair and vice-chair respectively, and Bob Earll was the project coordinator. UCS started to hold dedicated conferences from 1979 onwards, which were hugely popular (see Timeline 3, page 59).

The provisions for marine reserves in the Wildlife and Countryside Act 1981 made NCC's backing for a marine NGO even more important. We, in NCC, certainly recognised that such NGOs were an essential part of the conservation scene, enthusing and motivating the public and lobbying government. UCS morphed into the Marine Conservation Society (MCS) in 1983, and with my support and that of my colleague Ted Hammond (NCC's grants officer), NCC awarded MCS a development grant that enabled it to employ a wider range of staff for six years. In addition, MCS also expanded community support for marine conservation and data collection through their citizen science projects with divers, especially the colour identification guides to marine organisms, the mini-print sets coordinated by Bob Earll, which were transformational at the time. An NCC grant to the Field Studies Council also pump-primed the beginning of their excellent AIDGAP keys, with Sue Hiscock's guide on brown algae being the first of these.

A Nature Conservation Review (NCR) and the criteria for marine reserves

This publication, managed and edited by Derek Ratcliffe and published in 1977, set out to identify the most important sites for nature conservation in Great Britain under six major habitat types, including coastal. Derek Ratcliffe had made his conservation mark by highlighting the impact of pesticides on peregrines and other raptors. He now took on the task of trying to formalise what in effect had been an informal 'expert eye' process of selecting the best conservation sites, the preparations for which had been in progress since the late 1960s and early 1970s, with many people contributing to the surveys and planning.

It seemed to me at the time that the ten criteria that Derek Ratcliffe had developed for the assessment and selection of terrestrial conservation sites could also be used in the marine environment, with some redefinition. I aired this idea amongst my peers in a few articles and presentations at meetings, and it gained acceptance and was then published in *Nature Conservation in the Marine Environment* (NCME) in 1979. Later, in 1987, these site assessment criteria were also published in a Council of Europe report, which I authored, to establish a framework for the establishment of a network of marine protected areas in the North Sea and Baltic (Mitchell 1987). This was then turned into practice with the initiation of a Marine Nature Conservation Review (MNCR) in 1988 by the NCC and continued by the Joint Nature Conservation Committee (JNCC).

The NCC/NERC Report, *Nature Conservation in the Marine Environment*

The Clark committee, convened by NERC, produced a report on marine conservation in 1973 which had put rather a dampener on progress by indicating that the information was not then available to demonstrate any urgency. But in 1979 an NCC/NERC committee produced the key report *Nature Conservation in the Marine Environment* (NCME), for which I was the co-author/editor with Tom Pritchard (Director Wales, NCC). This was an important document because, in addition to setting out the background and documenting who was involved at that time, it set out the main drivers, the public interest, the threats, and existing legislation. It also set out a strategy for progressing marine conservation, including outlining the criteria for the selection of marine protected areas and how science could support this process. One of the documents prepared when pulling together the NCME was an inventory of coastal SSSIs to determine just how much intertidal habitat was already being protected. At that time, we asked our lawyers why we couldn't simply extend SSSIs out from the coast to include the Crown's ownership of the seabed as well as the waters above it. However, whereas our lawyers considered this was possible under the enabling Act, government lawyers were adamant that new legislation would be required to extend the designation offshore. A key influencing event around that time, arising from the NCME work, was a trip I organised to the United States in 1978 with a senior group of NCC colleagues, Bob Boote (Director General), Tom Pritchard (Director Wales) and Fred Holliday (Chairman) to discuss marine conservation with US officials and see for ourselves some marine reserves and national parks. This shared experience ensured that NCC's top management were even more supportive of increasing the efforts of NCC on marine conservation.

The Wildlife and Countryside Act and Marine Nature Reserves

In 1979 a White Paper emerged for what was to become the Wildlife and Countryside Act, but there were no draft provisions for marine conservation in this despite my attempts to influence NCC and government officials. However, Tim Sands (then of RSNC) and Chris Tydeman (WWF) lobbied to get Marine Nature Reserves (MNRs) included in the legislation; the newly formed Wildlife Link got behind this too. Eventually this coordinated pressure resulted in me joining the DoE Bill Team, which included briefing Michael Heseltine and Tom King (respectively, the Secretary of State and Minister for the Environment at the Department for the Environment), and Lord Craigton, an enlightened conservationist who eventually tabled the new clauses to ensure that provisions for MNRs were added to the Bill as it passed through Parliament. However, these were made nearly unworkable because, in parliamentary debate, assurances were given that should there be any serious objections to an MNR by local interest groups, then designation would not proceed in that place. Consequently it was largely vociferous community objections to the proposed Isles of Scilly and Loch Sween MNRs that sank these proposals.

With the benefit of hindsight, some said that we should have waited before pushing for legislation for statutory MNRs, but these Parliamentary opportunities don't arise often and I still feel it was better to act then when we had the chance, than to have waited. As it was, these new provisions for MNRs provided the momentum for three major initiatives: the Marine Nature Conservation Review, the scientific fieldwork to describe marine habitats and species, and NCC's development funding for the Marine Conservation Society. Around the same time several voluntary marine nature reserves were being promoted by local organisations and groups – Skomer in 1976, Purbeck with its first marine warden, Sarah Welton, in 1978, Wembury in 1981 and St Abbs Head in 1984. These VMNRs were also greatly helping to build experience and confidence in site conservation.

Marine Nature Reserves
Having succeeded with inserting provisions for MNRs in the Wildlife and Countryside Act, we then set about to describe and promote individual sites from an initial list of known hotspots of marine interest and importance: Lundy, Skomer, Bardsey, Isles of Scilly, the Menai Strait and Loch Sween. Some surprising things started to emerge. On Lundy, for example, the National Trust and Landmark Trust were initially against the idea, which was a bit disconcerting given that they were both conservation organisations. While it was never quite clear why they took this initial stance, I suspect they didn't want to lose control of their patch – but they eventually supported the proposals.

Even though we realised the legislation was flawed, we knew it was an opportunity, and it was important that we tried. However, this was all very frustrating, as a process for engaging stakeholders at proposed MNRs was a new experience and had not been properly thought out. Consequently, the approach was somewhat naive and uncoordinated. A view that seemed to be held by some was that as NCC was still a Crown body, whatever they proposed would be implemented – well, that didn't work on the Isles of Scilly or Loch Sween. This resulted in some uncomfortable internal meetings to determine how to do better next time, and, eventually, Lundy and Skomer were designated. I had also been loaned to the Department of the Environment (Northern Ireland) for short periods, where I discussed marine matters with Joe Furphy (DoE-NI), David Erwin (Ulster Museum) and Pat Boaden (Portaferry Marine Lab), which eventually resulted in Strangford Lough becoming an MNR too.

Marine Nature Conservation Review (MNCR)
We had started work on surveying the intertidal zone of Great Britain with the MBA and SMBA in the early 1980s. The involvement of Norman Holme (MBA) and Harry Powell (SMBA) and their organisational colleagues was very helpful because their wide experience allowed us to quickly develop a survey methodology to describe the attributes of intertidal sites. DoE had also given us some extra funding to acquire boats, diving gear and cameras to boost sublittoral surveys, and an early focus was the rich waters of the Hebrides – with Bob Earll, Frances Dipper, Gill Bishop, Christine Maggs, Bernie Picton and Alan Sime as diving scientists.

Internal bids for funding work on a comprehensive Marine Nature Conservation Review (MNCR) were made annually from about 1982, but were unsuccessful because so much of NCC's budget was already tied up with the Geological Conservation Review (GCR), which was always just about to be completed. An influencing initiative around that time was the formation of several groups to write responses to the World Conservation Strategy (WCS) coordinated by the Royal Society of Arts. I was a member of the group who prepared a report on 'Conservation and development of marine and coastal resources' which appeared in *The Conservation and Development Programme for the UK*, published in 1983. I also contributed to NCC's own published response to the WCS in 1984, *Nature Conservation in Great Britain*, which stated that action on marine nature conservation had lagged seriously behind, that an MNCR should be launched immediately and MNR establishment should be vigorously pursued as well as promoting nature conservation in the wider marine environment outside statutory sites.

In the meantime, funded by WWF (Chris Tydeman) (see Sue Gubbay, Chapter 7), MCS produced the first version of the *Coastal Directory* in 1985 in a classic NGO ploy to lever government into funding the MNCR. With this, and the realisation that the GCR was nowhere near finished, we did eventually get the funding to initiate the MNCR in 1987, which was subsequently led by Keith Hiscock and completed under the JNCC in 1998 (see Keith Hiscock, Chapter 6).

DEVELOPMENT OF MARINE CONSERVATION, INCLUDING THE CASE WORK

Many of the issues which required case work in the early years still seem prevalent today. Then, background information on sites and species and the scientific evidence on which to base our impact assessments and mitigation advice was sparse. However, whilst working for the DoE, I had quickly learned that one hardly ever has enough information on which to base decisions absolutely; advice had to be framed on the best available evidence. Nevertheless, by the late 1980s we had sufficient experience on a wide range of marine issues and impacts for Clare Eno, in my team, to collate a *Marine Conservation Handbook* of case-work advice (Eno 1991).

Invasive non-native species

The non-native American hard-shell clam, *Mercenaria mercenaria*, in Southampton Water and the Solent was the subject of my PhD. I found that it had been deliberately introduced and successfully colonised through sporadic breeding, in part due to the cooling-water effluents from several power stations and an oil refinery, which release heated water. This study, and the later appearance of the invasive Japanese seaweed *Sargassum muticum* on the Isle of Wight in the early 1970s, aroused my interest and gave me insights into the ecology and impact of non-native species. The DoE invited me to join a group to determine how to tackle *Sargassum*, which was thought to have probably come in from France on imported oysters or boat fouling. We met for three years and initiated several dredging and harvesting trials (led by Bill Farnham at Portsmouth University); herbicide treatment was considered but not used. All to no avail, as uncoordinated local efforts to clear the weed from beaches and pontoons ensured it floated off and spread rapidly in both directions along southern shores.

When at NCC, I had to assess an application by MAFF to introduce the Japanese oyster (*Crassostrea gigas*) and the Manila clam (*Venerupis philippinarum*) into British waters for shellfish farming – though we later found out that they had already been holding experimental stock in the Menai Strait. From looking at temperature records and experience with these species, my strong advice was that they would breed at temperatures occurring in British waters and thus become invasive. MAFF's response was that they would 'put them under mesh' – as if this would make any difference to reproduction or the release of larvae. They also stated that the larvae wouldn't survive. The licence for their introduction was given against my advice, and consequently both species are now established as wild, breeding populations in quite a few localities (Figure 5.1). Later, I was an author of a government report which reviewed policy for combatting the introduction of invasive non-native species.

Basking sharks

My passionate interest in these enormous beasts was aroused whilst swimming with and photographing them when they occurred in abundance around Lundy in the hot,

Figure 5.1 *Crassostrea gigas* oyster reef, a non-native species in the Yealm, south Devon. Source: Keith Hiscock

dry summer of 1976 (Figure 5.2). We were aware that, on the west coast of Ireland at Achill, the fishery for their liver oil had collapsed due to overexploitation. There had also been fisheries on the west coast of Scotland, and one boat was still operating. The quota allocation by UK to the Norwegians for the fishery in 1986 brought matters into sharper focus, and I asked Sarah Fowler to prepare a case for listing. MAFF once again thought they knew best and said there was no problem. However, pressure from the NGOs and the MCS citizen science project on basking sharks raised the profile of the unsustainability of the fishery. Eventually, in the mid-1990s, the fishery ended, and in 1998 the DoE, probably now weary of the endless debates, listed it as a protected species under the Wildlife and Countryside Act 1981 (Speedie 2017).

Fish farming in Scotland
My concerns about the impact of fish farming was first sparked when I dived underneath a small fish-farm cage in Loch Sween to see the seabed covered in a white bacterial mat called *Beggiatoa* or 'sewage fungus' with bubbles of methane and hydrogen sulphide emerging. To see that damage in what had been a pristine environment and a candidate MNR was a shock. A shellfish farm was also started in the same loch, but the TBT antifouling in which the fish-farm nets were treated soon put that out of business as well as damaging recruitment to the native oysters and clams in the loch. This loch was further seriously damaged by scampi prawn (*Nephrops norvegicus*) dredging when improved navigation

Figure 5.2 Basking shark at the surface off the Isle of Man: work to protect this species began in the late 1980s in the UK. Source: Jeremy Stafford-Deitsch

aids allowed fishermen to avoid the seabed snags that had earlier restricted their activities.

At that time, the Highlands and Islands Development Board were handing out grants for fish farming on the Scottish west coast with what seemed very little planning, guidance or regulation. A whole raft of issues ensued, including illegal pre-emptive shooting of seals and herons, visual and physical disturbance, litter, organic pollution of the seabed, and chemical pollution by antifouling and sea-lice treatment chemicals. These chemicals not only also damaged and killed other invertebrates in the sea lochs, but also affected many of those working on the sea farms.

Inevitably, many 'cowboy' operators jumped on this bandwagon, but because of poor husbandry quite a few of the early enterprises quickly failed. I was consulted by insurers for my views on what was going on, because they were fed up with paying out claims for damages which, as I told them, were self-inflicted. Many of the environmental issues with fish farming haven't gone away, and it is still a major industry in Scotland, but at least we now have a better idea of the associated environmental issues, which has resulted in some improvement in regulation.

Citizen science
In the early days, species recording coordinated by the Biological Recording Centre (BRC) at Monks Wood Research Station with its published distribution dot maps was supported by many colleagues in NCC and had led the way with using volunteers to record information virtually for free. I'd tried this whilst at Southampton University, using divers to record the distribution of easily identifiable species in the Solent, and I was aware that David Bellamy had also initiated a similar scheme, Operation Kelp, around the same time. I was very pleased that the UCY initiative built on these initiatives, and in 1988 we worked with and funded MCS to develop Seasearch, which is still going strong. A similar Coastwatch project focused on the coast and intertidal zone and this was energetically coordinated at NCC by Teresa Bennett and supported by Earthwatch and WWF UK. At NCC, working with BRC, we also produced the first provisional distribution maps of marine molluscs and algae.

I've always been a strong supporter of citizen science, because it is not only a good way of collecting data, but it also spreads ownership of the information and engages people with the issues that may negatively affect their favourite taxonomic groups. I found, as Chief Scientist at Earthwatch, that if our scientist project leaders had not properly explained the conservation background and needs of the research to the volunteers, they usually demanded to know it and performed better if they did.

Conservation versus natural history
A knowledge of, and a passion for, natural history is often a precursor to an interest in nature conservation – this is the route I took. However, from my experience in and with professional conservation organisations, it has sometimes seemed to me that too much of the focus of often well-paid individuals is still on satisfying their passion for natural history in pilgrimages to known locations of national rarities, or ticking a sighting on a list, rather than getting on with the more difficult job of conservation. Once, when NCC science staff were subjected to the scrutiny of a management consultant, one observation was that it was difficult to determine where their hobbies ended and their professional, paid work began.

But perhaps this blurring has as many advantages as disadvantages. The same could possibly be said about the passion for, or possibly distraction of, surveying and monitoring, which was and is often far more rigorous than required to satisfy the requirements of informing conservation action and measuring its success or otherwise. To some extent monitoring has now been modified, as the realisation developed that one needed more than the expertise of a few expert eyes to detect damage or change. So methods are now a great deal more fit-for-purpose for collecting useful data and information and can be carried out not only by most field staff but by citizen scientists too – and, increasingly, by automatic and remote sensing.

But even so, many projects are still not sufficiently well designed to reliably collect significant evidence of the effects of conservation interventions. When the planned intervention doesn't result in a success, the results are often suppressed or not published, which is a significant loss of useful information. I have tried to do my part in helping to promote and publish good science through being founder joint chief editor of *Aquatic Conservation: Marine and Freshwater Ecosystems* and joint chief editor of *Conservation Evidence*, publishing a number or articles and peer-reviewed papers of my own, as well as teaching and assessing on a number of MSc courses in ecology and conservation, and examining PhDs.

Looking back, what do you make of it now? How has marine done?
I'm not as in touch with progress in marine conservation as I once was. However, I am very pleased that it is higher up the agenda and no longer a novelty. It is now accepted as normal and a recognised necessity on an international scale. Good evidence for this in the UK is that we now have a large number of Marine Conservation Zones, much more recognition of the need for controls to ensure sustainable fisheries and fish farming, and a very large number of staff in governmental and non-governmental organisations now employed in marine conservation. Looking back, it now seems remarkable what was achieved in those few years at the end of the 1970s which set the groundwork for statutory marine reserves and where the need for, and purpose of, marine conservation gained a much wider formal recognition.

FUTURE CHALLENGES

What are the most important threats to the marine environment, and what needs to be done about these?
Many of the threats cited in the NCC/NERC report (1979) are still current in UK waters, although some haven't proved to be quite such a problem as we thought then due to improved regulation and mitigation, and new ones have emerged. Climate change, which was not mentioned in 1979, is now the major global environmental threat, as are new diffuse pollutants and the increasing levels of marine plastic litter including nano-particles. The damage from fish farming and the deliberate introduction of non-native commercial shellfish was perceptively foreseen and the pressures on fish stocks and the collateral damage to other fish and marine organisms, and to the seabed habitat through dragged nets and dredges are still a global issue. The solutions to present impacts include continuing with actions such as communicating knowledge of the consequences, behaviour change, precaution, regulation, polluter pays and – probably most importantly – continued investment in and development of clean energy, including for sea transport.

What are the hidden problems and barriers that make progress harder than it should be?
When I started in the NCC, I felt empowered in large part because I was involved in everything from science to practice as well as communication and education. As organ-

isations have grown and specialisms have developed, the sort of broad-based jobs that I enjoyed now seem rare; there appears to be a tendency for silos and personal territory to emerge and stultify progress. I've found increasingly that it is difficult to know exactly who is supposed to be doing what. Modern job titles don't help, and in some organisations people seem to be tripping over each other. At one level this may be necessary because of the scale of present activities through increasing regulation, but at another it is bureaucratic and ponderous. This has also spread into some of the increasingly powerful NGOs, which once seemed so light of foot and quick to act.

I've described our early work on marine reserves and the tension and institutional resistance to the idea from the fisheries interests in MAFF. MAFF's Fisheries Director at the time, H. A. Cole, seemed to object 'on principle' to NCC's views on the growing negative impacts of fishing and fish farming. He and I corresponded once he retired, and he came to accept that the principles of nature conservation were indeed necessary to sustain fisheries. A later Scottish Fisheries Director, Alasdair Macintyre, was a real ally in marine conservation, and with his sharp mind he moved into an academic career on retirement – and was then even more supportive in overcoming historical barriers.

Innovation – what are the most interesting and promising new approaches?
There have been so many technical advances: remote sensing, acoustic tracking, sonar, GIS, modelling etc. are now all routine tools we take for granted. In the 1970s, simply knowing where you were and measuring location accurately from small boats was very difficult. Now the phone with GPS, camera and biological recording apps has transformed capturing natural history observations, citizen science and conservation. One of the changes that is very apparent now is the number of social scientists working in nature conservation, because engaging stakeholders and behaviour change is often central to achieving many conservation objectives.

When you retire, what would you most like to have achieved for marine conservation?
Retirement isn't something I'm contemplating, because I still get a great buzz out of my involvement in nature conservation at various levels in education, communication, science and practice. Looking back at the marine side of things I have a warm feeling about the principles I have helped to embed and the projects and initiatives I helped set up, like the Seabirds at Sea Programme, the MNCR, Coastwatch and Seasearch, the Review of Coastal Saline Lagoons and the first MNRs. I'm very pleased that I was able to influence and initiate the planning and resourcing of those and other projects that helped many more people to become involved in marine conservation. We needed more people pushing at the wheel, and we've certainly achieved that.

Are you optimistic or pessimistic about the future?
I'm concerned about the current and future effects of climate change, because of its impact on both the marine and terrestrial environments, which is increasingly evident. It is very important to determine and communicate what is wrong and act to put it right, or at least practise precaution while assessing the evidence. But I'm optimistic in so far as the solutions needed to resolve many of the negative impacts and processes are clear. It is also important to celebrate our successes. One of the programmes I was involved in funding at English Nature was the reintroduction of the red kite to reverse the effects of their persecution by gamekeepers, and I feel so good whenever I see these birds on my travels or flying over my house! My optimism still drives me, and I am fortunately still able to be part of the conservation movement.

TIMELINE 3

Marine conservation in the UK: first steps, 1960–1991

1965 John Lythgoe prepares a note for NERC/NCC on 'Underwater Conservation', signed by sixteen other scientists.

1965 Underwater Association of Malta formed, providing a forum for UK diving scientists to meet. Meetings act as a focus for discussions of sublittoral ecology and MPAs until the 1980s.

1965 The National Trust launches Project Neptune, also known as Enterprise Neptune, to buy coastal land of outstanding beauty. By 2015, almost 1,200 km of coast had been acquired.

1967 Oil tanker *Torrey Canyon* wrecks, causing major concerns over oil pollution and threats to the environment.

1968 David Bellamy organises the first citizen science project with sports divers in the UK, Operation Kelp.

1969 The Nature Conservancy Council (NCC) produces a paper on conservation policy in the shallow seas which highlights its lack of remit for shallow seas.

1971–1973 The Natural Environment Research Council (NERC) publishes *Marine Wildlife Conservation*, also known as the Clark Report, on marine conservation; it highlights the lack of information and downplays the need for marine conservation.

1973 Lundy Island declared the first voluntary marine reserve. Other subsequent voluntary sites included Skomer in Pembrokeshire, Kimmeridge in Dorset and St Abbs in Berwickshire.

1973 The first public meeting on marine conservation in the UK: A Seminar on Marine Conservation, run by the University of North Wales Department of Marine Biology. Organised by Keith Hiscock and Ivor Rees.

1974 Control of Pollution Act: landmark legislation that has been developed incrementally to control many forms of pollution in the UK.

1975 Roger Mitchell appointed to NCC Chief Scientist team, the first official to be appointed in the UK with 'marine conservation' in his job title.

1975 Bernard Eaton, editor of *Diver* magazine, convenes first meeting for Underwater Conservation Year.

1977 Nature Conservation Review published by NCC – the 'Ratcliffe review'. It does not include any sublittoral marine sites.

1977 Underwater Conservation Year (UCY) launched: citizen science projects for sports divers; first coordinator Charles Sheppard.

1978 Underwater Conservation Programme evolves from UCY; Bob Earll takes over as project coordinator: it develops into the Underwater Conservation Society in 1979.

1979 NCC starts the Seabirds at Sea surveys to investigate the distribution and behaviour of seabirds in the North Sea.

1979 NCC and NERC publish the report *Nature Conservation in the Marine Environment* which sets out plans for marine reserves.

1979 The first annual meeting of the Underwater Conservation Society in Manchester.

1981 The Wildlife and Countryside Act published, which includes provision for statutory Marine Nature Reserves.

1984 Marine Conservation Society formally constituted.

1985 Coastal Directory first published (MCS and WWF): author Sue Gubbay identifies UK sites for potential marine reserves.

1986 Lundy declared the first statutory Marine Nature Reserve in the UK.

1987 Marine Nature Conservation Review of Great Britain launched by NCC. This was a scientific assessment of potential UK marine conservation sites; it was eventually published in 1998.

NATURE
CONSERVANCY
COUNCIL

1989 Initial discussions of the European Habitats Directive (1992), eventually including some marine habitats and species

1991 The Nature Conservancy Council is split into country agencies: English Nature, the Countryside Council for Wales, Scottish Natural Heritage and the Joint Nature Conservation Committee.

CHAPTER 6

Keith Hiscock

Keith is currently an Associate Fellow of the Marine Biological Association of the UK (MBA) and has worked on a variety of publishing, photography and consultancy projects since he retired from leading the Marine Life Information Network (MarLIN) programme in 2007. He developed an interest in marine life growing up in Ilfracombe, and was fascinated by what could be found on the seashore, especially the anemones and corals. In the local library, he discovered the books published by Victorian naturalists, especially Philip Henry Gosse. Gosse was a great populariser of marine life in the middle of the nineteenth century, and the local shores of north Devon were one of his areas for collecting.

Once at university, Keith took up scuba diving as a route to pursuing his interests in marine life and, now, underwater photography. His first degree was at Westfield College, University of London, where he further developed his research into cup corals especially on field trips to the Isles of Scilly and to Lundy as well as his native north Devon. He did his PhD at the marine station at Menai Bridge, north Wales, looking at the effects of wave exposure and tidal currents on subtidal marine life. Leaving Bangor, he went to work for the Oil Pollution Research Unit (OPRU) at Orielton Field Centre, where his work focused on researching and monitoring marine life mainly in the context of oil pollution, but he also undertook diving studies for the Nature Conservancy Council (NCC).

In the early 1970s there was a growing interest in the development of marine nature reserves, and Keith became a leading champion of the idea of marine reserves in the UK – and for the island of Lundy in particular. Lundy became the first UK voluntary marine nature reserve (VMNR) in 1973 and statutory reserve in 1986. In the early 1980s, pressure had been mounting to have an effective inventory and description of the UK's marine species and habitats, and in 1987 he moved to Peterborough to lead the NCC's Marine Nature Conservation Review of Great Britain (MNCR), and then became Head of Marine Conservation Branch for JNCC.

This chapter focuses on Keith's work on the use and development of diving science to describe and understand the distribution of seabed species and habitats and how that work shaped the MNCR. It was the awareness of the need for protection of the wider environment and case work that led to the development of the Marine Life Information Network (MarLIN) programme, which has made information more accessible to stakeholders on the sensitivity of marine species and habitats to development. His work on the leading marine reserve at Lundy has been pioneering, and the chapter reflects on this and his work on marine protected areas and marine conservation.

MARINE CONSERVATION AND ITS DEVELOPMENT

How do you define marine conservation?

I use a definition that is from a time before 'biodiversity' had entered our vocabulary and become fashionable: *the regulation of human use of the global ecosystem to sustain its diversity of content indefinitely* (Nature Conservancy Council 1984). Conservation can also include restoration. The overarching goal needs to be broken down into specific objectives and actions – which will include ensuring that representative examples of habitats are protected and that features (species and habitats) that are rare, scarce, in decline or threatened with decline ('sensitive' to human activities) are safeguarded.

How is marine conservation different from terrestrial conservation?

I have compiled a list of these points for my book *Marine Biodiversity Conservation: a Practical Approach* (Hiscock 2014), but in summary they are:

- Our understanding of what is where – the basic distribution of marine species and habitats – is very poor and it is very expensive to fill in the information gaps.

- Likewise, our knowledge of change brought about by human activities and our understanding of long-term natural fluctuations in abundance and distribution of marine species and of natural change in the character of habitats is very poor compared to the land.

- The above two points have direct relevance to approaches like Red Listing, where 'data deficient' is the norm for marine species and leads one to question whether Red List criteria can be applied to marine species and habitats.

- There are still places which are essentially untouched by human activities or at least by physical damage – this is very different in comparison to land, where disruption of natural succession and alteration of landscape is almost ubiquitous.

- The ecological processes that shape and maintain biodiversity in the sea are very different from those on land.

- There is natural connectivity for migratory species and for larvae or other propagules provided by the water column in the sea, unlike on land, where physical measures – often called 'wildlife corridors' – often provide such links.

- Restoration and recovery in the marine environment are greatly dependent on natural processes and, unlike on land, practical measures – 'ecological gardening', including reintroductions – are much less useful.

What are the differences between when you started and now?

In the early 1960s and 1970s it was divers collecting sea urchins and sea fans for curios and spearfishing that were regarded as the main threats, together with pollution in general. Now the issues are rather different, and damaging effects of fishing, eutrophication and non-native species are much more apparent. There was no science of marine conservation then. There were academic studies of particular species and their distribution and biology but very little underpinning knowledge. We didn't have a description of the distribution of our main subtidal species and habitats around the shores of the UK. There has been lots of survey work since then that has filled that huge gap.

There is also a major difference in our ability to recognise and identify marine species. It is important to remember that even into the early 1980s there were very few guides or colour illustrations of what marine species looked like when they were alive, and many of the descriptions and guides dated from the nineteenth century, often written in French or German. In the early years there were just a few journals covering marine science; now the science of marine conservation and management is generating more than 1,000 relevant articles a year and it is difficult to see how practitioners can keep up to date, despite Google and electronic journals. Marine conservation has become professionalised, and the numbers of people employed has grown enormously.

THEMES

The development of diving science and conservation thinking

Like a lot of people I was inspired and fascinated by the Hans and Lotte Hass and Cousteau TV programmes, and this led me to take up diving, which I continued at university to do scientific research. The period from the 1960s saw an explosion of interest in diving and the wider availability of affordable equipment and expansion of diving clubs throughout the UK. At that time, the Underwater Association, which was a network of diving scientists formed in the mid-1960s, was also really important in providing encouragement and feedback to like-minded people. Their meetings showed that it was possible to use diving science in your career.

Many traditional marine scientists were very sceptical about diving, thinking of it as 'a bit of a jolly', but there were others who were very supportive. My PhD supervisor, Professor Dennis Crisp at University College of North Wales at Bangor, supported me in using diving to investigate the effects of wave exposure and tidal currents on sublittoral marine life. Rupert Riedl, who did a lot of work using diving in the Mediterranean, was an inspiration because his work began to explain and provide insights into the distribution of subtidal marine life. In 1973 the Clark report of the NERC Working party (Clark 1973) dismissed the idea of marine reserves and marine conservation, which prompted a reaction in many of us, and I organised a conference in Bangor to pull together mainly scientists to help develop a counter view. From 1975 onward I also took part in the formative meetings led by Bernard Eaton.

In the early 1980s, money became increasingly available to undertake descriptive diving surveys for nature conservation (Figure 6.1). The reality was that we had a very

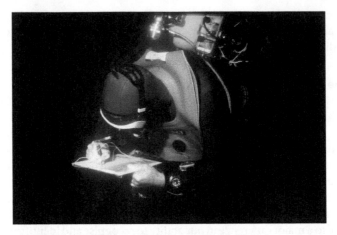

Figure 6.1 Diver recording marine life on a survey Source: Keith Hiscock

poor, almost non-existent description of the basic ecology of shallow waters (0–40 metres). Much of the work in the late 1970s and early 1980s started setting the scene for a much more systematic description of the marine habitats and their associated species present around our shores. In effect we were having to provide what the terrestrial ecologists and botanists had taken for granted for more than 100 years, because there was no effective catalogue of seabed habitats and associated species to provide the context for conservation or more specialist studies. I hope that my recent book, which illustrates the wide range of underwater habitats around the UK, begins to address the need to share our understanding of this environment more widely (Hiscock 2018).

The Marine Nature Conservation Review – providing the contextual information to inform MPA designation

The Nature Conservancy Council (NCC) had published their *Nature Conservation Review* (NCR), edited by their Chief Scientist Derek Ratcliffe, which provided the scientific basis for protected area designation on land (Ratcliffe 1977). There were very few references to marine sites in NCR apart from mudflats and estuaries, some intertidal seashores for geological interest but very few for their marine biological interest; there were no shallow sublittoral sites.

Roger Mitchell was appointed to the NCC in 1975 and, from the outset, wanted to start the marine version of the NCR. NCC had already commissioned a lot of work, notably on an intertidal survey of UK shores with experts at the MBA and SMBA and diving surveys. The Marine Nature Conservation Review (MNCR) began in 1987 with about nine staff in Peterborough and, later, there were regional teams. The aim of the MNCR was to undertake survey work to document the variety of marine habitats and associated species in the shallow seas around Great Britain. We did it using the established shore survey techniques, with diving techniques for the reefs and inshore areas and benthic grabs and dredges for the soft sediments. Not only did we do our own survey work but we pulled together a lot of existing work, both collecting and collating information to build into an overall picture.

We used an Advanced Revelation computer database from the start of the MNCR; at the time, it was novel and innovative. One of our key tasks was to identify recurrent habitat communities and to classify them. All the data in the literature and from new surveys was incorporated into the database and analysed to identify similar groupings of species from different locations. Those groupings were the foundation for identifying separate 'biotopes', a European term new to us in the 1980s.

In the early 1990s, an EU-funded programme – BIOMAR, initiated in Ireland – enabled us to work with European colleagues who were also exploring and developing marine habitat classifications. The MNCR biotope classification thus became the starting point for what is now the marine element of the European Union Nature Information System (EUNIS) biotope classification. David Connor and his team developed the classification. The first version was published in 1996, and it has continued to be kept up to date. This work was very complicated, not least in terms of naming the different biotopes, but it has stood the test of time. It gives us an important tool to compare biotopes on a like-for-like basis in relation to different locations in European seas. The biotopes classification is now relatively stable although, inevitably, ecologists are still discovering new assemblages which need to be included and there is an ongoing need to fine-tune descriptions as our knowledge improves. The classification has over 500 biotopes, and the current challenge is to make it as user-friendly as possible, especially to new people doing surveys.

The MNCR was finished in 1998 and there were two volumes published: the first explained the rationale and methodology and the second was the benthic marine ecosystems volume, explaining what we knew of the habitats and species in the fifteen coastal biogeographical sectors that the MNCR had used. JNCC also continued to bring together the results of surveys in a series of regional summaries. The MNCR was *finished but never complete*, and it was wound down in one of the inevitable reorganisations that occur in large organisations. In 1998 I went back to my parent agency, English Nature.

MarLIN – developing guidance for case work on the sensitivity of marine species
When I was made head of JNCC's Marine Conservation Branch I had to deal with applications for development and was given responsibility for providing advice on case work. We had to comment on all sorts of things, from applications for new pipelines to dredging and harbour works, to say whether such activities would cause unacceptable damage. One instruction was clear: we had to use 'available data'. Of course, very often there wasn't any, or if there was quite often the data couldn't be accessed because it was 'commercial in confidence'.

Not only did we need to do better in getting access to data, but it also became clear that we could add to the survey data information on sensitivity (intolerance and recovery potential based on knowledge of species life-history traits combined) of species and biotopes to provide more scientifically based advice in case work. I started to develop the idea at the MBA in 1998 with its then director, Professor Mike Whitfield, who was beginning to disseminate information using a new tool called 'the internet'. The project took off and, with the help of a steering committee, became the Marine Life Information Network (MarLIN) programme. The idea for the name arose because we thought that one computer couldn't hold all the data we would generate and a network would be required – but of course computing power has grown substantially. The programme's strapline was '*Information for marine environmental management, protection and education*'.

In those very early days of the internet, success was a case of having a good idea and being in the right place at the right time. In 1999 I was seconded from English Nature to lead the MarLIN team in Plymouth. We started with a small staff of four which also pursued educational activities. Our work was closely aligned with the developing National Biodiversity Network (NBN). In 1998 the NBN was being discussed as a way of collecting more data and doing more survey work, but today it is a repository for an enormous amount of biodiversity data which is displayed and accessible in a variety of helpful ways. The MBA, because of the MarLIN programme, became the marine node of the NBN. MarLIN is still going strong and concentrates on the sensitivity work and information to be used for environmental management applications and decisions. The data access activities have spun out to be a separate entity, as has the marine education part of the MBA work – and of course they are all dependent on external funding.

Marine protected areas – lessons from Lundy over forty years

In 1968 I was still an undergraduate but I'd taken an interest in marine conservation and had seen that marine parks seemed to be springing up all over the world, and I started thinking 'why not Britain as well?' At the same time in north Devon where I'd grown up, people like John Lamerton, Heather and Ron Machin from the Ilfracombe subaqua club were looking at the island of Lundy as a possible marine reserve, because of its small size and isolation and the variety of marine habitats and interesting marine life. They started making the case for Lundy to be a marine reserve, and I discussed this with them in 1969 and supported the idea. I was fortunate to be on the island when it was taken from private ownership into the National Trust and to be managed by the Landmark Trust. I spoke with John Smith, the founder of the Landmark Trust, and he was very favourable to the idea of a marine reserve, passing me on to his agent Ian Grainger to develop the proposal further – and it all happened from there. The history of the Lundy reserve, which has been at the forefront of MPA designation and management in the UK for over forty years, has been captured in the book that I prepared with Robert Irving (Hiscock & Irving 2012).

In the early 1970s we started to think about management plans and look at who we should be speaking to. We found that the organisation charged with fisheries protection, the Devon Sea Fisheries Committee (DSFC), was just not interested, especially if management meant curtailing fishing activities. It took a long time to engage with them but eventually they did contribute to the publication of a management plan in 1973. It is the publication of that management plan which we take as being the start of the voluntary marine nature reserve (VMNR) around Lundy.

Lundy was very popular with recreational divers, and it had a diving centre based on the island. Divers were collecting pink sea fans, spiny lobsters and urchins. It is interesting that marine conservation was then seen as necessary because of the 'awful' things being done by divers, but nobody was talking about fishing and especially mobile fishing gear. The voluntary agreement was adhered to by the dive centre and we hoped visiting divers, but the reserve also triggered an enormous amount of survey work. We obtained funding from all sorts of sources including the Royal Society and the Lundy Field Society and a lot of self-funding to undertake survey work over many years. We brought a wide range of marine scientists over to Lundy to investigate the subtidal marine life in particular and to catalogue the species present.

Lundy was well established as a VMNR, and after the Wildlife and Countryside Act (1981) made provisions for Marine Nature Reserves, in 1986 it became the UK's first statutory MNR. This designation didn't mean that it was suddenly well protected – that was not the case, because the legislation was very weak – but it did open the way to longer-term monitoring of change in marine communities of species which were fragile or rare. That work taught us a great deal about longevity, growth rates and persistence of species, which is the sort of information you need to develop management plans.

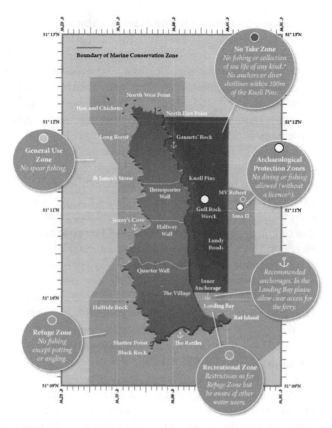

Figure 6.2 Lundy MCZ zoning scheme: *Lundy Management Plan* (2017). Source: Lundy Management Forum and The Landmark Trust

That monitoring essentially stopped when the Habitats Directive came along in the early 1990s and the work of the statutory nature conservation bodies had to be directed into the implementation of the Directive.

Lundy continued to tick over in terms of survey work and monitoring. There was a management plan, and by the 1990s DSFC had realised that marine conservation was here to stay and they played a role in management – but it was very much at their discretion and not supported by statutory backing. In 1994–95, with work done on zoning by Dan Laffoley at Natural England the idea of sanctuary and refuge zones was introduced, and this evolved in 2003 to become the first marine area in Britain to have a highly protected area or no-take zone (NTZ) where no extraction or deposition was permitted. It was only a small area, 4.8 square kilometres. The fishing community understood the idea and came on board very quickly, enabling DSFC to publish a bylaw to protect that area off the east coast of Lundy (Figure 6.2). It still remains an NTZ embodied in the European Special Area of Conservation (SAC) status and in the Marine Conservation Zone (MCZ) designation (2013). Lundy was also the first MCZ, but the only feature to be added to the MPA status was the spiny lobster, so this designation really didn't add much. The Lundy Management Forum (representatives of the island owners, management, statutory advisors and regulators and the Lundy Field Society) have published a management plan for the marine environment around Lundy, which is a pretty good document with goals, objectives and targets (Lundy Management Forum 2017).

What are the changes you've noticed to the marine habitats and species on Lundy over this time, and what are the lessons for the future?

Lundy has been very helpful as a source of information on natural variability. We now have a better idea of year-to-year variation for many species, but there are still many gaps, not least for decadal-scale changes. By their very nature it takes a long time to see what these events are driven by, whether they link to phenomena like the Atlantic Multidecadal Oscillation (AMO) or are simply one-off, episodic events such as good years for larval settlement or when diseases strike. We need to keep studying ecology to try and understand natural long-term cycles relative to those that might arise from human activities, pressures and issues like climate change. Such local studies need to be backed up by monitoring on a broad scale in terms of oceanographic information.

Since the mid-1980s we have noticed a downturn in many of the species that once made Lundy stand out. The nationally rare sunset coral *Leptopsammia pruvoti* is one of the most notable of Lundy's marine species and was recorded there for the first time in Britain in 1969 when it was abundant at a limited number of locations. In 1983 we started a monitoring programme and, since then, the population has declined so that, by 2014, they were at about a third of the 1973 levels, and one colony has completely disappeared. We noticed changes for other species as well. In the case of pink sea fans, they were being collected for the curio trade, and tea chests full of them were being sent to London in the late 1960s. That activity stopped and there was recovery, and then in the early 2000s we saw a *Vibrio* bacterial disease decimating the sea fans as well as at other sites in the southwest. This type of episodic event clearly has a major effect, but we understand very little about its causation or frequency. My view is that the *Vibrio* bacterium also affected other seabed species which we have noticed declining. We have seen recovery and recruitment, but for some of these species near the northern edges of their distribution recruitment doesn't seem to be keeping up with loss. The appearance and disappearance of the mud-dwelling red band fish off the east coast of Lundy seems a good example of natural change for which there is no really clear explanation.

There was a major issue with the level of potting in the 1990s and, when the NTZ was introduced, there was an opportunity to monitor the effects on lobster and crab populations. The work has demonstrated significant changes in the populations of lobsters, with seven times as many lobsters in the NTZ as in control areas which have continued to be fished (Wootton *et al.* 2012). The fishermen think that their catches outside the NTZ have increased, and they put this down to spill-out from the NTZ.

We have learnt not to expect changes in species which were unaffected by potting or other activities. Many of the seabed species simply haven't changed as a result of the NTZ. The spiny lobsters, a southern species, which were common in the late 1960s, have never really recovered from over-collection by divers, although there has been very noticeable recruitment in the last few years. Scallop dredging was not a frequent activity at Lundy and ceased after voluntary agreement in the early 1970s. The now undisturbed sediments off the east coast are rich in species. Lundy has become a major attraction for snorkellers and divers interested in swimming with seals. The seals were rarely seen underwater by divers in the 1970s, but since the early 2000s seals have been actively seeking interactions with humans in the water. It is notable that the population of seals at Lundy seems to have increased significantly over the last forty years, although any impact on inshore fish hasn't been monitored.

Another key lesson from Lundy has been the time it has taken for fishermen in particular to recognise that there are other legitimate uses of the marine environment.

Over the forty years of the reserve's existence, we have seen fishermen gradually becoming more accepting and supportive of the marine conservation work. Although stakeholder engagement has its pluses and minuses (as we saw in the MCZ process), bringing different stakeholders together means that people do get to see each other's views. The work of the various Lundy MPA advisory groups has demonstrated the value of bringing users together for over thirty years.

Marine protected areas and conservation in general – reflections and lessons

Forty years ago when we began to think that MPAs would be a good thing for protecting the marine environment, all the organisations, knowledge and systems were in place for terrestrial conservation but we didn't have any of that for marine. Now we are in a broadly comparable situation with terrestrial conservation. The Wildlife and Countryside Act (1981) provision for statutory marine reserves was very weak and didn't have much scope for providing statutory regulation, but the EU Habitats Directive in 1992 changed the effectiveness of MPAs for the better. Marine habitats were added to the Directive at a time when there was no clear idea of what marine habitats were threatened. The five Annex 1 marine habitats relevant to British waters were mostly very broad types in comparison with many other detailed categories of terrestrial habitats: there were, for example, seventeen sorts of sand-dune habitat. Eventually guidance was produced interpreting what these five marine categories included. Even though it was imperfect, the Habitats Directive dragged the UK government into effective marine conservation. There had to be a Marine Regulations Act and the ability to designate areas to be protected, which had to involve statutory authorities who needed to create regulations to protect designated habitats and their sub-features.

We are still living with the peculiar habitats listed in the Habitats Directive, and this has led to many difficulties, false starts, wasted money and unnecessary measures. Similarly, the Species of Conservation Importance listed in the Ecological Network Guidance for MCZs is a very restricted and distorted list and they are the only species that will be included in designation orders. We still seem to be missing the challenge of protecting vulnerable and sensitive species, and notice is not being taken of information which has been paid for by the public conservation bodies and which should be being used much more effectively. Across the board we need to continue to ask questions about what *really* needs to be protected from damaging human activities: MPAs have their role, but effective environmental management and regulation based on the knowledge that we have about rarity and sensitivity of species are required as well. One of the recent changes has been the good work of the Inshore Fisheries and Conservation Authorities (IFCAs) and MCZs – they've really stepped up to the mark. The MCZs might have these peculiar designated features and be missing important features, but the IFCAs are doing a good job with valuable bylaws. The Scottish government is also to be congratulated on the measures that it has introduced to protect marine habitats and species.

In 2003, the OSPAR Convention, Annex V, set up the requirement for an ecologically coherent network of well-managed marine protected areas. This new imperative was also seen as an opportunity in the UK to fill in some of the gaps of the Habitats Directive not covering certain marine habitats. How to do it became the challenge. This OSPAR work was incorporated into the thinking and legislation for MCZs in the Marine and Coastal Access Act (2009). However, OSPAR had to admit that defining what they meant by an ecologically coherent network was difficult, although they went on to produce guidance and tests. Their tests for ecological coherence included representativity and replication, which are very good and found their way into the Ecological Network Guidance (ENG) (Ashworth *et al.* 2010), which became the 'bible' for the MCZ designation process. There were other criteria on viability and how big a site has to be to be viable, which are very difficult to translate into practice. There were questions about adequacy, which concerns whether the number of sites we intend to introduce is adequate to be representative but a good report has been produced on the subject by Rondinini (2010).

Then there is whole stupidity around networks, which in my view – and I am unpopular for saying this – is a flawed concept because the sea is very well connected for species that are highly mobile or have widely dispersing larval and juvenile stages. There is good and long-standing science to show that for other species their life history does not have a stage where dispersal is very great at all, and those are the ones that need protecting: they will not come back if lost. For some reason, the scientists involved in applying the OSPAR goal had lost sight of information about dispersal distances, and a colossal amount of time and money has been wasted on identifying 'connectivity distances' between MCZs. At the core of the Ecological Network Guidance was a list of 'features of conservation importance' (species and habitats). The listed species were clearly not the only ones that were rare, scarce, in decline or that were in need of protection. There were many worthy species and biotopes which should have been taken account of in terms of the designation process of MCZs and for management. Once again, like the Habitats Directive before it, there was a poorly thought-through list of features to be protected, and we will no doubt have to live with its consequences for a long time. So the ENG process was flawed because it was too prescriptive, whilst the requirement for an 'ecologically coherent network' was linguistically inept and scientifically indefensible.

THE FUTURE

What are the most important future threats to the marine environment?
I see the current obsession with economic growth and that nothing should be done to stop it, even if this means damaging the environment, as the main problem. There are many threats from fishing still, and work needs to kept up on that front – and on eutrophication and non-native species. We will need to look very carefully at the impacts of Brexit, although the UK will still be party to legal drivers that stem from the OSPAR convention.

Getting MPAs and achieving conservation goals has been a long haul. What do you see as the major barriers to marine conservation?
Why has it taken so long? There has been resistance from the commercial fishing interests, but things do seem to take a long time from the grand pronouncements in conventions and laws and then the time it takes to implement them and getting people to take regulatory responsibility for the measures needed. Sometimes the long time is appropriate and other times not so. You could argue that the MCZ process and development of guidance in particular should have been slower, to get it right. We didn't for example look at biodiversity hotspots – where will we get the best bang for our buck? Bringing people along with you takes time, but the thinking general public does now believe that the seas should be protected, and that has translated into action. It is clear that if the public and MPs' constituents want to see protection then that makes a huge difference.

What are the most interesting and promising new approaches?

Survey is still important. Autonomous underwater vehicles (AUVs) could provide a major way forward. The developments in sonar are very important but we must be cautious about claims regarding remote sensing to identify biological features. Social media – in my case Facebook – is providing a really interesting and interactive way to communicate with a network of like-minded colleagues. It is proving useful in getting prompt responses to questions such as what were the effects of the 2013–14 storms, and then being able to collate the responses.

When you retire, what would you most like to have achieved for marine conservation?

I'm retired in the sense of not having a full-time paid job, but in many other respects I'm busier than ever on many of the issues covered in this chapter. I look back on a career that has, I believe, achieved many of the scientifically based foundations of knowledge for marine environmental protection and management. There is more to do. Retirement does allow increased freedom to express my views, and I will hope that those views will have been influential. I hope to see the science-based information on marine conservation that I have initiated or produced maintained, developed and used by those who make decisions.

Are you optimistic or pessimistic about the future?

Optimistic, because, although progress can be slow, it is happening.

CHAPTER 7

Sue Gubbay

Sue is an independent consultant who has worked on marine conservation issues since 1984 for a wide range of government and non-governmental organisations in the UK and internationally. Her work has focused on marine protected areas (MPAs), integrated coastal zone management (ICZM), marine planning, and the scientific descriptions of the habitats and ecosystems of the marine environment.

Sue was born in India and moved as a child with her parents to live in London, where an inspirational schoolteacher, Miss Fitz, fired her interest in biology. She did her first degree in ecology at Lancaster University, where she received her introduction to classic rocky shore ecology. Going on to York University to do a PhD on barnacles triggered a remarkable event. As someone who hated the cold, and avoided swimming in the sea because of the fish, much to her parents' astonishment she took up diving to support her PhD research. Experiencing British marine wildlife at first hand on her first snorkelling trip to the cold waters of north Wales caught her imagination, and she wrote in her log book 'hooked for life'. One outcome was a project recording barnacles with volunteer divers. This was supported by the active community of divers at York, which included other enthusiastic marine scientists such as Rupert Ormond and Lynne Barratt. It was Lynne who suggested she apply for the Coastal Directory job at the Marine Conservation Society (MCS).

The Coastal Directory project with MCS in 1984 was a one-year contract to produce the first UK-wide listing and description of sites of marine nature conservation importance in UK waters. This was expanded into an inventory of coastal conservation areas and descriptions of major coastal and marine resources around the UK, and was published in 1985. It went on to drive the start of the Nature Conservancy Council's Marine Nature Conservation Review in 1987 and a series of seventeen regional Coastal Directories for the whole of the UK produced by JNCC. Sue eventually got a permanent job at MCS, working for them until 1995 on the Coastal Directory, MPAs and ICZM. Since then she has been an advisor to parliamentary committees, a consultant working for a wide range of UK and international marine conservation NGOs, and a board member of government conservation agencies Natural England and the Countryside Commission Wales (now National Resources Wales).

This chapter describes her work on the Coastal Directory and inventories of marine conservation features, the development of MPAs, the development of the European Commission marine habitat 'Red List', and ICZM in the UK. Her work has broken new ground, helping to convert important ideas on MPAs into government thinking, but also illustrates how a relatively small amount of funding can kick-start big changes, the time it takes for new ideas to become mainstream, and the practical difficulties of working with organisations, even those with supposedly common aims.

MARINE CONSERVATION AND ITS DEVELOPMENT

How would you define or frame your approach to marine conservation?

I don't use a book definition of marine conservation. For me it is about safeguarding the future of marine wildlife and habitats in a way that people can enjoy and use without leading to its destruction. It is not about preservation – things change – but rather about protecting what we value for now and the future. It is about people and having a positive interaction with wildlife and the environment. Throughout my career I have worked collaboratively with various teams, and I have found that this provides a constructive approach to many projects. I am a scientist by training and believe in providing scientific evidence and making this accessible to inform decision making.

How is marine conservation different from terrestrial conservation?

There are the widely known and recognised differences. For example:

- It is a much more difficult environment to work in and collect scientific information

- Many people find it difficult to visualise the species and habitats

- The legal requirements are very different, as are the way different sectors operate

- It is not possible to control what goes on in areas of sea by buying them.

Despite this, I believe there is actually a lot of common ground with terrestrial conservation – similar difficulties (politics, funding, priorities), similar values to promote (the importance of wildlife and the environment), and conservation in both environments embodies the same ideas (taking a long-term view and safeguarding for the future). You could debate whether it was harder or easier than terrestrial conservation, but in reality I believe this depends on the level of motivation and support, not on any fundamental differences in the ideas.

What are the differences between when you started and now?

The biggest difference is that marine conservation is firmly on the agenda for NGOs, governments and industry. In the late 1980s that was a hard message to get across, and I'd often go to meetings simply to point out that projects and organisations had forgotten to consider marine issues; that is not the case now. One reflection of the change is that there are now many more people working in marine conservation, and it is also more diverse and more detailed in its scope. Another noticeable difference is that in the 1980s, from an NGO perspective, some of the main antagonists were the statutory nature conservation sector agencies and government. In our meetings with them there were always long and heated discussions about the low priority for marine work. Today, although there are still questions about the resources being put into marine work, the agencies are seen more as a partner in helping to deliver marine conservation. Finally, and perhaps most significantly, marine conservation is no longer seen as a fringe area of interest but as an accepted part of government and industry obligations. It is easy to take this for granted, but it was not always so.

On the ground things have been achieved, not least with marine protected areas. We have a good number in the UK, and they are considered a useful part of marine conservation. The case for having them has largely been accepted, and although there are always some dissenters their complaints are most typically aimed at the detail rather than the principle of MPAs. On the international level, another big difference is the status of NGOs. When I started going to international conferences it was difficult as an NGO to get observer status, and if you attended you could usually only speak when spoken to and were excluded from decision making. Now NGO delegates are present in many different influential forums, including OSPAR, ICES, IMO and the European Commission. They are seen as key stakeholders in many stages of decision making, but their involvement was hard won. These are very significant positive developments, so when we get demoralised about progress with marine conservation, it is important to remember these changes and achievements.

THEMES

The Coastal Directory – sites of marine nature conservation importance

In 1977 the Nature Conservancy Council (NCC) published a *Nature Conservation Review* (Ratcliffe 1977). This identified sites of importance for conservation around the UK. The only problem was it did not extend below the low water mark! At the time, there was also no conservation legislation equivalent to Sites of Special Scientific Interest or National Nature Reserves to protect such areas even if they had been listed. The means finally came in the 1981 Wildlife and Countryside Act, which made provision for Marine Nature Reserves. In light of that, the Nature Conservancy Council wanted to do a Marine Nature Conservation Review (MNCR) to provide the context for site selection in the marine environment. For various internal reasons, this kept being blocked, so in 1983 the Marine Conservation Society (MCS) (Bob Earll) and WWF (Chris Tydeman) worked together to set up a project to try and leverage action by NCC. It was fully funded by WWF UK, £10,000 for one year of work starting in 1984 to produce the *Coastal Directory*. This was the first piece of conservation work I did, and the end result was the first inventory of potential sites of marine nature conservation importance around the UK.

The *Coastal Directory* also described a range of marine features, and the threats they faced, along with information on 246 sites of marine nature conservation interest. These had been nominated by scientists, divers and a range of other stakeholders and were assessed using criteria that were standard at the time. There were two main reasons for producing the *Coastal Directory*: the first was to raise awareness of important marine habitats and threats, and the second was to try and persuade the NCC and government to take on their responsibilities to fund the Marine Nature Conservation Review (MNCR). The first edition of the *Coastal Directory* was published in 1985 and was very positively received. The Department of the Environment (DoE) representatives thought that it was very useful and wondered – out loud at a meeting – why NCC hadn't done this. I would like to have been a fly on the wall at the meeting where NCC eventually decided to proceed with the MNCR, which started in 1987 (see Roger Mitchell and Keith Hiscock, Chapters 5 and 6). Many of the sites highlighted in the *Coastal Directory* 1985 are those that are the centre of designation and management now, so we were certainly on the right track more than thirty years ago.

With further funding from WWF we produced a revised version of the MCS *Coastal Directory* (Gubbay 1986a), which is now the copy most people have. Shortly after this DoE through Chris Reid persuaded NCC that they should produce a series of what became known as JNCC Coastal Directories. This was a substantial project overseen by Pat Doody and Nick Davidson and which resulted in seventeen regional volumes describing habitats, species and sites of importance around the UK coast. The MNCR which followed was the 'official' starting point for the MPA programme in the UK. The lesson from all this history is that a very small initial investment from NGOs can lead to long-term gains and prompt government to do the right thing.

The development of MPAs and reflections on the progress made

International conferences on marine protected areas routinely get audiences of hundreds, sometime over a thousand, from across the globe (see Timeline 4, page 83). The essential idea derived from terrestrial reserves has stood the test of time and is now part of mainstream marine management and a device for promoting marine conservation. The build-up to marine reserve legislation in the UK had been under way from the early 1970s. The early debate focused on the questions of whether it was worth having marine protected areas – would they only be lines on charts? – and what they would achieve. That question has now mostly been answered, and there is an enormous and growing body of evidence about what they achieve in terms of both marine conservation and multiple benefits affecting many other sectors. Various themes relating to MPAs seem to come in and go out of fashion, including no-take zones, multiple-use areas, big sites, small sites, networks of sites and 'integrated' sites. Despite this the basic principle that it is a good idea to have marine protected areas to safeguard biodiversity has become mainstream and widely accepted.

In the UK there were a number of key steps. The Wildlife and Countryside Act (1981) made provision for Marine Nature Reserves (MNRs) and got the ball rolling but was fundamentally flawed in terms of the conditions for establishing MNRs. There was a killer clause about 'complete' agreement being needed before designation could proceed, and this led to several high-profile failures. The process of MNR designation and stakeholder engagement left much to be desired, and although eventually Lundy, Skomer and Strangford Lough became MNRs under the Wildlife and Countryside Act, the paucity of sites relative to the size and diversity of the UK marine environment – three sites with only a tiny proportion within that totally protected – was pretty shameful. This outcome and problems with implementation led to the whole concept of MNRs losing credibility.

In the mid-1970s to 1980s there were a number of voluntary marine nature reserves (VMNRs) – at Skomer, Lundy, St Abbs, Kimmeridge and Wembury, to name a few. I prepared a guide on VMNRs, an approach which was successful in getting stakeholders interested and together at a local scale, and often introduced codes of practice for different groups (Gubbay 1986b). Nevertheless, it was clear that there was a need for some formal support by government, even for voluntary reserves to work. Around this time there was also debate about what type of MPAs we needed and their objectives. In response, working with Lynda Warren and a joint NGO/statutory agency group, we published some proposals for a system of MPAs in the UK. This idea was to promote just two types of protected area: type I, where there should be strict protection, and type II, where the legal regime would permit unrestricted use unless activities were expressly prohibited or regulated (Warren & Gubbay 1991). Today we would refer to these as 'no-take' and 'multiple-use' MPAs, with debate on their scope and objectives still topical in the UK 25 years on. Working at MCS, I also explored the idea of having 'joint-interest' protected areas where more than one partner could get some benefits. This was primarily with the fishing and archaeology sectors, but we made very little progress with the concept in the late 1980s. This type of approach is much more acceptable these days.

The EU Habitats Directive was the big breakthrough for MPAs in the UK, as it included the marine habitats, required progress, and allowed the UK government to be called to account. The inclusion of marine habitats was a big positive step but there were also issues because there were very few marine habitat types, compared to the terrestrial list. They were also all coastal in nature and an odd mix of categories. This led to a liberal interpretation of the habitat types, with all sorts of biotopes covered in these limited categories because of the prospect of protection underpinned by a European Directive. The outcome has been the establishment of both Special Areas of Conservation (SACs) for benthic habitats and some marine Special Protection Areas (SPAs) for seabirds. The Directive also passed a recent 'fitness check' (2016–17) where its role and relevance was reviewed. More than half a million responses came in during the three-month public consultation period – a remarkable number, and the largest ever response to an EU public consultation.

There have been a number of notable legal challenges to the implementation of the Directive. For marine issues one of the most significant was in 1999 when Greenpeace took the UK government to court and clarified the geographic extent to which the Habitats Directive legislation applied. There was no issue with this with the European Commission, as the Directive states it applies to 'the jurisdiction of European states', which includes any exclusive economic zone (EEZ), but the UK thought otherwise. Greenpeace put forward the view that the Directive covered the whole of the UK 200 nautical mile zone, not just territorial waters, and that it should include deep-water reef habitats such as cold-water coral (*Lophelia*) beds. The UK government fought this interpretation, with their agencies' advisors, but lost – and today there are more than twenty SACs that include offshore waters.

At the end of the 1990s the lack of progress with marine reserves and the rather peculiar and constraining nature of the Habitats Directive marine categories led to a high-level review group set up by government which identified the problems with MPA designation and management in the UK. This led on in time to the national focus of the Marine Coastal and Access Act (2009), which included a major section on a new statutory designation, Marine Conservation Zones (MCZs). These recognised that a much fuller diversity of seabed features (biotopes) needed to be the clearly identified focus for site selection and designation. There was much more consultation than during previous attempts to get MNRs, but in the end the government managed to lose the goodwill of many who contributed time and resources to the process by unexpectedly adding another tier of internal review to the proposals which came forward.

More recently, the process has settled down and management measures are being put in place and managed mainly by the Inshore Fisheries and Conservation Authorities (IFCAs) in England. A different process in Scotland gained better support and is now being successfully implemented, whilst in Wales the whole MCZ initiative appears to be 'on hold'. Having designated some sites, the challenge of ongoing management has brought a huge number of different stakeholders back to the table to help with the management of the MCZs, with the IFCAs taking a strong lead to instigate and enforce bylaws for the sites in England. There are still the usual complaints about whether they work etc., but the reality is that these issues are exactly the same as those internationally, and the fast-growing weight of evidence from all over the world has helped support work programmes here.

Reflections on MPAs

I could never have believed when I started that something so obvious as MPAs would take thirty years to become mainstream in the UK. Even now (2017) we still don't have a programme of no-take zones, apart from a few 'dots' (Lundy, the Isle of Man, Arran), where we can fully explore the consequences for both biodiversity and fisheries. The evidence from international northern and temperate waters shows clearly what happens and the benefits of NTZs, not least in providing a scientific baseline for many other issues. The lack of progress is shocking, but it clearly reveals that both the need to gather scientific evidence and what evidence there is from long-running programmes does not carry enough weight to deliver MPA programmes. I have no doubt that the Habitats Directive provided a sea change in the approach adopted by the newly devolved conservation agencies in England, Scotland, Wales and Northern Ireland. They and government had clear duties to implement the Directive – and although, once again, this has been tortuously slow, with marine SPAs still be designated and declared 25 years on, it provided the statutory backing for a comprehensive suite of MPAs in the UK.

There is still much to be done on the MPA front, as progress so far has been very much about multiple use areas, and has focused on seabed habitats and species and inshore areas. The deep sea and pelagic environments, which can also benefit from this approach as part of a package of conservation measures, have hardly been touched in this regard.

Designating MPAs has implications for many marine interests, but it is the interaction with fisheries which is typically the most controversial. Fishing in MPAs has a direct impact, but adding further regulation or prohibiting certain types of fishing altogether can encroach on what fishermen feel are inalienable established 'rights'. This has made MPAs controversial, confrontational and difficult. These reactions do not seem to arise to the same fervour with other major use sectors like offshore wind, where for one reason or another the fishermen seem to accept the development and to come to the table more constructively. The other issue in dealing with fisheries is that it is such a diverse sector, with different types of fishermen (not least the static and mobile gear operators) often having conflicts over use among themselves. They do not necessarily have the spokes-people or representatives that you find with other sectors, and there are instances where they have chosen not to participate in the process of creating MPAs, and then enter in the final stages to complain. It always seems as if the MPA advocates are providing all the evidence, whilst the fishing community simply complains and provides no evidence. This is all rather ironic because even from the early 1980s it was clear that the fishing sector could directly benefit from MPAs. It is worth noting, however, that today there are examples of joint projects and positive collaborations with the fishing sector, so it is not all bad news. Perhaps we are moving very slowly away from the situation where, whenever there is an MPA proposal there is kneejerk negative response from fishing interests.

Integrated coastal zone management – the need for a broader management context

I started work on a coastal zone management project at MCS with WWF support in 1987, with a project team comprising Chris Tydeman, Lynda Warren, Hance Smith, Martin Angel and Bob Earll. We referred to it as coastal zone management then – the 'integrated' has come along subsequently. The USA had their Coastal Zone Management Act (1972) providing impetus but with individual states having the freedom to plan for their coasts and sea areas. At the same time, in the background people like Hance Smith at Cardiff University, who was a member of our project group, were thinking about sea-use planning as well (Smith & Vallega 1991). The reason for all this interest was that land-use planning stopped at the water's edge. There was no real context for managing the competing interests of all the different sectors that operated at the coast and in the adjoining marine environment and the pressures that arose from this, nor an obvious link between the terrestrial and marine planning issues in this zone. So whilst projects were considered from their terrestrial perspective, the impact on the adjoining marine environment was much less clear.

The ICZM project reviewed other approaches to coastal management (Gubbay 1989) and provided case studies (thirty or so) of examples from the around the UK illustrating the lack of coordination in decision making and planning at the coast. Some general principles for coastal zone management were identified, and ideas on how it might be taken forward. The report drew interesting comments. One of these was that it was commendable as it was short enough to get across the main points as a read on a thirty-minute train journey! Whilst people could see that there was a need for something like this, there was no existing structure to support the practicalities, no legal framework and no financial support to help deliver it. The report did however prompt a meeting with representatives of different departmental officials with responsibility for coast and marine, and one of them commented that it was the one of the largest cross-departmental meetings he had ever attended.

An important driver for progress was the 1991 House of Commons Inquiry into ICZM chaired by Sir Hugh Rossi, which I assisted as a specialist advisor (House of Commons 1992). At the same time, the RSPB had been working with NCC and later the country agencies on a major campaign on protection of estuaries and birds (RSPB 1990), which led on to the agencies setting up a network of estuarine and coastal partnerships (see Peter Barham Chapter 21). Many of these survive to this day, another very positive outcome of this initiative. There was widespread support for the ideas of coastal zone management across the NGO sector through Wildlife Link. The coastal local authorities also became very interested, and several local authority groups emerged and survived through the 1990s. The select committee response was positive and suggested that government should get on with it, but the government response was fairly non-committal. I remember an observation from Sir Hugh at the time that it usually takes government at least ten years to do something after an inquiry like this. His prediction proved to be accurate: the idea of marine spatial planning, including sections of the coast, eventually got statutory support in the Marine and Coastal Access Act (2009).

The DoE did start to move on this and set up a Coastal Forum to consider the issues that were arising, with a wide range of stakeholders, and later there was that work on the best practice guide (1994) – and so gradually things started to fall into place. At the same time there were European initiatives, meetings and a number of coastal groups covering the European coasts. After a major demonstration programme across Europe in the late 1990s and much wrangling, an ICZM Recommendation, not a Directive, was produced by the European Commission in 2007. The principles behind ICZM which were promoted bear a remarkable similarity to those stemming from sustainable development.

From the mid-2000s the idea of marine spatial planning began to take hold (see Bud Ehler, Chapter 22), going beyond the coastal strip to include the geographic area out to the edge of the EEZ. The UK was significantly ahead of the game as it developed legislation in the Marine and Coastal Access Act (2009) to set up the legal status of marine planning in the UK (it is a devolved power), but in essence this built on twenty years of thinking and work. A European Directive on maritime spatial planning followed in 2014 (there was no agreement on an ICZM Directive), which promotes the same key principles and the importance of considering land/sea interactions when planning. The remit of marine spatial planning in this legislation is the whole of the territorial waters out to the limits of the EEZ. In effect marine planning provides the context and framework for the management of all activities in the marine environment. This is important for MPAs because they are, in essence, a type of marine planning zone.

We are now well into the process of marine planning in England, with the Marine Management Organisation, overseen by a group of departmental representatives, taking the lead and producing regional marine plans. This is a step change because it means that regulation is happening along similar lines for all the sectors and moving away from silos for particular activities. These developments went along with a reform of regulation and licensing of marine activities, and so licensing and planning are included in the same approach. What this also means is that one can look at the combined and cumulative effects of all of these sectors for the national interest. Marine planning is now taking hold all over the world – it makes a great deal of sense. The fact that it is now mainstream and a major activity of government is a real achievement.

The European Red List – working together
From 2012 to 2017 the European Commission funded a project to assess the threatened status of all terrestrial and marine habitats in the European Union. It was known as the Red List, and I was involved in it as a leader of the marine assessment process. Standard criteria were used (across both terrestrial and marine realms) and terminology which has been developed by IUCN for Red Listing species, where the status ranges from 'critically endangered' to 'not threatened', as well as 'data deficient'. A habitat would be like the kelp forest illustrated in Figure 7.1.

The project was a first attempt at a standardised and comprehensive assessment of the habitats across Europe. The marine aspect covered the Baltic Sea, northeast Atlantic, Mediterranean Sea and Black Sea. We chose to do this at a particular level of detail, EUNIS level 3, which is a middle level, which is rather like saying you are looking at kelp forests rather than specialist biotopes within this. More than 350 marine scientists from all over Europe contributed to the project, providing the information, working through the criteria, carrying out the assessments and preparing the reviews (Gubbay et al. 2017).

Figure 7.1 Kelp forest and understorey of the soft coral dead men's fingers, St Abbs, UK. Source: Keith Hiscock

One of the clear observations from the work is that there are many more distinct marine habitats that are endangered and threatened than are listed in the Habitats Directive (Figure 7.2). Scientifically, another major outcome is that three-quarters of the marine habitats were 'data deficient'. People always say we do not know enough about the marine environment and so it is too soon to do this type of analysis. In one way, they were right. There was insufficient information to assess many marine habitats – but that certainly did not mean there was no knowledge about these habitats. We brought together a great deal of information even for the data-deficient habitats, and having done a first assessment this sets a benchmark for what we know as well as identifying significant gaps. Ten or more years down the line when this exercise is repeated there is a process in place to build upon, but as all the data are accessible on the European Communities website anyone can build on it through their own studies (Gubbay *et al.* 2017, http://ec.europa.eu/environment/nature/knowledge/redlist_en.htm).

Even with common objectives, the reality of working on this scale was challenging, not least because of different conservation politics, styles, agendas, priorities and ways of working among different groups, making it a complicated task. We were fortunate in some ways that this was 'a project', which meant it was time-bounded. There was a

Figure 7.2 Subtidal seagrass beds, a critically endangered habitat in the Northeast Atlantic, and an assessment of marine habitats (86 habitat types at EUNIS level 4) in the Northeast Atlantic for the European Red List (Gubbay *et al.* 2017). CR, critically endangered; EN, endangered; VU, vulnerable; NT, near threatened; LC, least concern; DD, data deficient. Source: European Union

deadline and a set budget which participants were contractually bound by, so it had to be finished. One can imagine that if this was attempted, say by a government without these clear boundaries, it could have taken forever. In that sense, the European Red List assessment was just like the Coastal Directory and Coastal Zone projects – we had limited money and time and the expectation that the work would be completed. Conservation is not an easy process, and these conditions produced a good outcome on a tight timescale.

THE FUTURE

What are the future challenges?
What are the most important threats to the marine environment? These would include:

- Climate change – this will change many things for us

- Getting the environment to have a higher profile – this is a priority

- The deterioration of marine habitats

- The simplification of food webs by fishing – fishing down the food chain

- The time taken between observing problems and getting some action

- Limited funding for marine conservation – like the environment sector in general it is typically regarded as a luxury rather than a necessity

- Legal challenges to what is being done

- Changes in fisheries – the ownership regime and more work on MPAs.

What are the barriers to achieving marine conservation?
People often cite the main barrier as lack of funding, but whilst this is relevant I am not sure it is the most significant barrier. The areas where I have worked reveal that a huge amount can be achieved with seed funding, not least to leverage other funds. Rather, the issue is people's perceptions, a desire to work together, and seeing environment as valuable in their world view. I believe the main barrier to achieving marine conservation therefore centres around the psychology behind change and understanding this. Finding common aims and objectives is particularly helpful, as is exploring how others can benefit from your aims. All standard thinking in negotiations.

Having good statutory backing for conservation is also important, and politics plays its part. Politicians can be very supportive but still not stick their necks out, so a great deal of work is needed, often behind the scenes, to try and change things. There is a lot of discussion currently about whether we have sufficient scientific knowledge and evidence to make decisions, for example about good places to have MPAs – and in particular, do we know enough? Well, we will never know enough, and the reality is that we need to make decisions anyway. I see reluctance to act, rather than lack of knowledge, as the greatest barrier – but it is often the latter which is cited as the reason for not acting.

What are the most interesting and promising new approaches?
Although I'm not a fan, or an expert on the economics and valuation of the environment, we do need to pay significant attention to this. I have come to conservation from a scientific background but realise that knowing more about the economic arguments for

conservation and the economic drivers of the other sectors is important.

Raising people's awareness about marine conservation remains fundamentally important, and today there is huge potential through the digital world. For example, Google Seaview, where you can go for a swim around a marine park, brings that environment to people. Spreading the word and making it fun is really important. I always remember a colleague, Bob Foster-Smith, with whom I did a seashore wardening course, saying you need to raise awareness of the marine environment and conservation issues. From this can stem appreciation and finally concern. If you built on the first two, the third would follow.

An example of a different type of positive approach is in policy, and it has been a long time coming. The most recent revision of the Common Fisheries Policy finally acknowledged, up front, that fisheries need to work with the environment. This has many implications. One is that the European Marine Fisheries Fund (EMFF) can be used to support the management of MPAs as it relates to fisheries. Who would have thought that funding for marine nature conservation would emerge from that source? The big charitable funds and trusts, Pew and so on, helping to organise major programmes to effect change is also a big difference, as is the number of new people coming into the field.

When you retire, what would you most like to have achieved for marine conservation?

For the UK to have a significant area of its seas as fully protected marine reserves. I believe these can deliver for nature conservation at the same time as acting as reference areas for science and properly demonstrating the benefits of MPAs in UK waters. Multiple-use MPAs have their place, but somewhere along the line we need to have the courage of our convictions and get some highly protected areas. This is considered acceptable on land, where extractive activities are certainly not seen as compatible with nature reserves. Why not in the marine environment?

Are you optimistic or pessimistic about the future?

My view changes, although I am generally optimistic. We have made great strides over marine water quality, for example. But it is the fishing sector that gives me the most cause for concern. It should thrive, but not at huge cost to the environment, as that does not help anyone in the long run. Humans are very adaptable, and they'll need to be to get us through a very uncertain future. However, I don't think we will go back to the early days when you'd be the only person in the room saying that the marine environment was important.

TIMELINE 4

The development of marine protected areas (MPAs)
Compiled by Sue Wells, National Trust, UK, based on Wells et al. (2016)

1879	Designation of Royal National Park, New South Wales, Australia. First known protected area to include subtidal and intertidal areas as well as terrestrial habitat.
1958	Designation of Exuma Cays Land-and-Sea Park, Bahamas. First underwater park in its own right.
1962	First World Congress on National Parks, Seattle. Resolution 15 encourages the establishment of marine or underwater parks. First recognition that MPAs were needed globally.
1966	Special Symposium on Marine Parks, Tokyo. First international meeting about MPAs; recommended the development of a 'systems approach to preservation of marine environments'.
1971	New Zealand Marine Reserves Act. First national legislation for MPAs.
1972	US National Marine Sanctuaries Act. Legislation for MPAs.
1974	UN Environment Programme Regional Seas Programme launched. The idea that marine conservation needs international/ regional collaboration and the governance framework of regional conventions, which are now key drivers for MPA establishment.
1975	Great Barrier Reef Marine Park, Australia, established: significant at the time for its size, zoning and multiple-use approach.
1982	Publication of first guide to MPA management (Clark *et al.* 1982, revised in 2000). First global guidance on establishment and management of MPAs.
1991	International meeting on Wild Ocean Reserves. First consideration of MPAs on the High Seas.
1994	COP1 – First Conference of the Parties of the Convention on Biological Diversity (CBD): Jakarta Mandate, providing a rationale and framework for MPA establishment and management, adopted by all parties.
1994	IUCN Guidelines for Protected Area Management Categories.
1998	OSPAR ministerial meeting in Sintra, Portugal, agrees to promote the establishment of a network of MPAs.
1999	Pelagos Sanctuary, Mediterranean: a transboundary cetacean MPA.
2000	Locally Managed Marine Area (LMMA) Network established. Global recognition of need for full stakeholder participation in MPA establishment and management.
2004	World Bank score card to assess progress in achieving management effectiveness goals for MPAs.
2005	First International MPA Congress (IMPAC1), Geelong, Australia.
2005	World Heritage Marine Programme established.
2006	Papahānaumokuākea Marine National Monument, Hawaii: one of the first 'very large' MPAs.
2010	CBD – Aichi Target 11, a network of comprehensive and ecologically coherent protected marine areas effectively managed and covering at least 10% of the oceans to be established by 2020.
2010	First OSPAR High Seas MPAs: six sea mounts and areas of sea underpinned by regulations from a fisheries organisation (NEAFC), as OSPAR has no regulatory powers in that regard.

CHAPTER 8

Joan Edwards

Joan is currently the Head of Living Seas for the Royal Society of Wildlife Trusts (RSWT) and is based in Plymouth, where she lives with her family. She was inspired to study marine biology at Bangor both by summers on the coast at Barmouth, north Wales, and also by watching the exploits of Jacques Cousteau and BBC wildlife programmes. She took up diving at university, and it has played a key part in developing her interest in the marine environment.

After graduation she moved to Plymouth, where she volunteered at the Marine Biological Association. In 1987, she moved to the Devon Wildlife Trust as their first marine conservation officer. There are 47 Wildlife Trusts that operate at a county level, and Joan was the first marine conservation officer to work for a Trust. Since then, her work has developed to supporting national marine conservation programmes for the Trust movement as a whole, and she now leads the marine programme at RSWT (the national umbrella body for all the local Wildlife Trusts). In 1995 Joan oversaw a major capacity-building programme in the Wildlife Trusts – the Joint Marine Programme. This was supported by WWF, and led to a broadening of marine conservation commitment and expertise in many coastal Trusts.

Joan played a key part in the development and promotion of a Marine Bill from 2000 onward, which nine years later became the Marine and Coastal Access Act (2009). This involved working in effective partnership with NGO colleagues, as well with government, in what has been a landmark piece of legislation. It introduced marine planning to UK waters, Marine Conservation Zones and many other reforms that have mainstreamed marine conservation practice into government and its agencies. The implementation of the Act has proved to be no less challenging, with a series of changing governments and, most recently, the forthcoming exit of Britain from the European Union.

MARINE CONSERVATION AND ITS DEVELOPMENT

What is marine conservation? How do you define it and what are its key ideas?

It seems to me that it's a vocation. The people that do it well – that are successful – are passionate about their subject; they want to see the marine environment managed in the right way, not only because it's the right thing to do, but for the benefit of humans as well. On land, conservation is easier because you can buy land, which you can't do in the marine environment. This means a wider range of people need to be involved and there needs to be a common desire to want to protect it; it's not just about yourself. People need to be inspired, which is more than just a methodology, about being a marine biologist or knowing the regulations. Passion is important, and marine conservation is certainly more than just being able to do an environmental impact assessment. This message is something we have to ensure is ingrained in our own staff within the movement.

What are the differences between our current situation and when you started?

When I started there were just a handful of people; now there are hundreds! When I started at the Wildlife Trusts in Devon (1987), I was the lone voice for marine conservation in the Trusts. Today there are over forty dedicated marine officers. There is little doubt that people have a much better idea of the marine environment today than when I started. We still have an issue, in that because of programmes like *Blue Planet*, people in the UK know about coral reefs or tropical sharks, but don't realise what wonderful marine life we have around our own shores as well. In the Wildlife Trusts, the Living Seas Programme has equal billing, although not equal resources, with the major Living Landscape programme we run for land-based conservation.

At the outset protecting the marine environment was a novelty and we were considered out on a limb. When I started we only had the *Coastal Directory*, written by Sue Gubbay, but the knowledge we have about the marine environment and its ecosystems has increased a hundredfold over this period. Nevertheless, as the Marine Conservation Zone (MCZ) process has recently shown, we still need more. We are in a better place – we have the people, there is greater political will and we have legislation – but progress on various fronts is still very slow. The legal foundation we have now, as a result of the Marine and Coastal Access Act (2009), together with the Marine Strategy Framework Directive (MSFD), is much more coherent, but there is still a major job to be done to translate this into practice (see Timeline 5, page 93). For example, we still do not have an ecologically coherent network of marine protected areas and we are still to fully realise the potential of marine planning.

So the foundations have been laid over the last 20 years and now we have to start making it happen. The implications of Brexit are potentially very serious for marine conservation. We will be fighting for the momentum to better protect our marine environment to continue and improve, rather than being stalled at this critical junction.

How is marine conservation different from terrestrial conservation?

On land there is more certainty. You can see what you're doing. Lots of people can get physically involved and you can easily see what measures such as restoration look like. It's much easier to convey messages of a healthy environment to the public and give them a feeling of ownership about it, for example by showcasing local nature sites on TV. Clearly the marine environment is much less accessible by comparison, although diving has been particularly important in making people more aware. In the marine environment, it is harder to mobilise people and inspire them. That's why photographers like Paul Naylor and TV film makers have such an important role in marine conservation; their images played a key role with politicians and the Marine Bill (Figure 8.1). For my terrestrial colleagues in the Trusts, many of their activities are now well established, even traditional, whereas in marine conservation, techniques are often viewed as being slightly mad, and there is a constant need for new, innovative and inspirational approaches to take people on the journey. With marine you can't just take people to a reserve.

Figure 8.1 The tompot blenny was an iconic species used to promote the UK Marine Bill. Source: Keith Hiscock

THEMES

What aspects of marine conservation are of most interest to you?

Marine conservation is seldom dull. One aspect of this job I find really interesting is working on emergencies and emergency planning. If I was to get another job, I think it would be in this area. I work well under stress, and I'm often called in to deal with emergencies in all aspects of the Trusts' work, not just marine. There is an energy to it as it often requires different, novel and practical approaches. During a poly-isobuty-lene (PIB) pollution incident in 2013, people laughed when I went out collecting 'oiled' birds on my canoe in Plymouth Sound, but that got picked up by the TV, which led to meetings with the Secretary of State and Prime Minister. Although disasters are, of course, a bad thing, and you certainly don't want them too often, they can provide a major opportunity. They have the ability to raise the public's awareness of the marine environment. During the bird wreck caused by the storms in 2014, in which over 32,000 seabirds died, even my mum rang me up to talk about it! The events provide a hook to put the marine environment in the news, and remind people that it should not be taken for granted. Subsequently it enables us to talk about a wide range of other things we are doing to protect it. Responding to unforeseen events requires a mix of talents: you need experienced people, to understand what is needed, but bringing in new people can also be a major advantage, in that they bring fresh eyes and innovation to the response. We often deliberately look for this mix when we develop new campaigns.

Often really interesting outcomes emerge from taking a completely new approach and thinking well outside the box. In 2004 I got involved in a project called Invest in Fish South West; it lasted three years and was really satisfying. The idea was to prepare a plan that would look to ensure that there were enough fish to catch in twenty years' time, based on the idea that there was a healthy environment to sustain this. It involved all sorts of approaches and it was facilitated innovatively and carefully. One of the things stakeholders did was to stay with one another for a weekend in their homes. We spent a lot of time together and got to know each other much better – for example, how our families view what we do and what else we are involved with in our local communities. This built trust between us so that we could have sensible and frank discussions about the future. We discussed the challenges of reducing fishing effort, talked about '*no* fishing areas' and '*go* fishing areas' (where scallops should or should not be fished) and a whole range of issues that would have been difficult in other contexts. I'd like to be able to do more of this in the future.

Capacity building

From the outset of my time at the Wildlife Trusts I've been involved with capacity building, not perhaps a very obvious direct marine conservation task, but actually essential in taking things forward. The need to engage more people became very clear to me from when I started with the Devon Wildlife Trust and then within three months working with RSWT. Staying the only person in the Trusts on marine conservation wasn't really an option. My thought was that what individual Trusts could do well was to give a strong local focus to support national campaigns. Early on I was encouraged to prepare an EU grant bid, which made me think hard about what we could achieve and what it would cost. I put forward the idea of four or five 'regional' staff – a team based around the country. Initially this didn't go down very well with the RSWT powers on high! But with funding and a great deal of support from WWF UK, we got the Joint Marine Programme (JMP) off the ground.

For over ten years the JMP built the capacity of the Trusts to work on marine conservation issues. One of the tasks of these new marine conservation officers was to spread the message that marine conservation was important *within* the Trusts and their staff. There are about 2,000 people working for the Trusts currently and, of these, over forty are dedicated marine officers. Although we always had five-year development programmes, it wasn't until Stephanie Hilborne took over as CEO of RSWT in 2005 that marine really went beyond just an afterthought. Consequently, we developed the Living Seas Programme within the Trusts which is equal partners with the main terrestrial theme, Living Landscapes. At chief executive and chairman level, all of the Trusts now have a clear idea of marine conservation, and even inland Trusts are doing their bit where they can. The JMP succeeded in making the Trusts much better prepared for future challenges. We adopted a similar approach to working with the wider conservation movement in relation to the Marine Bill.

The Marine and Coastal Access Act

The work on what became the Marine and Coastal Access Act (2009) had a very long and tortuous gestation, but it too required trust and partnership working by a very wide range of people. An extremely succinct version of its development goes something like this. By the mid-1990s it became obvious that the legislation governing the marine environment was ineffective, unclear and not really fit for purpose. This was true for marine conservation and many other activities. Some of the issues included: the lack of effective ways of protecting areas of sea for marine conservation; the role of the Crown Estate in planning;

the Coastal Protection Act (1949); the lack of transparency in licensing, for example, harbour dredging; the management of aggregates; and the outdated legal basis for the Sea Fisheries Committees. Multiple pieces of legislation had arisen over many years in an ad hoc and piecemeal fashion; in short, it didn't fit together and was a mess.

In 1998/89 the Countryside and Rights of Way Bill (CRoW Act 2000) came up for discussion to address a number of terrestrial issues. A bolt-on paragraph covering marine protected areas was put forward but eventually withdrawn. However, various discussions emerged from this between wildlife NGOs, government and industry stakeholders which clearly highlighted the issues. As a result, the Review of Marine Nature Conservation process was set in place (not to be confused with the earlier Marine Nature Conservation Review work on marine habitats around the UK). Representatives from many interest groups worked together, and although the process dragged on, eventually a report was produced in 1999. During this period there was really good liaison between industry and the wildlife NGOs.

Matters were brought to a head when Sian Pullen (of WWF) and I rather quickly put together a paper on a White Paper for UK Seas and presented it at Coastal Futures in January 2001. We'd not consulted on it and just did it pretty much overnight, but we got backing from our bosses, Chris Tydeman and Simon Lyster. We produced some text for *Marine Update* which prompted some very heated arguments. There were various other steps – for example the RSPB went for a Private Member's Bill which just covered marine conservation – but many of us wanted something much broader. All this time momentum was building, and in 2004 Defra announced they would develop a Marine Bill. Also during this period, the Seabed User and Developer group (SUDG) was set up by the Crown Estate, to provide a coherent voice for marine industry, with the leadership of Frank Parrish, Carolyn Heeps and Bob Earll, modelled on Bob's experience of Wildlife Link (see Peter Barham, Chapter 21).

The Marine Bill became the Marine and Coastal Access Act (2009) and became the largest single piece of legislation to be passed by UK government. It set out some very major steps, including: reform of the marine licensing regime; the establishment of Inshore Fisheries and Conservation Authorities (IFCAs); the statutory basis of marine planning; the creation of a new Marine Management Organisation (MMO) in England; and long-awaited marine conservation legislation that created Marine Conservation Zones (MCZs) in England and a suite of marine protected areas in Scotland.

What really characterised the process of the Bill through its various stages was:

- All-party support by politicians

- Highly effective leadership and consultation by Defra's officials (John Roberts, Dave Bench and Diana Linskey)

- Considerable buy-in from industry, led by SUDG

- Highly focused and coordinated efforts by the NGOs through Wildlife and Countryside Link

Once again it was our ability to sit down face to face and discuss issues sensibly, across a wide range of sectors, that enabled this work. To my mind, the support of politicians, Ben Bradshaw for example, and industry, notably Peter Barham (then ABP, and more recently SUDG) and Mark Russell (BMAPA), played a huge role in this process.

I have been chair of the marine group of Wildlife and Countryside Link since 2002 (see Box 2, page 32); I resign on a regular basis but they keep on asking me back. I have no doubt that Link is very powerful *when it works* because it channels the views of organisations representing 8 million members. It is a very effective partnership and it works when there is trust. We saw it at its best throughout the Marine Bill process as there was clear trust and shared goals amongst the members of our team. Each organisation had the confidence to undertake its own campaigns – like the Wildlife Trusts Petition Fish project – whilst having the trust to enable individuals to speak to the media representing a collective view. This trust issue is critical. In the PIB incident described earlier, I was able to represent multiple groups' views – organisations like RSPB, MCS and the Trusts – on the *Today* radio programme, a situation difficult to imagine without an effective Link organisation in place.

Getting an Act was the first step. Reflecting on this eight years after the Act came into force, my main feeling is one of frustration. The NGOs and politicians still aspire to see progress made, but for various reasons progress has been very slow. There are probably many reasons for this. We in the NGO movement were mainly geared up for campaigning rather than the implementation phase. The reality is that reorganisations of this scale do take time. The creation of the MMO and IFCAs and the subsequent need to get new processes off the ground, such as marine planning, are no easy task. But there seems to be a lack of ambition which is difficult to pin down.

The MCZ process in England has progressed, but nobody would say that it has proceeded well – in many ways exactly the opposite. The devil of legal detail in the way the Act is drafted has also been a problem, despite our seeking very expensive legal advice during the Bill process. All sorts of issues seem to have conspired to make it more complicated and expensive than is actually necessary. Once again we discovered that we need even more information on the marine environment, and this is still a huge challenge. We have also been held back, to some extent, by the statutory nature conservation bodies, prompted by some major changes in the last parliament (2010–2015) in the way they work, as they are no longer able to promote policy and are effectively gagged. They have also suffered and will continue to suffer more financial cuts. An obsession with measuring every feature has become counterproductive and slowed down the designation process. Marine planning has fallen into a similar trap. In summary, one could say we have the foundations and the tools. What we need now is the people who want to and are allowed to make a difference.

I did marine biology at university, but I hope you can see that this didn't really prepare me for the range of challenges that I've come across in my marine conservation career. The constant change is actually one of the attractions of the job. Trust is an essential ingredient in achieving anything, both with colleagues in the nature conservation movement and when working with a huge range of different interests. The ability to work with a wide variety of people, face to face, is key to success.

THE FUTURE

What are the most important threats to the marine environment? What needs to be done about these?

The two issues I'm most worried about at present are fishing and development. Fishing, despite the reform of the Common Fisheries Policy, still poses huge challenges; there are simply still too many fishermen chasing too few fish. Nomadic fleets for scallops and *Nephrops* pose particular problems, as they can cause massive environmental damage. It is not that there shouldn't be fisheries for these species, but *where* this takes place should be controlled. Scallop ranching should be explored in the UK as it has been in Brittany, with considerable success. There are many improvements that can and are being done to help mitigate the worst effects of fishing. Again, the role of the Wildlife Trusts at a local level is to provide insights, raise awareness and promote initiatives which make a real difference to developing sustainable practices; this is where we are seeing some progress.

For example, the Cornwall Wildlife Trust worked with local fishermen and technologists to develop what is known as the 'banana pinger'. Their work on cetacean and seal strandings has shown that porpoises often die when caught in inshore set nets. The banana pinger is a device which emits sounds that deter porpoises from coming near the nets, to prevent them becoming entangled. It's a brilliant example of local action, working with the fishing sector, gathering and reporting the evidence, helping develop new technology products and working with regulators to get the devices deployed. Even in this project quite how you describe and encapsulate the skills needed defies simple description. It shows just what marine conservation involves and how working *with* people from a wide range of interests is essential. The unique setup of the Wildlife Trusts has meant that the value of this project has been seen beyond just Cornwall. Sharing knowledge and expertise has been shared with colleagues in the northeast, and the banana pinger has since gone on to be trialled in the North Sea.

The local sustainable fish guides are also an important step towards reminding the supply chain, producers and consumers of the issues that fishing is posing. As long as the information in these guides is honest and up to date they can provide a valuable prompt to action. The Cornwall Wildlife Trust's recent guide provides another example of how this idea, developed nationally by organisations like MCS and Greenpeace, can be given a local face. Local sustainable fish guides have also been produced in Dorset and Wales.

Another example of how local expertise can be brought to bear on fisheries issues relates to the major breakthrough that was achieved in 2013, when UK government recognised the need to control damaging fishing practices in the UK's marine protected areas – both in Special Areas of Conservation (SACs) and in MCZs. Although the final step in this process was the Client Earth–MCS complaint, an enormous amount of work had been done in Lyme Bay, Strangford Lough (taken forward by the Trusts), the Firth of Lorne, Scotland, and the Fal SAC in Cornwall (highlighting the damaging impact of scallop dredging on long-lived seabed species). Fifteen years or more since this problem was first highlighted, with the passing of the Marine and Coastal Access Act we are now seeing the IFCAs enacting nature conservation bylaws to protect marine areas from damaging fisheries. As you can see, marine conservation can take a long time and involve many different players and strands of work.

In relation to fisheries in general I feel there is still a fundamental need for an effective long-term vision. Using the jargon of today, we need to create a safe space where all the parties interested in this, including industry, can come together and work through the issues. Although *Fishing News* vilifies and tars all wildlife NGOs with the same brush, we need to move beyond stereotypes and megaphone diplomacy. We don't want to close fishing down; rather, we need an industry which is sustainable but doesn't do enormous amounts of environmental damage.

Development of renewable energy is currently posing us big questions, and over half of our officers are currently involved with this work. With offshore wind, the increase in sheer geographic scale, with the footprint of the third round of developments, dwarfs anything that has gone before. We know little of the cumulative impacts of wind-farm development on this scale and I'm worried that we will be left with a legacy that nobody has foreseen. We are working with Natural England on ways of mitigating the impact of these developments, as well as with the industry, who also want to see issues resolved. The industry representatives are open to these collaborations, because they don't want to make a mistake either. One particular issue is the impact on porpoises of the noise of piling during the installation phase. With the smaller sites it was possible to envisage that animals would simply move away to avoid this, but with the larger sites this could be less of an option. We simply don't understand what the population impacts of disturbance on this scale would be. Underwater noise is a particular problem because there seems to be a wide level of disagreement among experts about the nature of the threat. The wind farms are getting development consents even though the developers, regulators and government know that development has an effect. We are discovering that the way the planning system works regarding these developments requires very particular knowledge, and we are actively engaged with all concerned to make sure our views are heard and listened to. Having a presence at planning hearings is very important in this regard, and we strive to attend as many as possible.

Until recently, the lack of SACs for harbour porpoise has meant that, in reality, there has been little spatial protection for the species from these huge developments. We were delighted that the European complaint by WWF UK has resulted in the UK finally taking forward designations for harbour porpoise SACs, something which we also have been advocating for many years. The management and control of development within these SACs is still a huge challenge. Over a year on from the close of public consultation, we still do not know how wind-farm construction will be managed in the southern North Sea SAC and what further mitigation of noise will be required. We continue to be in direct talks with developers and JNCC about how this will work.

What are the real and hidden problems or barriers that make progress harder than it should be?

For obvious reasons, there is no straightforward answer to this, but there are problems, and sometimes simply understanding what is happening at any one time can be difficult; old certainties are simply turned on their head and become unworkable. Two examples.

Firstly, we currently (2017) have a government that is pro-development. On land, even when we have protected sites, either Sites of Special Scientific Interest (SSSIs) or SACs, these designations are not stopping development. Despite three major complaints and two judicial reviews of the development of the high-speed rail link, HS2, these have made no difference at all to a project that looks set to destroy many sites of conservation importance. We are in what is one of the greatest surges of development since the war, and economic and business interests trump all opposition. Recognising this is very important. Judicial review (JR) can be successful in demonstrating the failure of process, but still make no material difference to the development outcome. Similarly, complaints to Europe can fail very easily for no apparent reason.

Secondly, evidence, proof and MCZs. I often summarise this by saying that government seem to want to map where every last seahorse is in an MCZ. For MCZs, an obsessive level of detail, evidence and proof is being sought by the nature conservation agencies, driven by their government departments. We currently find ourselves in a situation where if I want to prepare a report on MCZs, for it to be recognised it has to be peer reviewed. This is bizarre and conflicting on a variety of levels, not least since other government policies – over badgers, for example – clearly do not have scientific rationale or even the backing of past presidents of the Royal Society, many of whom were government chief scientists. It seems that Defra and their agencies, in this case Natural England, are petrified of JR by the fishing sector. This is in itself is odd, and perhaps a diversion, for two reasons: (a) the government has weakened JR over the last five years, and (b) the fishing industry is not organised enough to take this forward. It seems as if the fear of JR is being used as a deliberate excuse to justify the current MCZ designation approach. The levels of proof being sought for marine protected areas are obsessive, needlessly expensive and make very little scientific sense – yet this is the strange world in which we are trying to make progress.

What are the most interesting and promising new approaches to marine conservation?

There is a constant stream of developments that are difficult to judge. Whilst social media is all-pervasive and we, like all other NGOs, spend time and money using it, I'm not sure that it has really made any difference at all. You need new ideas and there is constant need to keep reinventing yourself. You have to use social media but on its own it doesn't make the difference. It has a place, you need a tool kit, including the 'old things' like petitions. One technological advance I do think is fantastic is the GoPro camera, as it raises the level of awareness among the population as a whole, rather than just 'preaching to the converted'. As often as once a week the *GoPro Video of the Day* is a marine video. Currently on YouTube the most popular videos often feature marine topics on their list of most watched – and over time the cumulative effect of vast numbers of people seeing these videos will make a huge difference to awareness.

But how do you really make a difference? For me, face-to-face meetings seem to count for a lot more with opinion formers and regulators. Building reputation and trust is an essential prerequisite before one can really achieve things. For example, with the PIB pollution incident, the International Maritime Organisation (IMO) regulations were changed at an almost unheard of speed on the back of working closely with people within government whom we knew.

Experience has also taught us very specific lessons about the way the system works and how to work with it to make a difference. Let me give you an example. With the MSFD consultation on 'measures' we didn't even do a Trust response and instead we did one response for all the collective NGOs through Wildlife and Countryside Link, as we were advised to submit just one response. We know that automatic computer systems are used to search for key topics and so we drafted our response in chapters that reflected key search words. This is not necessarily new, but it is insightful and essential in getting our voice heard.

When you retire, what would you most like to have achieved for marine conservation?

An ecologically coherent set of marine protected areas and the ability to say that we did it. That will be huge advance on having just one – Lundy – when I started. Although there is the challenge of making the aspiration and the reality meet, it will be a huge achievement. Oh yes, I would also like to have saved the manta rays in Indonesia, but I fear that I'm doomed on that front.

Are you optimistic or pessimistic about the future?

Optimistic! Twenty-five years ago all we had was Lundy, I was the only person in the Trusts and *marine* was 'out of site and out of mind'. Now there are more people, who are more aware and doing more things. But clearly we need to carry on building. We might need to stop fussing with the detail in this country and meet the global challenges facing marine conservation. This is where the biggest challenges really lie.

TIMELINE 5

Marine conservation: the European phase, 1970–2017

1973   UK joins the European Union. Now subject to Decisions, Recommendations and Directives, including what becomes the Common Fisheries Policy (CFP).

1979   Adoption of Council Directive 79/409/EEC, which provides the basis for the EU Birds Directive (2009/147/EC).

1980s   The precautionary principle emerges in relation to pollution.

1984   First North Sea Ministerial (NSM) meeting, Bremen, Germany, on the protection of the North Sea.

1985   Chris Rose and Adam Markham write in *ECOS* on the need for a Habitats Directive; Stanley Johnson from the UK produced an early draft of the Directive.

1985   The Environmental Impact Assessment (EIA) Directive (85/337/EEC) introduced; it applies to a wide range of public and private projects.

1987   Second NSM meeting, London. Agrees need for wide-ranging measures to help stop marine pollution, recognises the precautionary principle and the need for a scientific task force. A deadline for halting sewage sludge dumping at sea is agreed. A quality status report for the North Sea is produced.

1988   Phocine (seal) distemper virus kills a large number of common seals in the North Sea.

1991   EU Urban Wastewater Treatment Directive (Council Directive 91/271/EEC) puts in controls to stop the input of sewage into estuaries and the sea for conurbations of over 10,000 people.

1991   ASCOBANS introduced. It is a regional 'Agreement on the Conservation of Small Cetaceans of the Baltic and North Seas' under the auspices of the UNEP Convention on Migratory Species, or Bonn Convention, in September 1991, coming into force in March 1994. It covers all species of toothed whales (Odontoceti) in the Agreement Area, with the exception of the sperm whale.

1992   The EU Habitats Directive (Council Directive 92/43/EEC), which along with the Birds Directive gives member states responsibilities for declaring a network of protected areas of sea (Natura 2000), in terms of Special Conservation Areas (SACs) or – under the Birds Directive – Special Protection Areas (SPAs).

1995   Fourth NSM meeting, Esbjerg, Denmark. Fisheries included for the first time, especially sand eels. The Brent Spar is towed up the North Sea just prior to the meeting, causing a PR storm.

1996   London Dumping Convention protocol. In 1996, a special meeting of the Contracting Parties adopted a revised protocol to replace the 1972 Convention. This reflects the global trend towards precaution and prevention, with the parties agreeing to move from controlled dispersal at sea of a variety of land-generated wastes towards integrated land-based solutions for most, and controlled sea disposal of few remaining categories of wastes or other matter.

1997   Intermediate Ministerial Meeting on Integration of Fisheries and Environmental Issues (IMM 97), Bergen, Norway, to develop and apply an ecosystem approach to management of human activities and protection of the North Sea.

1998 OSPAR meeting in Sintra, Portugal. Ministers agree to set up a network of marine protected areas (MPAs), to be completed by 2016.

1998 *Marine Nature Conservation Review* published – JNCC.

1999 UK High Court finds in favour of Greenpeace in the case against the UK government regarding implementation of the Habitats Directive in offshore waters (CO/1336/99).

1999 Review of Marine Nature Conservation started.

2000 Water Framework Directive (2000/60/EC) commits EU member states to achieve good qualitative and quantitative status, including estuarine and coastal waters.

2000 OSPAR first *Quality Status Report*. Many things have been supported/developed with reference to this series of reports, from MPAs to improving water quality.

2002 EU ICZM Recommendation requires Member States to establish coastal management strategies; a weak agreement largely superseded by the Marine Planning Directive in 2014.

2004 Review of Marine Nature Conservation Defra report – 'strong arguments for a new approach to managing our seas, including a new Bill' (Tony Blair) – which was announced in 2005.

2004 Wadden Sea European Court of Justice (ECJ) judgment sets important case law that a fishing activity constitutes a 'plan or project' under Habitats Directive Article 6(3) and can therefore only proceed after an 'appropriate assessment'.

2005 *Charting Progress*, Defra report: an integrated assessment of the state of UK seas. Further reports in 2010.

2006 Sixth and final NSM Meeting, Gothenburg, Sweden. OSPAR and ICES take on the mantle.

2008 Marine Strategy Framework Directive (MSFD) (2008/56/EC) aims for Good Environmental Status (GES) of the EU's marine waters by 2020. It is the first EU legislative instrument with an explicit objective related to the protection of marine biodiversity.

2009 Marine and Coastal Access Act, UK. Informed by a comprehensive review of marine management, this includes provision for marine spatial planning, revised licensing arrangements, and provision for Marine Conservation Zones (MCZs) in England and Wales and MPAs in devolved countries.

2013–2014 In England, fishing in European marine sites addressed by the 'matrix' or the 'revised approach' after Client Earth and MCS threated UK government with legal action. Four high-profile cases of scallop dredging in SACs tip the balance.

2014 EU Marine Planning Directive (2014/89/EU), establishing a framework for maritime spatial planning.

2016 The UK in a referendum decides to leave the European Union. Reference to European Directives downplayed. The potential role of international conventions such as OSPAR and ICES comes to the fore once again.

CHAPTER 9

Dan Laffoley

Dan describes himself as an 'ocean conservationist, strategist, communicator and marine biologist' – which is rather shorter than two of his current roles, which are Principal Advisor, Marine Science and Conservation for the Global Marine and Polar Programme, and a global honorary role as Marine Vice Chair for the World Commission on Protected Areas, both for the International Union for the Conservation of Nature (IUCN).

Dan works on an enormously wide range of projects which take him all over the world, reinforcing the point that marine conservation and its ideas are truly global in their reach. This is a long way from growing up by the sea in Jersey, inspired by parents who were keen on the natural world and made wetsuits for their children long before they were commonplace in surf shops. Jersey has an enormous tidal range, and 'low watering', searching through deep rock pools, snorkelling and surfing were almost daily key ingredients in developing his early interest in the sea. By the time he was fourteen his interest in science was already strong, being described by his schoolteacher at Victoria College as 'extending well beyond the syllabus, and I am confident he will be successful in this field'. Further inspiration came from the Cousteau films, learning about island wildlife from leading local experts such as Frances Le Sueur and early natural history TV documentary maker Roderick Dobson, and having written to David Attenborough and got a handwritten reply encouraging him in his interests in natural history.

After a conventional university training at Exeter and during his PhD work on polychaete worms he got a job, as the 'young enthusiast with high potential', with the Marine Nature Conservation Review (MNCR) in 1987. By 1990 he had undertaken a massive amount of work and switched roles to become Head of Marine Conservation, leading a growing marine conservation team at English Nature, which later became Natural England in 2006. In 1992 a Winston Churchill fellowship took him around the world looking at marine protected areas (MPAs), a background that was immensely valuable in delivering the marine aspects of the Habitats Directive, and MPA management in England.

By the time he left Natural England in 2010, Dan had been involved in the Marine and Coastal Access Act, seeing Marine Conservation Zone (MCZs) and marine planning come to reality, seeing the ecosystem approach adopted in the Marine Strategy Framework Directive, the development of the Marine Climate Change Impacts Partnership and a host of other initiatives. He enjoys ideas but also has the tenacity to deliver them. Since he moved to the international stage after being unleashed from Natural England, his CV, publications and range of contacts have been remarkable. His interests have covered international MPA networks, promoting the ideas of blue carbon, co-originating and helping to create Google Ocean to illustrate ocean protection and assessments of the status of world's oceans (which expanded to become the 3D ocean you see when you do a Google Maps search), establishing programmes to make real progress on Marine World Heritage and on high seas conservation, and often facilitating, enabling and leveraging support to see progress made.

MARINE CONSERVATION AND ITS DEVELOPMENT

What is marine conservation? How do you frame or describe it? What are its key ideas?

The classical view would be that marine conservation is about protection and preservation of wildlife and biodiversity in the ocean and its ecosystems, and about lessening and removing impacts. The danger with the classical view is that its focus is often about describing the things we want to conserve. For me marine conservation is not just about that but much more about people, places, priorities and action, and in particular the reality and need to manage human behaviour. We need to allow breathing space in the ocean where recovery can take place. A lot of people want to debate and describe what we should do, in my mind seemingly endlessly, but I think marine conservation must be, especially now, about setting priorities, really getting on with it and delivering effective action and outcomes.

What are the differences between when you started and now?

When I started it in the late 1980s it was all about gaining knowledge and doing the right thing. Now that too has a cloak of process, procedures that are levelling the playing field around our activities. We were also lucky enough from those early days on diving field surveys throughout the UK to be able to see at first hand many of the species and habitats in situ that we went on to promote in discussions about marine protected areas (MPAs); we therefore usually knew first-hand the reasons why protection was required. Today I get the sense that there are fewer people in the system who are lucky enough to have such a volume of such widespread practical experience, and there is in reality less opportunity to gain the direct experience of the habitats and species through survey, diving etc., as part of the job. That really does make a difference when putting the case with officials. In spite of this we do have the victory in a sense of the scaling up of MPAs in the UK and elsewhere. We do have a healthy marine conservation community, who are more engaged, but we are also realising that the challenges are greater and more urgent than we previously understood. So in that sense things have changed a lot.

The status of the statutory nature conservation agencies in the UK has also changed. When the NCC was broken up, the country statutory nature conservation agencies were vested with a much wider role in society. The reason they were set up was respected, and they had a truly independent role and clear responsibilities, and there was an expectation in their reports to parliament that they would hold government to account; this was the independent advisory role set out in the founding legislation. In recent years a number of respected independent bodies have been axed by government supposedly on 'cost-saving' grounds, including the Royal Commission on Environmental Pollution, as well as widely regarded bodies like the Prime Minister's Strategy Unit. It is therefore debateable whether what is happening currently to the UK nature conservation agencies and the closer influence of government is in the spirit of the independent role that was vested in them by the original legislation. Over the last five years the status of the nature conservation agencies and the role of scientists and experts has in my view been severely undermined. It seems as if nature conservation is sometimes being directed more by career administrators and policy makers in government who have less direct experience or scientific understanding of the natural environment. In these circumstances it is quite remarkable that there has been the scaling up of MPAs despite what you sometimes think government might have hoped would rather have happened. One reason for this success has been the large groundswell of public support and community engagement; the ongoing all-party support for MPAs by members of parliament has also been very encouraging. How things will shape up under Brexit, however, remains to be seen.

The changes in technology have been remarkable too. I can remember in the early days of diving surveys the hours spent processing colour film in the evening after the survey to see if we had the pictures we needed. Now we have the expanded and instant capabilities and capacities of digital cameras. The advance in the use of satellite technology has also been invaluable, and we are now able to visualise the virtual oceans on the web – and that has massively enabled communication and ways of displaying information. All of this has been very helpful but altered the essence of the way we do conservation and made it much richer. I do get the feeling though that we are no longer prepared to invest in individual training for diving or field-work expeditions as we were 25 years ago, and gaining that experience and knowledge is just as critical now as when I started out. It has to be not just about eyes 'on' the ocean, but also eyes 'in' the ocean. Personal experience is essential.

On a global scale there is another way of putting the changes we have seen, which I term the 'three eras of arrogance', reflecting the disregard humans have shown for the ocean over the years – seems we still don't realise it is a finite resource that should be cherished. The three sequential but overlapping eras are:

1.   We can't deplete the oceans' renewable natural resources – wrong.

2.   We can't change the oceans' processes and underlying chemistry – wrong.

3.   The third era we entering ... we can't exhaust the oceans' non-renewable resources, e.g. the rare earth metals – we will be wrong, and we will wish we had focused on reducing and recycling much earlier than we did.

In taking these short-sighted approaches, humans have unfortunately significantly damaged the blue heart of the planet that keeps us alive.

How is marine conservation different from terrestrial conservation?
Most terrestrial ecologists would only get a sense of the difference by working in a dense fog where you can't see much around you. That is the reality of diving work in the marine environment, and although technology has helped us, getting our information base is much less easy than for those on land. The processes in the sea that regulate ecology are much less predictable. On land you can simply see the effects of particular actions. The fluid dynamics and 3D structure of the oceans means that connectivity is much more important than on land. Many marine species move very significant distances and across many administrative boundaries in their routine behaviour. Things can affect places from a much greater distance. Toxic chemicals exemplify this, with the polychlorinated biphenyls (PCBs) turning up in the very extremities of the Arctic and Antarctic many thousands of miles away from points of origin, and years after some may have thought we had 'solved' this problem.

Perhaps the most significant difference is that because nobody owns the marine environment action has to be based around working with people's activities – and this requires important interpersonal skill sets. Not doing the obvious can be helpful at times. In my global work, for example, in relation to the marine–terrestrial question, I have been very careful not to automatically resort to special pleading for marine. In the protected area sphere, whilst there are clear physical and locational differences, the principles are similar on land and in the sea. With the whole conservation community behind one, an attack on MPAs can be seen as an attack on the idea of protected areas and the system in general, and that can be a great strength. In other situations, and we are currently scaling up work on world heritage and high seas conservation, the differences are so great that special pleading has been helpful.

Using a terrestrial model was also helpful in the approach we took to 'blue carbon'. The importance of carbon sinks on land is well established, and we simply applied that to the marine environment, highlighting the value of marine habitats and ecosystems such as mangroves, seagrass and kelp as carbon sinks (Figure 9.1). In considerations of carbon management systems we pointed out the gap in coverage when people hadn't taken marine habitats into account. We were transferring a powerful idea that people understood in a terrestrial context to the ocean environment and highlighting a real gap – and people hate having this type of thing pointed out, as it showed that their policies were not complete. Logic followed, and action has been and is being taken to better protect some of our amazing marine carbon sinks (Laffoley & Grimsditch 2009). We used the blue carbon arguments in the debate over the Severn estuary, where understanding the value of mud has always been a challenge. However, once we had data and information on this, there was another powerful argument that could be used for conservation. So there are differences and contrasts between conservation on land and with marine but they are more similar than we think, and whilst as a community it is easy to highlight the differences, I often think together is better.

THEMES, STYLES AND APPROACHES

How have the styles and approaches you've taken changed over your career? What works?
There is no simple answer or style that will deliver the responses we need to meet the huge challenges we face. What interests me is that whilst we may be in a world of 7

Figure 9.1 Seagrass beds are one of the major stores of blue carbon. Source: Keith Hiscock

billion people, individuals can still make a difference. I have a global role, a figurehead for the ocean for IUCN's second-largest commission, where the challenge is how we marshal the global effort of over 11,000 marine specialists and over 900 NGOs. I see my role as supporting and creating the freedom and opportunities to meet the challenges we face. There is a spectrum of approaches used by the NGOs and the way they relate to society and develop support for issues; they all have their place. Organisations like Greenpeace have been instrumental in raising the profile of issues and getting change on the back of this. What was particularly effective about the NGOs' Marine Bill campaign led by Joan Edwards and Wildlife and Countryside Link was the way they developed and gave societal representation a voice for MPAs. NGOs need to stand up and be visible. Governments have different ways of coping with the NGOs. By actively including them in processes it can attempt to buy them off, and it often comprises their ability to speak out. I have seen this happen. I think it is for this reason NGOs like Greenpeace and Friends of the Earth (FoE) often deliberately stay outside government processes. Either way NGOs have to be engaged inside or outside the process because silence is taken as tacit support for the government position.

I work a great deal on strategy now, and the ideas of Adam Kahane have influenced my thinking; marine conservation is what he would call a 'tough problem' (Kahane 2008). Kahane was a strategist with Shell, and in his book he describes the example of the birth of modern South Africa and how it emerged from a deliberative strategy. They worked through various scenarios to finally choose one called Flight of the Flamingos, and it is this that has helped take the country forward. So, if you can do this with South Africa then you can do it for the ocean's challenges, which some might say are rather easier. It is essential to think through the threats and the critical timescales of the emerging issues like ocean acidification.

On reflection, there has been a sequence to what I've done. It's been messy, but there has been a logical progression. It started with a scientific and marine biological foundation and describing marine habitats and species using diving, then progressed to looking at the range of measures that could be used to advance MPAs. Increasingly it became clear that there were major issues that needed to be addressed for which MPAs weren't appropriate, and I worked on blue carbon and on climate change report cards in the early 2000s with the Marine Climate Change Impacts Partnership (MCCIP). My work on the ecosystem approach also highlighted the importance of the major sustainability principles in marine conservation, not least in the way we drafted our input into the Marine Strategy Framework Directive (MSFD) and marine planning in the UK. Promoting MPAs is the essential core of my international work with IUCN, but dealing with global issues affecting the oceans and facing up to the challenges of scaling up global awareness and putting oceans at the centre of global debate is my current challenge. Some of the key moments and lessons from this progression are highlighted below.

My first proper job was with the Marine Nature Conservation Review (MNCR), which was a major survey of seabed habitats and species around the UK. It was a fantastic opportunity because diving all around the UK provided the knowledge that helped me enormously with the Habitats Directive. First-hand experience of the marine species and habitats that we were trying to protect made it much easier to justify the selection of sites. By the early 1990s I'd switched jobs to English Nature with responsibilities for MPAs, and we worked on a range of approaches including with Sue Gubbay on voluntary marine reserves (see Sue Gubbay, Chapter 7). Developing technologies also played their part in the 1990s, and we were able to deploy remotely operated vehicles and Roxanne (a seabed imaging and classification system) for the first time in inshore sites to supplement diving studies. I was awarded a Winston Churchill Fellowship in 1992, and that opened my eyes and had a powerful impact on my thinking because what people were doing in New Zealand and Australia was far ahead of the UK in both spatial scale and ambition. One of the ideas I could bring back was to have a zoning plan for Lundy, one of the very few marine protected areas at the time, with a coloured map which eventually led to the implementation of a no-take zone (NTZ). The Lundy NTZ has been a good example of recovery. The fishermen were relatively happy to allow 3.3 square kilometres to be assigned to this category because they thought it had no value; in fact only a small proportion of it is rocky reef (0.3 square kilometres); however, as the lobster populations have recovered and built there have been occasions when there have been attempts for it to be 'poached' by local fishermen.

With MPAs, governments embrace the basic idea, but what they don't 'get' is recovery and the need for full protection. This is even though there is a host of peer-reviewed papers, including over 200 documented cases, which show they work with increases on average of 446% in biomass inside such strictly protected places that allow the ocean to (Lester *et al.* 2009). This accumulating evidence base makes it difficult to understand the continued reticence of governments to fully support strictly protected MPAs when they could provide for a more productive and plentiful future.

The Habitats Directive introduced in the early 1990s had a huge influence on the way we thought about marine protection. One key point in 1996 was when we secured a big sum, €4.9 million, of which 50% was funded by the EU, to look at how we would implement the Habitats Directive. Nobody had managed one of these funds before, and it produced a range of outcomes including a generic management model for marine sites that set the basis for how Natura 2000 sites are now managed at sea. This was an important step because it signalled that MPAs could be delivered in the UK and elsewhere in Europe on a large scale. In 2006–2007 I was seconded to the European Commission to help with the development of the MSFD to help bring the thinking we'd developed with the Habitats Directive to the MSFD.

One experience from this time illustrated that there is a limit at any time to what government will allow. English Nature had the power under the Habitats Regulations to implement the Habitats Directive in England, to issue so-called 'stop orders' if they felt nature conservation interests were being impacted by different activities. We used this successfully on several occasions to stop fishing for razor clams in the Wash, cockles in the Solent seagrass beds, scallop dredging on maerl beds in the Fal estuary and scallop dredging in Lyme Bay that was destroying the reefs there. On the first occasion when I asked for one of these orders I remember being summoned to Defra to explain what we were doing to the Head of Fisheries and various officials. At that time they were committed to this process as the last stop-gap to protect sites. When Natural England was formed we were no longer allowed to pursue stop orders, and I am unsure if that power was ever used again to protect marine conservation sites. As we are seeing now with the nature conservation agencies in the UK, government will only allow so much, and that is why the NGOs and others have a critical role to play in holding the government to account over its duties. When I reflect on what I've done and ask why, the reality is that there has been core work on MPAs which we need to do, and do increasingly well. With MPAs we also need to be honest with people over our expectations and our knowledge, recognising that that there are things we don't know and we need to discover ways of managing the oceans together.

By the end of 1990s the Joint Nature Conservation Committee (JNCC) was actively working on providing examples of how the ecosystem approach could be put into practice. I then worked with Ed Maltby on 'seven areas of coherence' as a way of illustrating how the ecosystem approach could be applied in the marine environment (Laffoley *et al.* 2004). As the thinking on the MSFD and marine spatial planning began in the mid-2000s we were able to introduce many of the ecosystem-approach principles into the way these programmes were written. The ideas from the ecosystem approach are still developing and evolving, not least in the area of natural capital and the development of the ideas around ecosystem goods and services. You might say that the ecosystem approach is still alive and well but in different guises. The MSFD, drawing on a wide variety of influences, has sought to provide an integrated framework of regulations to protect the marine environment in European seas, including many issues like litter and underwater noise that had previously not been covered.

I also recognised in the early 2000s that there are limits to what one can do by promoting ideas like marine protected areas through conventional Government and NGO routes since only part of the community gets this connection and they are not appropriate for other issues. We needed other ways of engaging with a bigger and wider audience, which is what has driven my work and thinking on blue carbon, and with the MCCIP report cards. With the report cards we had the opportunity to show a much bigger audience, in a way which ministers love and support, that there are real issues at stake, and that we need to do more to manage the oceans.

It has been interesting reflecting on the path I've taken and its progression. The interesting start in life which gave me the passion to do this job, and understanding the context, the way the planet works, gives you the ability to try and change things. When you have this experience, as Sylvia Earle puts it, 'you know too much', and this drives you in two ways. It highlights the need to do more, and do it right now, but you can also see quite clearly that we are not doing enough, or the right things, and that is frustrating – and that also drives even more action. It's a feedback loop, it drives the passion and you can't let it go. There is a theme to this, and I've been quoted as saying 'We need it all, we need it now and we need more.' You also realise you must put the ocean back into the debate. It is unbelievable, for example, that until the discussions at the Paris meeting

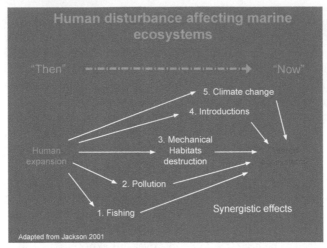

Figure 9.2 Threats to the marine ecosystem. Source: redrawn by Dan Laffoley from Jackson *et al.* (2001)

of the United Nations Convention on Climate Change in 2015 the ocean really wasn't factored in. The world had been missing the point about where key things come from, the weather, rain, fish and oxygen and carbon processes that support us. You end up going through these types of processes on that journey and you find ways of bringing people together differently.

THE FUTURE

What do you see as the future challenges?

There is the classical view, that it's fishing and it's getting worse, and pollution has long been a major problem. Jeremy Jackson and his colleagues produced a really helpful paper looking at the way the major threats have developed and work synergistically (Figure 9.2).

A number of the most pressing problem, such as plastics in the marine environment and ocean acidification, have, however, taken us by surprise in ways that hadn't been anticipated. The discovery of a 'soup' of small plastic fragments from all our rubbish in the largest five gyres of the ocean has come as a huge surprise. We now realise that that was the tip of the proverbial iceberg. Microplastics affect the whole ocean system and not just those gyres. These plastic particles, which we think harbour harmful chemicals, are also accumulating in marine sediments and entering the food chain via that route. Richard Thompson from Plymouth University has been looking at how these plastic particles impact on the biology of marine species for 25 years and has highlighted these relationships (Thompson *et al.* 2009). Describing this has been an important first step, but what can we do to address it? This is a massive challenge without many options at present. I'm sceptical about the large-scale booms technology being put forward by Boyan Slat's Ocean Cleanup operation (www.theoceancleanup.com), not least because there are lots of questions about the marine species caught up in the boom collections. We are looking to get an X Prize, which is a widely recognised mechanism for this kind of environmental project where a major donor puts up the fund to take projects forward. One of the ideas we are exploring is replacements for plastics that degrade quickly, or to convince people not to use them at all. We've seen progress with charging for plastic bags and the use of cellulose packaging, but there is a long way to go in this area. Partnership working will be particularly important between agencies, industry and the public.

With ocean acidification, Carol Turley introduced me to the problem in Plymouth in 2004. Ocean acidification is a very serious problem, and the first step is to plan for what is coming, since the world hasn't fully appreciated the scale of what is in store. We are changing ocean chemistry, and the timescales for this are very worrying. Even if we cut emissions using the low choice RCP2.6 scenario it may take forty years for the situation to stabilise. The latency in the system – the time lag – before action cuts in means that it will still be between 2035 and 2065 before we see change for the better. We have seen recently that scientists are reporting that we are currently living through the world's sixth mass extinction (Ceballos *et al.* 2015), and ocean acidification is going to contribute massively to this as many marine species depend on a delicate balance of seawater chemistry to maintain their calcium metabolism in their shells, for example.

We are used to being able to buy our way out of big problems such as the BP oil spill in the Gulf of Mexico or superstorm Sandy, but with ocean acidification it just isn't that simple. We have started to think about forecasting systems that highlight where the most vulnerable places will be, which will also help to identify places that because of their chemistry and physical features are likely to be more resilient in the long term. We are looking to see if we can get groups of nations to invest in programmes that will highlight where best to base shellfisheries and MPAs.

The additional challenge now is that we know ocean acidification has been joined by ocean warming and ocean deoxygenation as two more critical overarching issues we now must tackle. When we realise the enormity of what we face and when we have hindsight we might and probably will wish we had acted much earlier. Current political squabbles will be seen much more as they are – political posturing of the time – while we will realise that all along we were watching 'Rome burning' to use that classical phrase to describe what is happening our world. Chillingly, we already know that the combination of ocean acidification, ocean warming and deoxygenation accompanied the last five major extinction events in earth's history. It is just now we are creating the conditions that lead to extinction perhaps ten times faster than in past events – and we ignore that fact at our peril.

What are the hidden, real problems and barriers that make progress harder than it should be?

This is an interesting question. There are a number of barriers that can be characterised and that pose real challenges; they come up frequently, often in combination, and they include:

1. We know too much, and people who are not environmentalists can't reach across the gap in understanding between different viewpoints – and this make it difficult to sell them ideas.

2. Marine conservation has its strengths but also its weaknesses, one of which is the use of language. Conservationists are seemingly constantly renaming things like MPAs, which simply leads to more confusion both within the community and, more importantly, among non-conservation stakeholders.

3. The NGOs by their very nature claim issues as their own and understandably want to be seen in the best light, which reads well to their supporters and internal managers but doesn't always lead to effective collaboration, especially if they can't

agree between themselves. People can, and do, then say, well, if they can't agree why should we act? A key action should be to continually look to how they can collaborate for even more effective positive outcomes

4. Political barriers. Wildcard decision making by governments often undoes years of work on programmes which have been well funded and are delivering, and then they just decide not to carry on. We saw this with the axing of the estuary partnerships in the mid-1990s, but this has also been evident with the back-tracking on MCZs more recently. This is a challenge, but multifaceted approaches can provide resilience to deal with this.

5. Timescale mismatch. Ultimately some measures happen, but with the benefit of hindsight you think if only we'd done it earlier it would have made an ever bigger and real difference.

6. Money issues. There are clearly money issues. It is possible to raise money to do good things, but sustaining funding is a huge issue. With ocean acidification we saw the UK and Europe put much money in early on and then an argument forms that 'we've done our bit', and funding is reduced significantly or not renewed. So what is the scientist supposed to do? How do you keep some of these things running over the years that are not measured in political appointments? We have to be able to sustain our approach to these critical issues, as it is the long-term sustained view that matters and not just shorter-term observation.

7. Being perfect! The aphorism originally from Voltaire that 'perfect is the enemy of the good' has never been truer. I have my own saying that '70% is better than 0%'. So what would you have, MPAs that are imperfect and have their faults but that deliver 70% effectiveness and benefits, or a world of 'paper parks' where degradation continues? The realisation must be that we will never be perfect in safeguarding the ocean, but that all efforts must count. The reality is that as climate impacts take hold and grow so we will need to rethink how we can best protect the ocean, so with that realisation perhaps we should try all we can to protect resilience and restore and recover it wherever we can.

What are the most promising, interesting and innovative new approaches?

This question permeates everything I've been trying to do. Ideas and innovation are really important, whether it's working with the X Prize or leading climate scientists, or looking at new partnerships between NGOs to solve global problems. I think the issue here is that we are constrained by how we think, how we view things, and there is a world of opportunities out there. That's why, despite the doom and gloom, this is one of the most exciting times to be around. There are very real issues to deal with, and you can create good fortune in the middle of this – you can make things happen and change things. Individuals still count, and can make a difference, even in this highly populated world.

Innovation is really important, and although you know not everything will work it is vital to try new things and to have even more new ideas waiting to be tried. Recently, we had some success with the America's Cup, one of the most prestigious yachting races in the world, in encouraging them to use the yachts as a platform for marine conser-vation. Now it only worked for part of the series and it was a small step, but it was a unique experience. It then fuels this innovation engine that encourages you do more with different people.

I'm constantly thinking about new ways we can do things, and I get encouragement from a whole range of different things that are currently under way. A few examples include:

- Recently I've been really encouraged by the High Seas Alliance, a partnership of over thirty leading NGOs who are working to change the Law of the Sea to get effective conservation for what is 50% of the world's oceans. This collective effort is productive and impressive to see in action, and even now we have been pushing the boundaries to encourage others such as World Heritage to become involved (Laffoley & Freestone 2016).

- At the moment there is major programme putting up 648 cuboid satellites to provide internet access to the world including the half that has yet to be connected (OneWeb; http://oneweb.world/#solution), providing levels of high-speed connectivity that were unimaginable in the past. If we look at the ability to see the earth's surface at higher resolution and the challenges around that, a number of companies such as Earth-i (see for example http://earthi.space) are now chasing the associated business opportunities that will come with providing up to one metre and perhaps daily resolution, and that may well soon become the norm. Tracking the activities of fishing vessels by satellite has been a possible for some time now, but it is now happening en masse too, and both Google's Global Fishing Watch and Pew are developing projects on this. The Pew Trusts have supported the development of a Virtual Watch Room where one person can sit in a room and track fishing vessels over vast MPAs and detect where fishermen are behaving badly (Pew Charitable Trusts 2015). Most recently one of the biggest fishing boats in the world was tracked across the Southern Ocean and prosecuted when it tried to land its catch, and fined millions of dollars for fishing illegally; it has now been successfully blacklisted. Not a single fishing patrol vessel was involved with this. Remote monitoring will become more and more part of normal practice.

So, have we reached a peak of innovation? No. Have we done everything we could? No. Have we made all the links yet? No. Are there new partners out there? Certainly. Is there a world of innovative opportunity awaiting us? Absolutely. That is why right now is one of the most exciting times for marine conservation.

When you retire, what would you like to have achieved?
We are in the middle of this 'silent storm' sweeping through the ocean, where major issues such as ocean acidification, ocean warming and ocean deoxygenation are catching up with us, and that provides every reason to keep trying to resolve these issues and act. I like to think that my epitaph might say something like 'he tried' or 'he made a difference', set against the rising odds stacked against us as a species. Every generation knows more than the next and is in a better place to judge, but I'd like to think that our efforts have made a difference and there is still everything – quite literally 'our world' – to play for.

Optimistic or pessimistic?
Despite the challenges, optimistic.

CHAPTER 10

Callum Roberts

Professor Callum Roberts is a marine conservation biologist in the Environment Department at the University of York. Before university he lived in Wick in the northeast of Scotland and was fascinated by the sea. Like many, he was inspired by the Cousteau films, but it wasn't until he went to university in York that he learnt to dive. As an undergraduate it was an expedition to the Red Sea in 1982 with his lecturer, Rupert Ormond, to study reef fishes that kindled a lifelong love of coral reefs and began a career in marine science. In the early 1990s his interests in fish behaviour gave way to concern about the deteriorating condition of coral reefs and what could be done about it. He began research on marine protected areas in 1991, exploring how protection from fishing affected fish populations. The study of marine reserves is still a central pillar of his work on marine conservation.

Currently, Callum's research focuses on human impacts on marine ecosystems. While his interests in marine conservation have blossomed over the years, his field research remains firmly rooted in coral reefs. Away from the coral, he has also taken a lead on promoting marine reserves, working with many organisations including the United Nations, governments and conservation organisations all over the world. Callum has served on a US National Research Council Committee on Marine Protected Areas, and has been a member of the Marine Reserves Working Group headed up by Jane Lubchenco at the National Center for Ecological Analysis and Synthesis in Santa Barbara, California. He was awarded a Pew Fellowship in Marine Conservation in 2000 to tackle obstacles to implementing marine reserves, and in 2001 was awarded a Hrdy Professorship in Conservation Biology at Harvard University. This chapter outlines his work on marine protected areas and historical ecology, but it also reflects on how marine conservation science interacts with other interests, often challenging the status quo.

MARINE CONSERVATION AND ITS DEVELOPMENT

How do you define marine conservation, and what are its key ideas?

Defining things is always hard. How I feel about marine conservation is easier to explain. Marine conservation is about giving nature relief from the adverse effects of human influence. It's all about revitalising the oceans, and for that you need to appreciate history. Digging deeper into the history of exploitation of the sea – both hunting and fishing – made me realise that the oceans have been profoundly altered over hundreds of years, in some places more than a thousand years. Species and habitats that were formerly abundant have undergone dramatic depletions, sometimes to the point of complete elimination. Thinking about marine conservation is all about not accepting the status quo. It's an exploration of how you can effect recovery. This is not to say that conservationists should try to rewind the clock to some halcyon prior state. Ecosystems are affected by climate and people, and people cannot be disentangled from the world we inhabit. We can only go forward.

Recovery is about allowing the sea to achieve its potential in the present frame of current conditions. You protect places to create refuges in which nature can flourish, but you cannot dictate the direction of recovery. With sufficient protection, nature determines the endpoint, not us.

My thinking on that was shaped in the Caribbean. Areas fully protected from fishing supported much more abundant fish stocks than the fishing grounds around them. With time, the big fish like cubera snappers and tiger groupers began to return. Marine reserves gave species time and space to achieve the full capacity that the environment offered; that for me was incredibly telling. Ever since, I've promoted the introduction of protected areas that offer the conditions for a transformation, a comeback, in whatever form that takes. It is not about specifying this kind of reef and these sorts of fish. It's all about removing human pressures to the extent we can and allowing nature to take its course.

I think some people, and government agencies in particular, get far too hung up on controlling the outcomes. They say 'We must specify goals for protection and measure ecosystems to see when they are achieved.' That's a recipe for failure, since recovery rarely follows a carefully planned trajectory. What is more productive, I feel, is to make sure we remove or reduce the activities causing depletion and damage. The rest will take care of itself. What you can be sure of is that there will soon be more wildlife – more species, more biomass, greater numbers, more complexity. To me, that is the best measure of success.

How do you see the differences between marine and terrestrial conservation?

The thinking in terrestrial conservation, especially in the UK, is very different to what I've just described. On land, the goal is usually to protect what we have here and now, whether that is a natural landscape or not. We have come to value species and habitats in human-altered states and seek to maintain them in a particular condition, rather than letting them achieve their own balance ecologically in the absence of human intervention and tinkering. We British have always been much more wildlife gardeners than embracing the idea of wilderness unfettered from human influence. Things are very different in North America, where the Wilderness Act of the early 1960s codified that there would be areas where nature took its course with minimum human influence. This has never been on the cards in the UK, probably because we did not think there was enough space left by the time we realised nature needed help. But there has recently been a vociferous questioning of this approach, led by people like George Monbiot, who has raised the option of rewilding in his book *Feral* (2013). He questioned why we spend so much time keeping large areas of countryside in ecologically impoverished states, much of the uplands for example. These landscapes were once far more diverse and often forested. They supported an abundance of animals and plants, including predators like lynx, bears and wolves where now they are bland landscapes dominated by sheep or grouse. There are other reasons for questioning this model, not least because by doing things differently, such as afforesting uplands, we can restore other ecosystem functions like reducing the rates of rainfall runoff and downstream flooding.

In the sea we don't have the option to wildlife garden as we do on land; we don't have the tools for it. You can't unload a flock of sheep to trim seagrasses, or prune sea fans to foster corals. It's that inherent wildness, a refusal to be tamed, that led me to conclude it's better to simply remove or reduce adverse human pressures and let nature take its course. Unfortunately, what we see today with the approach to Marine Conservation Zones (MCZs) in the UK is an extension of the wildlife gardening approach, the application of terrestrial thinking to the sea. The attempt is flawed on several levels. By putting the focus on particular 'features', or habitats, you first have to find and map them, which delayed and nearly derailed the process of establishing MCZs. Our seabed maps are often rudimentary, and the positions of features uncertain. Managers spent a lot of time deciding what activities each of the dozens of features could cope with so they could then be subject to separate, tailored management measures. Aside from the impossible management complexity this patchwork approach would entail in diverse seascapes, if you don't have precise maps, you can't easily apply piecemeal management. The approach also ignored a huge body of historical evidence that the features present today have been transformed over decades and centuries by industrial fishing and other human pressures. Taking the maintenance of present conditions as a desirable conservation goal – 90% of conservation goals set for MCZs were to 'maintain', just 10% to 'recover' – institutionalises past degradation within the conservation bureaucracy.

Marine protected areas (MPAs) fail for many reasons, lack of management, lack of funds or staff, poor planning, but ours fail on the most fundamental level of all: lack of ambition. We are dreadfully off course at present, and this terrestrial model – which is also flawed on land for other reasons – is a non-starter in the sea. We can't exert that kind of control. We need far greater ambition, and we should learn from experiences in other countries. There is no free lunch. High-level protection of whole ecosystems is necessary for high levels of benefit. Carrying on as we are will produce no tangible benefits at all.

What are the differences between when you started and now?

People know a lot more about what goes on underwater now. When I started it was very much a black box; the sea hides its secrets well. People didn't think about ocean life much; they didn't know what to expect. When Cousteau and others brought the oceans into the living room, it was a revelation. Now such scenes are the staple of wildlife documentaries and there is a much greater awareness. Creatures from the sea have taken on new lives as characters in advertising, literature and movies. People have always been drawn to the water, but that fascination and love goes deeper now, far into the realm of fish and whales.

People are much more aware of the problems of the oceans too, which are all of our making. In part this is because the problems have grown so large in the last thirty years as to be unavoidable. The price of wild fish has outpaced the rate of inflation for decades as stocks have fallen; beach holidays are marred by the scum-line of plastic and trash;

coastal waters are declared unfit for bathing. This awareness of the need to protect and conserve the ocean goes right to the top, to national governments and the United Nations. For example, at the UN, when I began my career the talk was about how to extract more wealth from the sea. Protection was hardly mentioned. Now the whole debate has changed – we recognise the need for marine conservation, even on the high seas, beyond the limits of national sovereignty. So we no longer have to make the case for conservation from first principles. People understand the issues; the whole debate has changed.

Awareness of climate change has grown massively over the last 25 years and is now being closely linked to the oceans too. People are aware and concerned about the ice caps, polar bears and sea-level rise in ways that they weren't when I started out. That has come about because so many more people are engaging with the sea, studying it, filming it and talking about it. Getting into the water is so much easier too. In the 1980s it would take many days' travel to reach exotic and remote places. Now globetrotting is routine. The remote places of yesteryear are the fashionable resorts of today. Spectacular marine environments and wildlife have never been closer. The number of people who have experienced coral reefs has soared, way beyond the hard core of divers of the early days; it's commonplace.

THEMES

Marine conservation as an applied science

My style is very heavily data-driven, and as a scientist I like to go to the field to collect evidence and use that to figure out what it can tell me about how the world works. I've spent thousands of hours counting fish underwater in hundreds of different places around the world. That can't help but give you a good understanding of what makes the seas tick. I started out as a behavioural ecologist and spent years watching the daily lives of fish in their natural surroundings. For me they are not the abstractions of mathematical fisheries science, but real beings looking for food and mates, seeing off rivals, avoiding predators. That understanding has coloured my approach to marine conservation. Habitat quality, distribution and connectivity cannot just be taken for granted, as they generally are by fisheries managers, they must be built into management. Alongside fishing, habitat loss and degradation explain the declines of many once abundant species like halibut, turbot, skates and cod. Long-lived animals thrive best when their populations include lots of big, old, fertile and experienced fish.

The many hours underwater also gave me a keen understanding of the dominant role played by fishing in shaping the species you see, their sizes and the state of the seabed; the human imprint is all around. Over the years I have gained a good understanding of the vulnerabilities of species to human impacts. This is where I part ways with certain fisheries scientists who have spent their lives in front of a computer screen. Experience indicates just how easy it is for species and habitats to get knocked down in abundance by human impact, and when they are down populations don't necessarily bounce back when you relieve the fishing impact. Fishing can have a long-lasting legacy. This background of fieldwork has also inspired my interest in MPAs and historical ecology.

Marine protected areas

My interest in marine protected areas (MPAs) was born in the Red Sea in the 1980s. My PhD supervisor, Rupert Ormond, led a marine conservation project on the coast of Saudi

Figure 10.1 A variety of fish swimming in a no-take zone in one of the world's most successful MPAs, Cabo Pulmo, Mexico. Source: Octavio Aburto-Oropeza

Arabia to map marine habitats and identify important areas for marine parks. We were there to both describe the marine habitats and help drive marine conservation. I was mapping the distribution of fish. It was a voyage of discovery for everyone. We reached many unexplored places on the long coast between Jordan and Yemen. Many places we jumped in the water had never been dived before. Then in the mid-1990s I worked on the islands of St Lucia and Saba in the Caribbean, looking at the effects of marine reserves that had been closed to all fishing. Those studies revealed both the huge scale of human impacts on the sea, and the means of protecting marine ecosystems from such effects.

About this time, I came across the work of Bill Ballantine, who studied a marine reserve at Leigh in New Zealand. Bill is often referred to as the father of marine reserves. By the time I came across his work, he'd written many influential and wonderfully articulate and principled papers; he summarised his ideas in clear and simple language, not least in advocating the use of no-take zones (NTZs) in which no fishing was allowed. His thesis was simple, and it is one I have taken forward. When the activities of people, most notably fishing, are removed from an area of sea it is possible to see major changes take place – recovery – in the fish populations and seabed habitats (Figure 10.1). If you don't fish an area, the animals live longer, grow larger, become more numerous and produce many times more offspring. Seabed habitats benefit from the exclusion of destructive fishing methods like trawls and dredges, and from the re-establishment of more intact food webs that favour structural species like kelp and coral.

It is not very easy to predict exactly what will happen after protection, as I've already said, since nature follows its own course. But today this experiment has been run hundreds of times around the world, in tropical and temperate seas and for a very wide range of geographic sites and scales. It is easier to see and measure the changes of fish and seabed species at locations where underwater visibility is clear, which is why marine reserve science blossomed first in the tropics. But the same effects are also seen in murkier temperate seas (PISCO n.d.). The adoption of NTZs is an idea that challenges the status quo, not least of major fishing interests (who are often backed by government), and it has led me into many heated debates.

In the late 1990s, I worked with the US National Research Council Committee on Marine Protected Areas, and as a member of the Marine Reserves Working Group, headed up by Jane Lubchenco, Steve Gaines and Steve Palumbi at the National Center for Ecological Analysis and Synthesis in Santa Barbara. We sought to develop a more robust theoretical underpinning for the design and implementation of marine reserves (Roberts *et al.* 2003a, 2003b), doing spadework on which MPA science has since built.

I think one of the most important things I've done in conservation is to study some really highly protected places and demonstrate how the effects of protection play out over time, both for wildlife and for fishers. Watching populations rebuilding, often very quickly, you couldn't fail to be impressed at how much more highly protected areas deliver for conservation than more weakly protected places. I've taken that insight and tried to spread the word through scientific publications, but also by writing books in which people can read about the ideas without being scientists or professionals.

Historical ecology

Jeremy Jackson is an American scientist who has been a big influence on me in terms of thinking about the importance of history to understanding the environment. There was one particular talk I went to which he gave in 1995 at the International Coral Reef Symposium in Panama. He called it 'Reefs since Columbus' (Jackson 1997) and in it he pieced together a compelling rewriting of the history of Caribbean coral reefs, gleaned from information in old books and manuscripts; I was spellbound. Jeremy was really the first person to highlight how what we think of as the 'natural world' has changed hugely since the early 'pristine baseline'. Ecology is, he said, a historical science. What happened in the past is key to understanding the present. Jeremy's talk and a paper written by Daniel Pauly about the same time (Pauly 1995) led to the emergence of a key idea, 'shifting baselines', that has played an important part in reframing how we view the environment. Shifting baselines are intergenerational shifts in how we perceive the environment. Where older generations may have experienced steep declines in the abundance of wildlife during their lives, younger generations looking at the same scenes may see a world of abundance they think is natural.

I quickly immersed myself in old books, finding an entire library full of them in the attics at Harvard, where I spent a wonderful year as a visiting professor in 2001. It was among these dusty shelves that I began my book *The Unnatural History of the Sea* (Roberts 2007), which sets out how centuries of hunting and fishing have had profound influences on the populations and habitats in the sea.

Like Jeremy Jackson, Daniel Pauly has been a massive influence as one of the most brilliant, prolific, creative minds in fisheries science (among his many other interests). Daniel's great skill is to think very broadly and come up with neat concepts and ways of looking at familiar and unfamiliar ideas. His ideas like fishing down the food chain (Pauly *et al.* 1998) are, unfortunately, particularly evident and relevant today. The concept ties in closely to what I discovered writing *The Unnatural History of the Sea*. In the past we targeted big fish high up in food webs, because they were easy to catch, tasty and valuable. With time and growing effort, we depleted them, progressively fishing further afield and deeper. As big fish dwindled, we moved on to other species, generally smaller and lower down the food web (Figure 10.2). We invented better, more destructive ways of catching them, like bottom trawling, which led to habitat loss, reinforcing fishing-in-duced shifts in the structure of ocean food webs.

One of the main insights gained by a long historical view is that it is easy to do lots of damage to fish stocks and the environment very quickly. Even in the early days of sailing trawlers in the first half of the nineteenth century, you can see this new fishing method quickly supplant longline fishers and netters. Trawlers changed the rules of the fishing

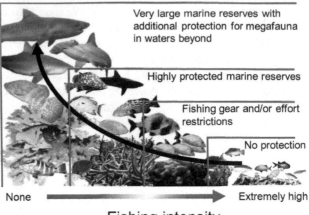

Very large marine reserves with
additional protection for megafauna
in waters beyond

Highly protected marine reserves

Fishing gear and/or effort
restrictions

No protection

None   Extremely high

Fishing intensity

Figure 10.2 Fishing down the food chain. Source: Callum Roberts

game, catching more for less effort. They depleted stocks, and the line and static-net fishers lost out as fish declined. Again and again we see the same pattern repeated in different places. What the sum of those experiences says is that a small amount of fishing is capable of removing a large fraction of biomass. The impact happens fast, and recovery takes a long time. The corollary is that if we want those vulnerable species back, we have to supply a high level of protection. These facts are often unappreciated by those tasked with creating MPAs. UK government rhetoric, for example, is that the creation of MCZs will bring back life in the sea and create resilience. But the rhetoric is not matched by the tools. Hardly any of the protected areas established so far have any new protection at all. Most remain open to fishing, including destructive bottom trawls and dredges. Business as usual won't change anything.

Fisheries scientists and marine protected areas

Some colleagues – lots of them fisheries scientists – have never got in the water and watched the lives of fish unfold in their natural habitats, or seen the impacts of fishing at first hand. They view fish as numbers on spreadsheets or particles at large in homogeneous seas. Such people often don't appreciate how ecology shapes populations in the wild and how fisheries management approaches work in the real world. I've butted heads with fisheries scientists over the years who deny that MPAs work, especially for mobile species and at small scales; they say you need to protect vast areas to have any benefit. Yet when you go into a small MPA and see supposedly mobile fish species stay put and grow large and old you can see that their world view is flawed. Spatial protection works – which, as it happens, many fisheries managers know from long experience using closed areas. My world view has come from under water, watching life respond to protection in dramatic ways. I'd never have formed this understanding if I'd only worked with numbers on spreadsheets or been sat in a lab. Most of the people I come across who are switched on to marine conservation have direct experience in the water and have spent long hours watching lives lived in the sea. Those who spend lots of time modelling need to be alert to the limitations of their models in representing the realities under water.

Advocacy and science

Lots of scientists agonise about 'advocacy versus science'. Some think that if you become an advocate for any particular policy you lose your objectivity and become unscientific; you sacrifice your right to the title of scientist. I think that is just plain wrong. What you do is science according to the best principles of scientific enquiry, with objectivity, replication of results and sufficient numbers of the examples of the phenomenon investigated to be sure what you have seen is real. Having done good science that is robust, and having come to a conclusion – for example that our seas have seen some species of megafauna reduced to a few percent of their former abundance – that opens a whole raft of other questions. Is that a good thing? What is the impact of the change on ecosystems and society? Having looked deeper into the ramifications, I think it is legitimate to argue for policies that will bring back some of those animals. For me, conservation science is a package that includes both doing science that is relevant to public policy, and then acting on the results. You have to promote the findings to have any impact in the real world; if you don't the research is worthless.

Conservation science by its very nature is applied science. That includes using the media (digital, TV and newspapers) and campaigning, where needed, for policy change; you must persuade people of the relevance of the findings to them. For me there is no conflict. But you have to make sure that the science is sound, and given the human foible of believing only what you want to, you have to remain open to challenges to your ideas.

Conservation organisations and styles

I've worked with lots of conservation organisations and I've always tried to give them a solid foundation of good science on which to develop their campaigns. The conservation world is an ecosystem in itself, each organisation filling a different niche. Greenpeace, for example, has a very different approach to WWF. I admire both and am an ambassador with WWF in the UK. WWF tries to effect change *within* the establishment. Greenpeace on the other hand tries to push for change from outside; the moment Greenpeace staff start to dress in suits and become parliamentary lobbyists will be the moment their power is lost. Sea Shepherd takes a confrontational approach, and in doing so makes negotiating space for others who work in their own ways towards the same ends. There is room for all sorts of styles and organisations, from those who work directly with governments to those who chain themselves to oil rigs.

There is a risk, however, in getting too cosy with your adversaries, that you will end up diluting your approach. The more closely you work with regulatory bodies the more likely you are to come to a negotiated compromise which goes too far, undermining your principles. There is a well-known phenomenon of government departments being *captured* by industry, and this leads to a very dysfunctional relationship which serves the interests of a minority of people in industry but not the public; we've seen that with energy companies and with financial services where there has been a lack of effective regulation. I think the same is true in fisheries. Over time, rather than regulating for the public good, government fisheries management bodies have come to serve the fishing industry, giving them what they demand in the short term but at the cost of the long-term viability of the industry and the environment. You see the same risk of NGOs being captured by government. NGOs working closely with government can be flattered by their inclusion, but can end up compromising their principles. I've seen them accept outcomes insufficient to achieve their objectives, and which they wouldn't have accepted at the start. There is no point, for example, in celebrating the establishment of MPAs that don't offer any real protection.

How do you see the intervention of the large international conservation organisations in Europe?

Lots of people talk about this, the power of the Pew Charitable Trusts and other big organisations like Oceana. They all have their different ways of doing what they think works for conservation. There is space for all of them. The injection of large amounts of funding is certainly necessary to challenge the power of vested interests and to break through in places like the European Parliament. To have a chance of swaying opinion in major debates, you must have a weight of power behind you and savvy people who know how to run excellent campaigns; it's good to see them wade in.

What lessons can we draw from different styles of working?

There has been a fairly steady increase in the number of people working in marine conservation, which has to be good for enhancing impact, especially as governments cut expert staff. Some argue that this causes problems, governments playing one organisation off against another. But I see the multiple approaches as complementary: sheer force of effort will enhance our ability to get better protection. Time has taught me that science alone has limited power to effect change. I used to think that if I could collect the right evidence – demonstrate a problem and its solution – the arguments would win themselves. There would be logical outcomes with policies enacted that, for instance, protected nature and fisheries. But that isn't the case. Knowledge is not necessarily power, and although important, knowledge alone won't win the arguments. Vested interests run deep and reach far into the past; the relationships between government and business are more baroque than you might ever suspect, and unravelling these interests is very difficult.

One thing that does have great power though is good stories and images. Stories can often win conservation arguments better than detailed data, but only if they have a clear message. So when at meetings and events I show pictures of docksides stacked with huge fish from the last century, we know and feel immediately that the experience people had in the past was fundamentally different from today (Figure 10.3). That captures people's imaginations and starts conversations about how to change things. Storytelling is powerful, which is why I took up writing books to make science more accessible to a broader audience. That way I hope to be more persuasive, showing how the ocean has been compromised by human activities, but, crucially, what we can do about this.

Figure 10.3 Large fish on quayside in the early years of trawling. Source: Callum Roberts

MARINE CONSERVATION – WHAT HAS BEEN ACHIEVED?

I've seen wonderful achievements. Take protected areas, for instance: in Pacific islands, where people have established networks of protected areas off-limits to fishing, you can see the comeback of fish populations, which then spill over to surrounding fishing grounds, reseeding them. Also encouraging is to see places like California roll out protected areas state-wide so their benefits are multiplied and enhanced. There have been gains at a species level as well; the whaling moratorium has been a great success for many. The eastern Pacific grey whale has come back from a few hundreds in the 1930s to over 20,000 animals today, a fantastic comeback. The grey seal in Britain has bounced back from persecution in the seventeenth and eighteenth centuries. The surge in numbers of seabirds that have recovered from hunting and egg collecting is a result of legislative protection. There is currently an increasing number and scale of shark fishing bans around the world, which again is very important and good conservation. The high seas drift-net ban in the early 1990s had a big impact, although illegal fishing continues; but this was major global milestone. Good as they are, however, these conservation successes are still too fragmentary and isolated.

In the UK, we have been successful up to a point. The arguments for marine conservation have been made strongly and have been heard. Legislation has been passed which enables us to create MCZs, complemented by European measures like SACs and SPAs. On paper at least, we have a good legal framework that could facilitate very strong conservation outcomes. But we are only halfway there. We have to match designation with strong protection to turn around the state of life in the sea. The management half of the argument has still to be won. Life in the sea needs strong protection from fishing, as well as from other impacts. If you don't deal with fishing, conservation just isn't going to happen.

I think the arguments can be won, but to do so we have to focus on public education around the lack of effective protection. People are being misled. They have been told that things are protected that in reality are not. When that shortcoming becomes widely known, I hope we'll see genuine protection put in place. We need large areas with no fishing at all. And no MPA that aims to achieve broad conservation goals should be fished with destructive gear like bottom trawls and dredges. To carry on with such activities in protected areas makes a mockery of conservation. It's scandalous.

On the plus side, it should be easier to fold strong protection into the existing protected areas than to start again from scratch. Some countries are far ahead of us. In Palau, for example, a tiny group of islands in the western Pacific, they have protected spawning aggregations of reef fish, established a network of fully protected areas, and banned shark fishing and bottom trawling from all of their waters. They are putting in place a plan for sustainable management of their rich marine resources which by 2020 will see 80% of their waters fully protected from fishing and other extractive uses. Chile has recently declared 46% of its waters as protected areas, the great majority of it no-take. More and more, we see countries stepping up with bold protection plans, inspiring others to act.

At the international level there has been progress too. The UN has been negotiating towards an agreement to enable conservation on the high seas. It has taken several years, but finally the process has reached the General Assembly level. There has long been recognition of the need for protected areas on the high seas, where the fishing industry is rapidly creating a conservation crisis and piling up problems for itself as stocks decline. Although some nations remain opposed to high seas protected areas, I'm optimistic we'll find a way forward.

Frustrated by the lack of high seas protection, some regional bodies have gone it alone. For example, OSPAR, a collective of fifteen northeast Atlantic nations and the EU, has established a series of high seas MPAs in their waters, although they still need to implement effective management. But it is a major step in the right direction. I'm proud that it was my research team that did the science that underpins the OSPAR high seas protected areas; it's a fantastic feeling to draw lines on a map around areas of high biological value, and a few years later they are made real.

THE FUTURE

What are the future challenges? What are the most important threats to the marine environment?

Five years ago, I would have said without hesitation that the biggest threat was overfishing and the package of collateral damage to species and habitats that goes along with it, certainly more important than pollution or climate change. For the sheer scale of its historical impact on the destruction of wildlife, fish stocks and habitats – fishing is still a major force that continues to shape life in the sea. But it is joined now by the utterly pervasive impacts of climate change. With warming and acidification, we see changes that are unprecedented on timescales of millions of years and, even during the mass extinction events, the changes have never been as fast. If that were not bad enough, the effects of climate change interact with other pressures like fishing and pollution.

Alongside greenhouse gases, there are other pollutants, one being what you might call the 'biological pollution' that comes from the movement of species from place to place, such as the steady trickle of species from the Red Sea into the Mediterranean (Galil & Goren 2014). This planetary diaspora is breaking down evolutionary barriers that have existed for hundreds of thousands, often millions, of years. The mixing and blending of biotas is going to cause profound change. We can see it happening in the Caribbean now, where lion fish have swept in, altering the whole reef system in unprecedented ways. This biotic blending is an unstoppable force accompanying globalisation. We are homogenising the planet's wildlife, and the outcomes are totally unpredictable. The directional creep of climate change is mixing with the wildcard of accidental introductions to provoke ecosystem reorganisation. Change is the only thing we can take for granted any longer. What we have seen so far is going to be dwarfed by what will come. The only rational option is to build up the resilience of life in the sea.

What needs to be done – what actions would you take?

What would I do? We need to reduce the pressures that are under our immediate control, which is where those fully and strongly protected areas come in. Reducing impacts of fishing is one of the most important ways to create breathing space for ocean life. Such action is practical and its benefits almost immediately tangible. Rebuilding populations and the integrity of their habitats will reintroduce much more complexity in habitat structure and the variety of creatures present. It can restore wildlife phenomena, like some of the great fish migrations and spawning aggregations. In time, the larger, older animals that are the engines of reproduction in populations and the stocks will come back. These changes will give life greater resilience to cope with increased human pressure and environmental stress. Reducing fishing and its collateral impacts is the most important dial we can turn in the short to medium term to buy us time while we get to grips with moving the planetary levers of climate change and global pollution.

Of course, there are many other pressures to alleviate in the short term. What they are depends on the location. The Bohai Sea of China is today swamped by aquaculture on a monumental scale. In 1976, two-thirds of this coast was natural habitat, today it us just 15%. The rest has been turned into fish farms or built over. Just outside the Bohai Sea is Qingdao, which was the Beijing Olympics yachting venue. They have a horrific problem with 'green tides' which smother the coast with a metre-thick layer of seaweed washed up onshore. Cutting green seaweeds from aquaculture ropes further south releases millions of fragments of weed that drift north in a seawater broth enriched by agricultural runoff and fish-farm inputs. By the time they arrive at Qingdao they have grown into a massive problem. Chinese aquaculture has far outstripped the carrying capacity of their seas. The Chinese must quickly get to grips with reining it in and reducing nutrient inputs. Fixing ocean problems isn't just about fishing.

What are the most promising new developments?
The challenge is to get people to change the ways they interact with nature, to their own and nature's benefit. We need to see more direct connections made between actions and outcomes and change the incentives. A lot of progress will be made in the social sciences where we need to understand better how to motivate people to think and reduce their impact in order to nurture and cherish their environment for the long term. That goes against the grain of human nature because we have a strong short-term bias. We like rewards to come early rather than defer gratification. I worry that it is going to be hard to counter that hard-wired bias; for me this is the big challenge.

In a nutshell, just one sentence, what we have to do to reinvigorate life in the sea is to fish less, using less destructive means, waste less, pollute less and protect more. The tools are straightforward. The knotty part of the problem is getting people to agree. I've seen protected areas that work well for years and then the manager is replaced, protection slips and the whole thing goes to hell; that vulnerability is a big problem. You need to maintain protection in the long term so you don't win little battles only to lose the war.

Are you optimistic or pessimistic about the future?
Optimistic – my glass is half full. I've seen great progress over the years, especially with MPAs and the engagement of environmental issues in fisheries management. Whether action is fast enough to counter the scale of change under way, I don't know. But you've got to try.

BOX 5. BILL BALLANTINE AND NO-TAKE MARINE PROTECTED AREAS

Bob Earll, Keith Hiscock and Keith Probert
Dr Bill Ballantine QSO MBE (1937–2015), originally from the UK, was an outstanding scientist and marine biologist whose early major contribution to rocky shore ecology resulted in a 'biologically defined exposure scale', often known as the Ballantine exposure scale, for rocky shores. This described how the distribution of marine life was affected by different levels of exposure to wave action. In 1964 he left the UK to become the Director of the University of Auckland's Leigh Marine Laboratory on New Zealand's North Island.

In 1971 the New Zealand Government passed the New Zealand Marine Reserves Act (1971) 'to provide for the setting up and management of areas of the sea and foreshore as marine reserves for the purpose of preserving them in their natural state as the habitat of marine life for scientific study'. This set the scene for the first in what has become a series of over forty no-take MPAs around the New Zealand coast. The New Zealand marine reserves legislation took its lead from science and the need to know how natural systems worked free from human activities. This was in stark contrast to the model adopted in the UK, derived from terrestrial conservation and placing emphasis on selecting special places and criteria.

The first marine reserve in New Zealand was set up adjacent to the marine laboratory at Leigh in 1975 after fierce battles with local fishermen about the nature of the no-take idea. Bill Ballantine described the Leigh site as an unremarkable piece of coastline – not special in any way – with no criteria applied to its selection. The overriding purpose was to use science to explore the changes that took place once extractive activities ceased. As the science undertaken in the reserve progressed over the years it became apparent that major changes had occurred (Shears & Babcock 2003). As fish predators grew larger and more numerous, the urchin barrens declined, seabed communities were transformed, rock lobster populations grew in number and size of individuals and didn't migrate as many people had predicted. The fishermen discovered they could 'fish the line' – the boundary of the reserve – and catch larger fish.

Today this reserve attracts thousands of visitors a year who come to see the abundant marine life and is a major boost to the local economy. With this strong grounding in the science of how MPAs actually work free from human disturbance, a programme covering a wider range of habitats and species has now been put in place in New Zealand.

Known as the 'father of no-take marine protected areas', Bill Ballantine has strongly advocated the need for MPAs at numerous meetings around the world, and his pioneering work has inspired many working in this field. His paper, 'Fifty years on: lessons from marine reserves in New Zealand and principles for a worldwide network' (Ballantine 2014) summarises his huge contribution. Keith Hiscock wrote on his death, 'Those of us who had the privilege to know him, will remember an individual passionate about conservation who did not suffer bureaucrats gladly, who "managed" opponents, who could be cantankerous and seemingly impossible, but who made things happen.'

CHAPTER 11

Jon C. Day

Jon Day has been closely associated with planning and managing the Great Barrier Reef for over 30 years. Based on his experience in this globally iconic MPA, Jon is today widely recognised by his peers as a leader in marine planning and management. Jon's efforts, and the efforts undertaken by others learning from his work, have resulted in more effective marine planning and management around the world. Jon's work with the rezoning of the Great Barrier Reef Marine Park has set many precedents that have been applied elsewhere, including the representative areas approach, the effective use of community and stakeholder engagement, and comprehensive multiple-use management – in effect, marine spatial planning. This chapter reflects on many of the lessons from Jon's work that together have set the benchmark for an expanding global network of MPAs.

HOW HAS MARINE CONSERVATION DEVELOPED IN AUSTRALIA, WITH REFERENCE TO THE GREAT BARRIER REEF?

When the Great Barrier Reef Marine Park (GBRMP) was declared in 1975, the marine park was the largest marine protected area (MPA) in the world, covering an area about the same size as Italy, Japan or Vietnam. More importantly, it was set up as a large-scale multiple-use area where most reasonable activities could occur, provided they were managed to minimise conflicts while allowing for the protection of the area. The reason the marine park was declared were the threats in the early 1970s to the Great Barrier Reef (GBR) from proposals for both limestone mining and oil drilling (Bowen & Bowen 2002). The resulting public outcry led to a Royal Commission and then to the Great Barrier Reef Marine Park Act 1975 (Australian Government, 1975). The Act provided the basis for the creation of the marine park and the managing agency, the Great Barrier Reef Marine Park Authority (GBRMPA).

Many people think that the GBRMP was zoned from the day it was declared in 1975, but the Act only specified the external park boundary and set out the zoning process. In reality, it took thirteen years of sequential zoning of smaller sections of the park before zoning was completed for the entire GBR in 1988. Also for the first 29 years after the GBR was declared, less than 5% of the entire marine park was in no-take zones. Only since mid-2004, when the current zoning came into effect, was one-third of the marine park declared in no-take zones, with a further one-third zoned to protect the seabed from damaging activities like trawling. I started with the GBRMPA in 1986, but prior to that I had worked for eleven years in terrestrial park planning and management, mainly in Victoria and in the Northern Territory. I was always keen to work with the GBRMPA because they were widely regarded as world leaders in marine park planning and management. On my third attempt, I was offered a job as a park management officer and planner with the Authority. Among the reasons I was employed at GBRMPA was my experience with park planning and public engagement from my work in Victoria; I was able to adapt a lot of what I'd learnt from my terrestrial park planning experience and apply the relevant bits in the GBR.

Complementary management and the Emerald Agreement

There are a number of reasons for the success of the GBRMP; a key one is the fact that the State of Queensland and the Commonwealth of Australia (the federal government) work effectively together to jointly manage the entire area. Consequently, from high water mark, all the way out to the seaward edge of the marine park, which is 250 kilometres offshore, and including most of the islands, there is what is formally known as *complementary management*. This is a joint management agreement between the state and federal authorities that means the zoning rules and regulations are virtually the same across all the various marine areas irrespective of the jurisdictional boundaries.

This complementarity has occurred for various reasons, the primary one being that the federal and state governments cannot agree where the 'low water boundary' between the two jurisdictions actually is, and it moves! The Emerald Agreement in 1979 formalised a cooperative approach between the two governments, which means the complex mix of marine, coastal and island issues throughout the GBR is all managed in a complementary way. For example, within five months of the federal zoning plan coming into effect, Queensland effectively 'mirrored' or duplicated the federal zoning provisions in the adjoining state waters. Today a series of collaborative arrangements have evolved over time between the state and federal agencies addressing aspects such as complementary legislation, joint-permits and enforcement (Day 2016).

Rarely do you see a close working relationship between differing levels of government, and the only other MPA in the world where I am aware this occurs is in the Florida Keys where good cooperation exists between the State of Florida and NOAA's National Marine Sanctuary programme. Such a complementary approach to me seems fundamental and very important because the jurisdictional lines on the maps mean nothing for marine species or processes like connectivity. Such collaboration is also one of the key reasons for success in the GBR. Many other significant precedents have been set over the decades by the management of the GBR. For example, in 1990, the GBR was declared the world's first Particularly Sensitive Sea Area; that was a significant development in managing shipping in the GBR, and the concept has now been applied by the International Maritime Organisation (IMO) in other significant maritime areas around the world.

Early involvement in GBR planning and management

One of the zoning processes in the GBR in which I was heavily involved was the rezoning of the Cairns Section between 1988 and 1990. As the lead planner for this project, I proposed changing a number of zoning aspects, some of which were accepted (e.g. zone names that remain in place today) and some were not (e.g. an attempt to use zoning to regulate tourism), but that is just part of the adaptive and learning process. In 1990, an opportunity enabled me to move into the field management side of the GBR, coordinating the day-to-day management activities with the field staff including the park rangers. This required me to transfer from the GBRMPA, a federal management agency, to the Queensland Parks and Wildlife Service, a state agency; however, I was still involved in a lot of interaction with GBRMPA given the complementary management approach I mentioned previously. Initially I was appointed as the Area Manager for part of the GBR and then subsequently I was promoted to Regional Manager, responsible for the middle third of the GBR, including all the islands within that area. I remained with QPWS for seven years mainly because I recognised the value of having a better understanding of practical on-ground, and on-water, management experience; I am convinced that helped me later to become a better planner and park manager for the GBR.

The national context

In the late 1990s there was a lot happening on the federal marine scene in Australia, including an Oceans Policy that was launched in 1998 (Commonwealth of Australia 1998). This national policy provided an overarching guiding document for much of our work. Around the same time, Australia was developing a national marine bioregionalisation, the Interim Marine and Coastal Regionalisation of Australia (IMCRA Technical Group 1998). Subsequent versions of IMCRA provided a regionalisation for the entire exclusive economic zone (EEZ) around Australia including the benthos and pelagic waters. The federal government also played a coordinating role, along with input from a group of Australian and New Zealand marine managers, developing guidelines for a national representative system of MPAs (ANZECC Task Force on Marine Protected Areas 1998). Collectively these three initiatives – the Oceans Policy, the IMCRA approach and the National Representative System of Marine Protected Areas (NRSMPA) guidelines – all provided a good basis for managing Australia's marine environment.

Each state within Australia has had its own approach and marine legislation for MPAs, and while there was some cooperation, each state largely operated in isolation when it came to marine planning and management. Even though some states have greatly expanded their MPA networks in recent decades, and some (e.g. Victoria and South Australia) consider they have produced representative networks of MPAs, most MPAs remain relatively small and confined to the inshore (i.e. state) or internal waters. The GBR model, with complementary federal/state management from the high water mark out to the deep offshore waters, has yet to be replicated with any other Australian state. So while the states and the federal agencies have been doing different things in their various jurisdictions, it is usually the same group of stakeholders who are being engaged in two separate and unrelated marine planning processes. The federal government is still working on developing a national network of marine reserves in the Commonwealth waters (i.e. outside the state waters, and outside the GBR, but to the edge of the EEZ). The outer boundaries of a network of marine reserves were declared in 2012 covering around 36% (or 2.4 million square kilometres) of the entire EEZ of Australia. However, today those areas remain without finalised management plans; in fact, the most recent review of the draft plans released for public comment is proposing a 30% reduction in no-take zones compared to the previously proposed network, so in many people's eyes it appears as a backward step.

The representative areas approach, bioregions and rezoning

In 1997, I took a sabbatical to spend a year in Canada, and fortuitously ended up working with WWF Canada and with Professor John Roff at Guelph. I helped John Roff complete a report for WWF called *Planning for Representative Marine Protected Areas* (Day *et al.* 2000), but John also helped me develop my thinking about systematically protecting representative areas instead of just protecting important or significant places. John got me thinking about mapping the bioregions of the entire GBR and how a representative areas approach might be applied.

When I returned to Australia in 1998, I was appointed as one of the directors at GBRMPA, responsible for conservation, biodiversity and world heritage matters. Along with a small team of colleagues, we were tasked with progressing the Representative Areas Programme (RAP), which was effectively a rezoning of the entire GBR (Figure 11.1). RAP started small but grew and grew, and by the time we submitted the final plan to parliament for approval at the end of 2003, virtually the entire agency had been involved. It became far more complex and contentious than anyone had envisaged.

There are a number of reasons why RAP was successfully completed, with a key reason being the outstanding leadership shown by Virginia Chadwick, the CEO of GBRMPA. Virginia was really the one who negotiated the complex social and political 'minefields', and enabled GBRMPA to undertake this extremely contentious planning task, and complete what many thought was impossible. Essentially, we took a 'whole of GBR ecosystem approach' and used expert knowledge to help us map seventy different bioregions as the ecological underpinning for the rezoning. These forty reef and thirty non-reef bioregions were very broad-scale habitat types, such as mid-shelf reefs or inter-reef seagrass areas. They were mapped by scientific experts in a series of workshops where they combined their knowledge with the data sets that GBRMPA had compiled. A scientific advisory group also recommended eleven biophysical principles and another four socioeconomic managerial principles (Fernandes *et al.* 2009). Several

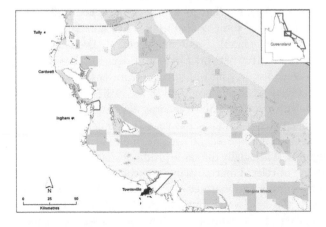

Figure 11.1 Great Barrier Reef Marine Park: a portion of the zoning plan; the different colours relate to different management objectives. Source: © Commonwealth of Australia (Australian Government 2017)

of the biophysical principles turned out to be very significant, particularly the one that recommended protecting a minimum of 20% of each of the seventy bioregions in the new no-take zoning network. There is no right or wrong reason for the 20%, but the advisory group recommended a minimum of 20% per bioregion based on a variety of reasons – not the least being the fact the GBR was a significant ecosystem and a World Heritage Area. These principles were not targets or rules, however; rather they were planning principles, so in some bioregions we went way beyond 20% but in other more contentious areas, it was very hard to find 20% without many compromises being made.

Rezoning and the Outlook Reports
The rezoning was completed in 2003 and the revised plan was put to the federal parliament, where it was approved and then became law in mid-2004. Around this time, there was an election and the Prime Minister promised a review of the GBRMP Act, which in effect was an undertaking to review how the rezoning had been done. This was an independent review led by another arm of government, but when it was finished it effectively concluded that GBRMPA had done the best we could with the resources available and in the time frame allowed, so that was a positive tick for the zoning. In addition, the review suggested the need for a new process to trigger any rezoning, plus a requirement for a periodic Outlook Report to assess the state of the GBR. As a result of this review, the GBRMP Act has now been amended to include a legal obligation to produce an Outlook Report every five years, requiring an assessment of nine specific thematic topics, including biodiversity, health, and the commercial and non-commercial uses of the GBR. This concept of a systematic assessment of values and risks has now been adopted elsewhere; for example, the national environment report has adopted a similar approach, and IUCN now undertake a similar review of their natural world heritage sites too, so the Outlook Report is something of which GBRMPA should be very proud.

To date, there have been two GBR Outlook Reports, in 2009 and 2014 (GBRMPA 2014). The conclusions of both these reports have been stark: in essence, they both conclude that the long-term outlook for the GBR is poor and likely to get worse in the future. Both reports highlight the four big issues facing the GBR as climate change, water quality from land-sourced runoff, impacts from coastal development and unsustainable fishing practices. However, they also showed 41 different pressures in a risk matrix highlighting the wide range of threats, with all but three of these 41 pressures happening now. The real concern is the cumulative impacts of many of these pressures. The number of very high risks that are certain to occur and are of major consequence, increased from three to eight in the five years between the two Outlook Reports, so the GBR definitely remains under pressure.

Community input and engaging stakeholders, politicians and the public in the process
One of the major components of the rezoning process was the huge amount of public input. Over 31,000 written submissions were provided during the two formal phases of the rezoning, with 10,000 submissions before we had produced a draft plan and over 21,000 after the draft was released. One of the reasons for such a huge response after the draft was released was that many fishers complained that the proposed no-take zones were a disaster. However, when pressed, many of them admitted that they had not made a submission before the publication of the draft, or had not mentioned any of their important fishing places ('We don't tell anyone where these are'). We pointed out that unless they informed us about their special spots, how could they expect us

to consider them? So many fishers recognised the value in helping inform us and so grudgingly made a submission (Day 2017). During this process, it became even more obvious to GBRMPA just how much the community knew about their part of the GBR. The researchers in the GBR are well informed but only intermittently access the GBR, whereas the fishers, tourist operators and locals are out on the water almost daily and therefore have a great deal of knowledge to contribute. In order to tap into this local knowledge, GBRMPA opened three regional offices along the GBR coast.

We learnt a lot about how to communicate effectively. Soon after I commenced the RAP with our small team, Bruce Kingston, who was an expert in communication, was specifically employed within GBRMPA. Bruce said to me, 'You're developing a new zoning plan, but the public doesn't understand there is a problem with the old one; I suggest you back off until we can help them understand the pressures facing the reef and then they might accept that the old plan isn't working'. The resulting public education programme (*Under Pressure*) was one of a number of such programmes that played a major role in educating the public during the overall planning process.

We also learnt a lot about how to effectively engage with the public. One of the main lessons was that public meetings are rarely an effective way to engage local communities. We had to attend some when other people organised them, but whenever possible we steered clear of public meetings, because more often than not they provide an opportunity for a noisy minority to protest and are not a good way to exchange information with the wider community. Instead, we undertook what we called *community information sessions*, which were much more labour-intensive but more effective and enabled us to have much closer interaction with the community. We would set out a programme of attending particular locations (like a community hall) at particular times (usually 3 p.m. to 7 p.m.), with an invitation to come along anytime in that period and 'meet the GBR planners who are doing the rezoning, who are keen to talk to you and hear your views'. We also made sure we had experts on hand to talk to say, fishers or tourism operators, so they realised GBRMPA had expertise in all these sectors. Having a naive young planner trying to engage an old experienced fisherman can be a disaster! However, if you have an experienced fisheries manager on staff, who can recognise and call out points when fishers start being *creative* with the truth, then that can gain respect and can make a real difference.

How we framed our messages for particular stakeholders also was targeted – so while we had the same overall message, just how we said it to, say, researchers, indigenous people, fishers or politicians may have differed. It was the same message but it was tailored for different groups. We also learnt that it was very important to engage the politicians from the start of the planning process so they were well aware of the efforts made to seek the best possible outcomes (Day 2017). Effective and timely communication was one of the key lessons of the RAP, but we were lucky we had help from communication experts who really understood how to do this well.

Structural adjustment and compensation for fishers – how not to do it

One of the least successful aspects to emerge after the rezoning concerned a programme of 'structural adjustment' for fishers affected by the new zones. We were not allowed to mention compensation throughout the entire rezoning process, but as soon as the final plan went to parliament it became clear that there were implications for fishers. The politicians therefore approved the concept of a structural adjustment programme (SAP). This had an interesting evolution, and there is an excellent paper reviewing the SAP which effectively documents an example of how not to do it (Macintosh *et al.* 2010)!

The fishers convinced the government that GBRMPA should be precluded from being involved in the SAP, so instead the programme was coordinated by another department within government with far less understanding of the issues or who really was involved. The whole process became caught up in political trade-offs, marginal electoral seats and inefficient procedures, which resulted in the government expending some AU$250 million in structural adjustment – which was more than the commercial fishery was worth. If GBRMPA had been provided with that sort of funding up front at the start of the rezoning, we could have bought every fishing licence in the GBR and allowed them back under careful control. All the money did was allow many fishers to buy bigger boats or better gear, and carry on fishing, albeit in a reduced area because of the new no-take zones. In hindsight, the SAP was badly handled; however, subsequent state and federal MPA programmes around Australia learnt valuable lessons, so the SAP is an example of how a bad outcome can actually lead to better processes elsewhere.

The Reef Guardians programme
In 2003, GBRMPA also started a Reef Guardians programme, which has become extremely successful. The programme began with a few schools and one teacher seconded to the GBRMPA working with those schools to get the kids aware and involved. Today the latest available figures show some 275 schools throughout Queensland, involving more than 120,000 students of all ages and about 7,000 teachers, all now have a far greater awareness of the need for greater protection for the GBR and its connected ecosystems. The Reefs Guardian programme has continued to grow and grow, and there are some amazing students who have come through the programme from its start, so I think the future is in good hands. The concept has also spread further and now engages local councils, farmers and fishers, and continues to increase public awareness of the issues facing the reef. The concept has now been copied by many other agencies, in Australia and around the world.

Lessons on zoning – zoning by objectives and using multiple layers
Over the years, we have learnt a lot about zoning. In the GBR, we don't zone by activity, but rather we zone by objective. To explain, I'll use trawling as an example. If you are looking to protect the seabed, then the objective should focus on seabed protection, irrespective of whether it is trawling or some other activity that disturbs the seabed. So instead of a no-trawling zone, we have a habitat protection zone with the broad objective to protect the benthos from anything that might affect the seabed. The zoning plan lists clearly what activities can happen in each zone ('as of right' or without a permit) as well as the activities that may allowed to occur provided a permit is granted, after being assessed against a number of specified criteria (Figure 11.2). Within the seven marine zones in the GBR, we are able to cater for all reasonable maritime activities. The zones become more restrictive on extractive activities as you move through the zoning spectrum. At the least restrictive end of the spectrum (*general use zone*), all reasonable activities are allowed; you can trawl in this zone, but trawling is prohibited in all other zones. As you move along the zoning spectrum, there is a *limited fishing zone* where only one line and one hook per person is allowed, irrespective of whether it is for commercial or recreational use. Further along the spectrum is a no-take (i.e. no fishing) zone called the *marine national park zone*, and, at the far end of the spectrum, is a no-go, no-access zone (*preservation zone*) which acts like a reference area or scientific baseline.

Many people think that our GBR management occurs just within a single two-dimensional spatial layer (i.e. the zoning plan), but in reality there are multiple layers of management (Day 2015). Therefore, the seven marine zones make up one layer, which is of prime importance for the fishers. However, sitting on top of the underlying zoning is another statutory layer that deals with designated shipping routes, which cut across some of the underlying zones. Then there is another statutory layer which recognises indigenous 'sea country' irrespective of the underlying zones, and in some locations there is yet another layer showing statutory defence training areas, some of which are temporal closures only and not always closed to the public (Day 2016). Some of these multilevel activities are not conflicting and can overlap, so what we have is an integrated and multi-layered 'system' of management, rather than just a single layer. This 3D multilevel marine spatial planning approach would appear too complex if every spatial aspect were depicted in only two dimensions; so particular spatial layers are made obvious only to particular user groups, so they specifically know what they can do where.

Our management approach also incorporates more than just ocean zoning and multi-layered marine spatial planning. Many of the factors affecting the GBR come from the surrounding area – for example, water-quality issues from adjoining coastal areas or from higher in the catchments or else further out to sea. Many MPA managers do not think outside their MPA when it comes to management. What I try to stress to MPA managers is that you need to 'think outside the MPA box', given that is where many of the impacts are originating. This means you have to work with industry, with other sectors or other levels of government as to how they can help to mitigate damaging actions. Today the GBRMPA spends a lot of its time (and effort) in the catchment working on the land-based issues that affect the GBR. We also know that global issues like climate change are affecting MPAs, and while a single agency or even a single country cannot address climate change alone, you can reduce some of the other cumulative impacts that are also affecting your MPA, thereby allowing the area to cope better with the remaining impacts (i.e. build resilience).

ACTIVITIES GUIDE (see Zoning Plan for details)	General Use Zone	Habitat Protection Zone	Conservation Park Zone	Buffer Zone	Scientific Research Zone	Marine National Park Zone	Preservation Zone
Aquaculture	Permit	Permit	Permit 1	×	×	×	×
Bait netting	✓	✓	✓	×	×	×	×
Boating, diving, photography	✓	✓	✓	✓	✓ 2	✓	×
Crabbing	✓	✓	✓ 3	×	×	×	×
Harvest fishing for aquarium fish, coral and beachworm	Permit	Permit	Permit 1	×	×	×	×
Harvest fishing for sea cucumber, trochus, tropical rock lobster	Permit	Permit	×	×	×	×	×
Limited collecting	✓ 4	✓ 4	✓ 4	×	×	×	×
Limited impact research	✓	✓	✓	✓ 5	✓	✓ 5	Permit
Limited spearfishing (snorkel only)	✓	✓	✓ 1	×	×	×	×
Line fishing	✓ 6	✓ 6	✓ 7	×	×	×	×
Netting (other than bait netting)	✓	✓	×	×	×	×	×
Research (other than limited impact)	Permit	Permit	Permit	Permit	Permit	Permit	Permit
Shipping (other than in a designated shipping area)	✓	Permit	Permit	Permit	Permit	Permit	×
Tourism program	Permit	Permit	Permit	Permit	Permit	Permit	×
Traditional use of marine resources	✓ 8	✓ 8	✓ 8	✓ 8	✓ 8	✓ 8	× 8
Trawling	✓	×	×	×	×	×	×
Trolling	✓ 6	✓ 6	✓ 6	✓ 6,9	×	×	×

Figure 11.2 The GBR activities matrix, indicating which activities can occur in which zone, which activities are not allowed and which activities need a permit. Source: © Commonwealth of Australia (GBRMPA, 2017)

Practical lessons – boundaries, adaptive management and information

One of the practical lessons we learnt from the early days of GBR zoning was the location of zone boundaries and lines on the map. In the early days, we used a specified distance from a natural feature (for example, '500 metres from the reef edge'), which looked fine on a map but was virtually impossible to locate while on the water. Along the coast some natural features may be used where they can be clearly identified, but further offshore using such a method of identifying zone boundaries by distances from largely unrecognisable features is virtually useless, and helps neither users nor enforcement officers. We therefore moved to coordinate-based zoning, using points of latitude and longitude and straight lines (primarily east–west or north–south) between such points. How this was implemented in the 2004 zoning network can be seen in the zoning maps (available on the web). With the advent of GPS technologies, the majority of users know exactly where they are on the water, and this is obvious when we see people 'fishing the line' along zone boundaries.

Adaptive management in relation to short-term changes is an essential way MPA managers need to work (Day 2008). The realities of changing patterns of use, changes in technology, and/or political or ecological changes, all mean that as managers we need the flexibility to respond to changes quickly without relying solely on legislation. So we have the ability to bring in other statutory layers of management (such as our Special Management Areas) which can be activated relatively quickly and can adapt or reinforce the underlying management. In the GBR, we don't have a statutory time limit stipulating how often a zoning plan must be reviewed. However, the law does stipulate that a new zoning plan must remain in place for a minimum of seven years, with the aim to give certainty to users and managers (that time frame is now well past for the zoning that came into effect in mid-2004).

Planners sometimes say to me, 'I don't yet have enough information to start planning.' My advice is 'Don't wait until you have perfect information, otherwise you'll never start.' Use the best available advice, but during your planning, be prepared to take on new information if it becomes available. During the GBR rezoning, and after we had released the draft plan, we obtained access to the Vessel Monitoring System (VMS) data showing the location of key trawling areas. We were able to consider this information when we revised the draft plan – and so it is important to realise that you need to adaptively plan, as well as adaptively manage.

Indigenous people and marine conservation

In Australia, we still have much to learn on how to engage effectively with indigenous people, especially when we look at the approaches used in Canada and New Zealand. Indigenous people have been in Australia far longer than any 'white fellas', and the GBR is considered by indigenous people to be their country. Whilst *native title* has been formally recognised on land, this is exceedingly difficult in marine areas. What we have done in the GBR is to work with indigenous traditional owners for them to develop 'traditional use of marine resource agreements' (TUMRAs). The TUMRA for a specified area of sea country may then be accredited by both state and federal governments (this effectively allows 'white fella law to be used to reinforce black fella hopes and aspirations'). While this approach is working effectively with some indigenous groups, some groups are not so organised and have yet to develop these sorts of agreements. Today more than 20% of the GBR coast is covered by a TUMRA, but we still have a way to go for co-management with indigenous people.

Figure 11.3 A bleached individual coral. Source: the Ocean Agency, XL Catlin Seaview Survey

MPA targets

Targets are a real double-edged sword (Agardy *et al.* 2016). They can be useful, and politicians like having something to aim for, but equally targets can be misused and misinterpreted by user groups. We ran into this with the 20% principle when people complained that we had protected more than 20% in some bioregions. We clarified that 20% was not a target but a minimum recommended amount, which explained why we went well beyond in some offshore bioregions where there was little interest or conflict so we were able to protect 80% or 90%. With the Aichi global targets, for example, governments are rushing to meet the target of 20% of oceans in MPAs by 2020, but how and where they are doing it is not in a very thoughtful way. So targets are a mixed blessing and everyone needs to be aware of their shortcomings.

CONCLUDING THOUGHTS

Although I no longer work for the GBRMPA, I am still proud of what that agency achieved during the 28 years I was there. Today the GBRMP is far from being the largest MPA in the world any more, but it is still widely regarded as the grandfather of modern MPAs (Day 2016), and people still look to the GBR for guidance. Some management approaches have worked really well, others not so well, but that is what adaptive management is about. Today the GBR has helped to shape marine conservation around the world, in particular the application of the multiple-use approach, through zoning and multi-layered marine spatial planning. There were high hopes that our efforts would help to build resilience in key parts of the Great Barrier Reef, particularly in the face of climate change. However, the lessons of the recent massive coral bleaching events in 2016 and 2017 (Figure 11.3) have shown that Australia needs to greatly increase its efforts to address issues like climate change that are exerting massive impacts despite our best efforts for place-based conservation. What is needed today are both local and global efforts if we are to ensure the future of our MPAs as well as our planet.

CHAPTER 12

Keith Probert

Keith Probert, now retired, was an associate professor in the Department of Marine Science at the University of Otago, New Zealand, where for nearly thirty years he taught marine biology and ecology. In 2014 he was presented with the New Zealand Marine Sciences Society Award for his contribution to marine science. He had previously worked as a marine environmental scientist in government and industry organisations in New Zealand and the UK, with involvement in a number of marine environmental impact studies. His main area of interest has been benthic ecology, particularly the biology and ecology of sediment macrofauna and meiofauna, and with studies that have ranged from coastal to deep-sea habitats. The impact of human activities on benthic systems and the implications for marine management and conservation have been important aspects of much of this work. His book *Marine Conservation* (2017), is a global synthesis of marine conservation, covering the major issues and ways to address them.

WHAT ARE THE KEY DEVELOPMENTS AND LESSONS OF MARINE CONSERVATION IN NEW ZEALAND?

The New Zealand marine environment

It is important to explain some of the key aspects of the New Zealand marine environment, to provide a context for developments that have taken place. New Zealand is a very maritime country in the southwest Pacific, and in fact roughly in the centre of the earth's water hemisphere. It has a huge exclusive economic zone (EEZ) of over 4 million square kilometres and a very diverse oceanography and seabed topography. It's a long country with two main islands and stretches from a subtropical region with mangroves in the north to the sub-Antarctic and fiordic systems in the south. It also straddles the Subtropical Front, a major circumglobal frontal system separating surface subtropical water in the north from surface sub-Antarctic water in the south, which is important for biogeography. New Zealand also has a long coastline of some 15,000 kilometres. The continental shelf tends to be quite narrow by world standards, but the EEZ includes very extensive deep-sea habitats with large plateaus at slope depths, numerous submarine canyons and seamounts and a range of chemosynthetic habitats, and as the country lies along a plate boundary we have deep ocean trenches to the north and south. All of this means it is an area of high marine biodiversity.

The human impacts on the marine environment

Exploitation of living resources has been the major impact. The New Zealand landmass was the most recently settled by humans – the early Polynesians from the thirteenth century and the Europeans only from the mid-nineteenth century – so its colonisation is very recent. The Polynesians had a major impact on megafauna including on fur seals and sea lions, some populations of which are only just recovering now. With European colonisation there has been the increasing pressure on fish stocks and in particular on deep-sea stocks in recent decades. In relation to pollutants we don't have the large-scale industrial pollution of many densely populated locations elsewhere, but increased nutrients and sedimentation in coastal areas are having clear effects, and ocean warming and acidification will affect us too.

The New Zealand inventory of marine species

A few years ago New Zealand completed an inventory of its entire known living biota, and in fact was the first country to produce such a catalogue (Gordon 2009–2012). This checklist shows that we have about 17,000 known eukaryote marine species in New Zealand. But it is estimated that there is a similar number yet to be discovered, so the country does have a rich biodiversity. Indeed, if you look at the best-known phyla and compare these with the European Register of Marine Species, which covers an area more than five times that of the New Zealand EEZ, we have a roughly similar numbers of species (Gordon *et al.* 2010).

A challenge then for marine conservation in New Zealand is the comparative dearth of information on species in the region. One of the lessons relates to our knowledge of the region's marine biota, which is nothing like as thorough as for many locations in the northern hemisphere. So we have this urgency to protect areas but have limited biological knowledge on which to select representative MPAs. An option is to depend more on proxy methods to address this problem, such as using abiotic surrogates to identify areas that are biologically meaningful. This approach has, for instance, been used to identify groupings of the more than 800 seamounts that we have, and habitat modelling has been used to predict the distribution of frame-building bryozoans in New Zealand waters. Bryozoans can form significant biogenic habitat on the New Zealand shelf, comparable to that of coral reefs in many ways.

Marine protected areas

The initial focus of marine conservation in New Zealand was very much on establishing marine protected areas (MPAs) under the Marine Reserves Act 1971. The first reserve was established at Leigh, adjacent to Auckland University's marine lab in the north of the country (Figure 12.1). This was due in large part to the efforts of Bill Ballantine, who is known here as the 'father of marine conservation' (see Box 5, page 117). Progress since then has been somewhat patchy, often with opposition from fishers, but there has been strong public support and we now have over forty no-take reserves. MPAs are all well and good but they do need to be embedded in wider management to help address

Figure 12.1 Snappers and snorkellers at Marine Reserve, Cape Rodney – Oakari Point, New Zealand. Source: Department of Conservation, New Zealand

the effect of more pervasive stressors. A significant step in this regard was a legislative overhaul when the Resource Management Act 1991 (RMA) was passed.

This sorted out a lot of miscellaneous bits of legislation and provided a more coherent framework with sustainable management as its focus. In this regard New Zealand was following international trends, as we'd had for instance the World Conservation Strategy (1980) and the Brundtland Commission (1987) stressing sustainability. There has been lots of discussion about what sustainable management and development actually mean and what they involve. But without getting bogged down in semantics this debate has been important in efforts to reconcile societal needs and wellbeing with maintaining healthy ecosystems. With the RMA, New Zealand was one of the first countries to enact environmental legislation with sustainable management as its focus. The act covered both land and our territorial seas and so it was important in a move towards more integrated management of the marine environment, at least for coastal waters. The New Zealand Coastal Policy Statement of 1994 (now superseded by the NZCPS of 2010) was drawn up to provide guidance on achieving the aims of the RMA for the coastal environment. The RMA was also precautionary in its ideas, although it didn't invoke the precautionary principle as such. This has been made explicit in later legislation, such as the Fisheries Act (1996) to help promote more sustainable fisheries.

New Zealand was quick off the mark with MPAs and the Marine Reserves Act, and particularly good evidence about how no-take MPAs work comes from New Zealand, notably from the Leigh reserve. Here there is very clear evidence of the increased size and abundance of exploited species within the reserve and of the importance of safeguarding large mature females, as they make a disproportionate contribution in supplying eggs and larvae. Studies on rock lobster and snapper show these MPA effects particularly clearly (Shears & Babcock 2003). Some of these areas are hugely popular with visitors. The small Leigh reserve, for example, attracts many thousands of visitors a year (Figure 12.2). There is a lot of grass roots support for these places and managing visitors is a developing issue as tourism increases.

MPAs in New Zealand are mostly small, nearshore areas, and selection has been somewhat ad hoc. Also we are still falling short of the expected UN Convention on Biological Diversity (UNCBD) target of 10% set for 2020. Whilst we might meet this target for coastal waters, less than 1% of the EEZ is protected, although in this regard we

Figure 12.2 Visitors at the Leigh reserve. Source: Tony and Jenny Enderby

are no different from many countries around the world. Efforts are now under way to set up a network of MPAs that is more representative in terms of biogeography and habitats. I think one of the important lessons is having a more structured community involvement in this MPA process. In this regard New Zealand's Department of Conservation is working with community groups with the aim of setting up a representative network of MPAs by 2025. Input is being arranged by having marine protection forums based on biogeographic regions around the country. These forums contain a range of stakeholders who may have an interest in local MPAs. Another lesson we have learnt concerns the current MPA legislation. The Marine Reserves Act dates from 1971, and whilst it was very significant for its day, its somewhat narrow focus has become an increasing issue over time. In particular, the purpose of setting up reserves under the Act is specifically for scientific study rather than for biodiversity protection, and the legislation applies only to the territorial sea. A draft Marine Reserves Bill was introduced to parliament in 2002 to overhaul the legislation, but the original Act still hasn't been superseded.

The role of indigenous people and community engagement
Another key aspect has been the role of indigenous people in recognising the aspirations and approaches of Maori. This has included putting in place methods for the sustainable management of traditional fishing grounds. Notable here has been setting up reserves known as *mataitai* where fishing has been of customary significance and where commercial fishing is not normally allowed. These areas are managed by local tribes to promote sustainable management of fishing grounds, and there are now more than forty *mataitai*. Community involvement has been a significant development in New Zealand's coastal management, with, for example, marine protection forums in MPA planning and in managing areas of traditional importance to Maori. I think there is considerable potential in integrating customary and more typically Western styles of resource management. In Oceania as a whole there has been quite a resurgence of interest in these hybrid approaches (see Jon Day, Chapter 11).

Deep-water habitats
Another important development has been the increasing attention paid to deep-water habitats. This has been driven mainly by concern over seabed habitats that are very susceptible to damage by bottom trawling, and highly vulnerable species like the orange roughy being fished at unsustainable levels. These deep-sea fisheries took off in the late 1970s. The impacts are often worst where target species aggregate on seamounts. Here

the seabed epifauna often form reef-like structures, such as those produced by cold-water corals, greatly increasing habitat complexity. These habitats are very vulnerable to physical damage and can take centuries to millennia to recover, given estimates of coral growth rates.

Similarly, target species, such as orange roughy, have all the life-history characteristics which make them highly susceptible to overexploitation. They mature at over thirty years, can live for over a hundred years, and have a relatively low fecundity. So the scope for sustainable fisheries in these habitats is pretty limited. Most of New Zealand's EEZ is deep water and encompasses a wide range of habitats, such as seamounts, canyons and chemosynthetic environments, so we need to include such deep-sea environments in a representative network of no-take MPAs. There are certain marine protection tools outside the Marine Reserves Act (1971) that have been used in the EEZ to protect areas from bottom fishing, for example the Benthic Protection Areas, but most of these have been established in areas beyond fishing depth so don't provide representative coverage.

Species-based marine conservation efforts

Species-based conservation efforts in New Zealand have often targeted seabirds and marine mammals. The region is something of a biodiversity hotspot for seabirds and cetaceans and there are concerns over mortalities from fisheries interactions. New Zealand's Marine Mammal Protection Act (1978) includes provision for establishing marine mammal sanctuaries, and one of the best known is that for Hector's dolphin around Banks Peninsula on the east coast of South Island, set up to protect them from gill-netting. Hector's dolphin is a small coastal endemic species and has become something of a flagship for marine conservation in the country.

Another endemic species is the New Zealand sea lion, and a sanctuary was set up for these in 1993 around the Auckland Islands. The sanctuary extends to 12 nautical miles offshore and is to protect the species' main breeding population. The problem is that sea lions can forage much further offshore than that and are then vulnerable to being caught in squid trawls. The fishery uses exclusion devices in the trawl nets – escape hatches like those used for turtles – but there is concern that sea lions may be injured in the process of escaping. The Auckland Islands population is still declining. Fishing mortality looks to be significant, but other factors may be contributing.

Seabirds and marine mammals are among the stars of ecotourism in New Zealand, an activity that has grown dramatically over recent years. Tourism as whole is now on a par with the dairy industry in terms of its contribution to New Zealand's GDP. Ecotourism activities need careful management, and there has been much work in New Zealand on developing guidelines and codes of practice to try and minimise disturbance to target species.

With regard to marine mammal sanctuaries, and especially work on Hector's dolphins, one of the things that has emerged is that populations may be genetically distinct, so we need to be looking at management plans for these distinct populations and also protect corridors between them. The Hector's dolphins off the North Island west coast are in fact recognised as a separate subspecies, known as Maui's dolphin. Its total population is now only some tens of individuals and it faces a very uncertain future if gill-net mortalities continue. The Banks Peninsula sanctuary for Hector's dolphin was one of the first to demonstrate how MPAs can work for marine mammals (Gormley et al. 2012).

On the world stage New Zealand has, for a small nation, been an active supporter of marine conservation and a strong advocate of protected areas in the wider southern hemisphere, such as the Southern Ocean whale sanctuary and more recently the Ross Sea MPA.

In the wider context of your marine conservation book (Probert 2017), what struck you about the development of marine conservation?
Marine conservation has emerged much more recently than conservation in terrestrial environments. There are some obvious reasons: often a matter of the sea being out of sight out of mind, but also as a free-for-all, as a repository for waste and for plundering natural resources. But that approach has caught up with us now. And importantly too we have come to understand that the major natural systems are interlinked, that strong interactions occur between the land, ocean and atmosphere.

In the epilogue of my book I've outlined what seem to be the underlying social issues that have underpinned this ecological crisis and why marine conservation is so important in this context. Basic problems relate to population growth, the way economic and ecological thinking have developed separately, and unfortunately the way that economics has not taken account of the ecological externalities, such as ecosystem services. This is particularly difficult to address unless economic activities take full account of their ecological implications. What one does about human population growth is another matter. Although it is levelling out in Western cultures, in many other parts of the world population growth continues – and projections indicate that globally we are heading for more than 11 billion people by 2100. Developing countries are in a bind, as they are often forced to overlook or neglect environmental protection for their very survival, and many developed nations have an excessive appetite for natural resources but poor accounting for the ecological implications.

There are no easy answers. We do need a more widespread understanding of basic ecology and the teaching of this in schools. There is now perhaps no knowledge more important to society at large than a grasp of basic ecology and of the environmental implications of our actions, not just locally but globally. There are glimmers of hope in increased awareness of the marine environment, a greater appreciation that major systems are interlinked and that we need to look after our seas and oceans. Maintaining healthy marine ecosystems is essential to human wellbeing. We basically understand the problems, and in fact have the tools to more effectively advance marine conservation. Concerted action is needed at all levels, from local and national to regional and global.

Conclusions
To sum up, there has been a shift in New Zealand marine conservation from a focus on MPAs to more holistic approaches that seek to place area-protection measures within overarching frameworks; and also an attempt to move from just coastal waters to cover the previously neglected offshore areas in the EEZ. We still, however, have issues with somewhat restrictive legislation, limiting the scope and siting of reserves, and the promotion of sustainable management. The approach to ocean management in New Zealand is still largely sector-based, and I do think we need a National Oceans Policy to provide a comprehensive and integrated view on how we use ocean space. In 2000 the government did start work on this but then pulled the plug on it because of political issues. Unfortunately the initial momentum has now been lost, but the process needs to be revived. We need to see how we can best accommodate the different uses of ocean space with healthy ecosystems. We have the tools but it needs the political will and resolve to deliver this.

CHAPTER 13

Heather Koldewey

Heather Koldewey (right) with a team in the Philippines

Heather is currently the Head of Marine and Freshwater at the Zoological Society of London (ZSL). She combines this with living in Cornwall, being an honorary associate professor at the University of Exeter, Penryn Campus, and Field Conservation Manager for Project Seahorse. Her early interests involved working with animals and helping on the family smallholding in north Devon and focused on becoming a vet. She had a strong interest in the ocean and rock-pooling and a love of wildlife. At Plymouth University (1986–1989) she read biological sciences and did her degree project on trout in upland streams, which paved the way for a PhD at Swansea University on Welsh trout population genetics with the Environment Agency (National Rivers Authority at that time). This was formative because it was being undertaken at a difficult time for depleted trout populations and was designed to help inform restocking policies. Working with river bailiffs and managers and having to explain her work to highly informed and motivated local interests was both eye-opening and challenging.

This experience reinforced her interest not just in science but in its practical application. Her postdoctoral studies were at the Institute of Zoology, part of the ZSL, which is when she met Amanda Vincent and began their hugely productive collaboration on seahorse conservation, which has had many successes; together they founded Project Seahorse in 1996. In 1995 Heather became the Assistant Curator of Lower Vertebrates at ZSL, which was huge step out of her comfort zone of research. This involved being duty manager in charge of the zoo, where her smallholding animal husbandry experience stood her in good stead.

This chapter explores Heather's diverse career, covering seahorse conservation work on captive breeding, marine protected areas, trade in endangered species, mangrove recovery programmes and developing community networks – which also led to disaster relief after a major earthquake and typhoon. Other themes explored include the power of individual action, collaboration and the need for ocean optimism.

MARINE CONSERVATION AND ITS DEVELOPMENT

What is marine conservation? How do you frame or describe marine conservation, and what are its key ideas?

When people ask me to make sense of my work there are two elements which, simply put, are nature for nature and nature for people. I got involved in marine conservation because I find the ocean amazing with its magical and extraordinary species. Working on seahorses has been fascinating, and whilst they are just a small part of the ecosystem a world without them would be far less interesting – and for me this is the intrinsic value of nature. Similarly our work on the Chagos Archipelago project has been about nature for nature and the need to value and protect these last ocean wilderness areas.

The challenge of thinking about nature for people is how we find a balance between, for example, fishing, with its important cultural and economic values, and marine conservation. Framing that balance is about long-term security and persistence of species and ecosystems and defining what conservation is. This then takes us in a wide variety of directions with projects like *Project Ocean* with Selfridges, who embraced the challenge to engage new audiences in marine conservation in their store, or *One Less* to reduce single use plastic bottles, or the *Net-Works* project that converts discarded fishing nets into carpets. It's about how we rethink conservation. Conservation is moving beyond biologists: to leave everyone feeling like they can be a conservationist, you don't need a degree, or to be a scientist, to achieve a more sustainable future.

What are the differences between when you started and now?

Massive differences and in a positive way. I don't think the ocean used to feature much in the news, and it certainly didn't have A-list Hollywood celebrities shouting about the ocean from the rooftops. The marine conservation community was a lot smaller, and when we started our careers there were only a handful of people in conservation organisations. The ZSL conservation work was small and only covered terrestrial animals; there wasn't a marine conservation programme.

Marine conservation came out of the extraordinary passion of the aquarium team, which I was leading at that time. There is a great deal of marine heritage at ZSL: we created the first aquarium in the world, and invented the word 'aquarium', which was derived from 'aqua vivarium'. This was in the era of the great Victorian collectors like Philip Henry Gosse when the interest in marine animals and exploring rocky shores was expanding. We even had celebrity fish – Clarissa the record-sized carp was one of the most famous animals in 1952, not quite Guy the gorilla, but in angling heritage she's up there.

When I started in the aquarium I couldn't understand the logic of why you would have conservation breeding programmes for terrestrial species but not for fish, and that has now completely changed. My remit as curator of the Aquarium and Reptile House included conservation, and one of the advantages has been the freedom I've had to explore how to achieve that. That freedom has possibly gone slightly beyond its intended boundaries, but enabled a host of projects which started as small, focused research projects, like Project Seahorse, but then grew arms and legs.

Do you see marine conservation as being different from terrestrial conservation?
Yes and no. There is a big difference in connection and visibility, so whilst many people have an intrinsic love for the ocean it is a lot harder to see what's going on below the surface. I use the example of the people on the cross-Channel ferry – if you asked them what was swimming under the boat, most people wouldn't have thought about it or know what was there. By contrast, people love going to public aquariums or watching *Blue Planet*, so we know there is a fascination with the ocean and its wildlife. It is a challenge to engage people in marine conservation, both conceptually and practically. First they need to know what is there, the wildlife and the habitats, but getting people to care about it is much harder. The Thames estuary is a perfect example, teeming with wildlife – but most people think it's brown and dead.

THEMES

Seahorses and conservation
My work with seahorses has involved many different strands. Their iconic, flagship status has helped us with captive breeding, establishing MPAs and managing their trade internationally, but this all began with a meeting with Amanda Vincent. She had done her PhD on seahorse behaviour but had become aware of the trade in seahorses (for traditional Chinese medicine, souvenirs and aquarium fishes), which she then documented through a landmark study and report (Vincent 1996). Seahorse species are hard to identify, and in Hong Kong you can see hundreds in baskets, and Amanda was interested in whether they could be identified using genetic techniques (Figure 13.1). Our collaboration led to us co-supervising two PhD projects, one to work on genetics and another that used classical taxonomy, and this resulted in the publication of the first definitive seahorse identification guide (Lourie *et al.* 1999).

In the late 1990s ZSL had some success in captive breeding seahorses and there was a very passionate community working on seahorses both in the UK and globally, which provided a great opportunity to address this aspect of the seahorse trade. In 1998 with my mentor, Dr Gordon McGregor Reid from Chester Zoo, we set up the first formal European breeding programmes for fish within the zoo and aquarium community. It seemed a completely crazy thing to do, and I remember four of us – me, colleagues

Figure 13.1 Dried seahorses for sale in Hong Kong. Source: Heather Koldewey, Project Seahorse and ZSL

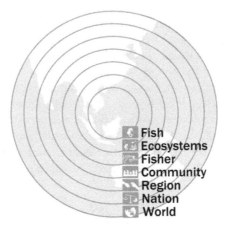

Figure 13.2 The world as an onion: addressing issues at all geographic scales. Source: Project Seahorse and Amanda Vincent (2008)

from Rotterdam Zoo, Burgers' Zoo in Arnhem and Gordon – presenting our ideas to the European Aquarium Community about why we should set up conservation programmes for fish championed by public aquariums. One response was 'Why do we need that? No fish are endangered.' Now, this is just an accepted normal part of what aquariums do, along with the very strong connection with active programmes in the field. There has been a huge positive transition since I've been involved in conservation. Seahorses were an ideal flagship species for this, and improvements in their captive breeding mean it's now very rare to see wild-caught seahorses in public aquariums.

Seahorses and marine protected areas

Amanda Vincent had discovered that the Philippines was the centre for the seahorse trade, centred on the island of Bohol. In 1995 Amanda went there and met a fantastic local NGO (the Haribon Foundation) which ensured she started by engaging a community organiser to talk to communities – experts with people rather than experts with fish. So Project Seahorse started with research interest, but then developed more widely than we could ever have imagined. As soon as we started working in the Philippines, questions emerged of how we could do something for seahorse conservation in a situation where people are very poor, earning an income from seahorses which could make the difference as to whether their children go to school or not. Going in as a Western scientist and banning fishing and trade was not an option. We came up with a helpful concept that Amanda crystallised perfectly by describing the world as an onion, with the seahorse in the centre, around which is habitat and environment, then the community, then national and global contexts (Vincent 2008, Figure 13.2). So in our work with Project Seahorse, we have tried to alleviate the pressure pushing down on that seahorse in the centre by work at every level.

Marine protected areas (MPAs) in the Philippines are a very familiar concept and are usually sanctuaries where there is no fishing activity at all. So it's very simple in terms of management guidance – here's the place, don't fish there, don't go there, don't touch it – which is why community acceptance is key. These MPAs were in their early days when we arrived, but it was part of an emerging marine conservation movement and now there are well over a thousand MPAs in the Philippines, so they are actually extraordinarily progressive globally. Driven by the communities from the bottom up – those are the ones that work the best, self-enforced, self-managed, often with training and guidance from NGOs. So there was already the framework we needed to provide the community support to MPAs for seahorses.

The first site was 100 × 150 metres, to protect Amanda's research site, but after that it was a case of listening and learning what the community wanted. The seahorses at these sites are night active, so a lot of the fishing was at night with lanterns, and catching anything – spear a squid, pick up a crab, and catch a seahorse – it's a complete multi-species fishery with no fisheries management. What gets caught dictates what people can eat or sell the next day, so it's a completely subsistence fishery. However, as in many fishing communities, there are lots of traditional methods of temporary closure of areas of the sea. Building on that tradition, we worked with communities to design MPAs from social principles without really putting any biological science in place to inform where they should go. Later on, when we did some studies to look at the ecological coverage of those MPAs, we found that it was pretty good – and it showed the value of progressing with MPA designation rather than waiting years while we did the science to decide where to put them, which may not ever be accepted by the community.

Global trade, CITES and seahorses

At the same time, we were looking at the global trade and realised that so much was coming in and out of Hong Kong. The Traditional Chinese Medicine (TCM) Association was a relatively small group and was controlling a huge percentage of the global market, so we started engaging with them directly, with a local Hong Kong Chinese team member who was based with WWF and Traffic in Hong Kong. We were clear we were not there to ban anything but to work out how to ensure there would be seahorses in the future, and obviously that is what that TCM community wanted as well, and so we had a common goal. The official records of the trade were very poor so we continued Amanda's pioneering work doing trade surveys, using enthusiastic early-career conservationists. These used a mix of methods, including questionnaires, looking at official records, counting animals, checking fishing ports and integrating all of those data. One option was to look at trade legislation, specifically CITES (the Convention on International Trade in Endangered Species). We ran a series of consultation workshops, and I remember one in the Philippines included Chinese medicine practitioners, which was one of the scariest and most exciting workshops I've ever been involved in. However, the TCM community were very supportive of CITES legislation as it could underpin a sustainable supply of seahorses, at a cost of additional paperwork. So we identified a common goal, even if our backgrounds and motivations were different.

Amanda led the process of securing the evidence and policy momentum to secure CITES legislation for all seahorse species in 2002. Other than beluga sturgeon it was really the first time that fish were on the CITES agenda, and there was a lot of resistance from fisheries interests to list fish species under wildlife legislation instead of fisheries legislation. Seahorses were non-controversial as they were not on the fisheries radar, in spite of the scale of the trade (millions of animals traded by over seventy countries), so we were able to secure this landmark step in legislation.

The challenge with the CITES listing of seahorses was that the Philippines had a strange national law that meant you couldn't do any trading at all of any CITES-listed species, including domestic catch and trade (CITES should only include international trade). So we were doing all of this great policy work internationally, knowing that the Philippines, which was where we were working hardest on the ground, could potentially ban all seahorse trade and impact many local fishers. This was one of the hardest decisions we made because we thought that globally CITES listing was the right way forward for seahorse conservation. We negotiated a two-year deferral of the implementation of the CITES listing to do preparation work in the Philippines (and other countries), to make sure we could find positive ways through this and prepare communities and our team for the ban in trade.

Sharks, rays and groupers have followed the seahorse listing, and CITES is definitely now an instrument for managing fish species that are under threat from trade, so the seahorse work really unblocked that. People said this last CITES meeting (in 2016) was an ocean meeting, and species that wouldn't have even been considered previously are now listed and protected, and countries have had to wake up and look at what management measures they can put in place, so that's a big win. For seahorses it has transformed the wild trade significantly, with a big drive to culturing seahorses instead of collecting them from the wild. The dried trade remains challenging because it is valuable bycatch and fishers on a shrimp boat select seahorses out of each trawl; cumulatively that's an awful lot of seahorses. So that's still the hardest nut to crack. There are terrifying reports of so-called 'biomass fisheries' where everything is caught and usually converted to fishmeal with no species discrimination. Destructive trawling remains the most significant challenge for seahorse conservation globally.

Mangroves, MPAs and communities

While working in the Philippines, we met an amazing Filipina scientist, Jurgenne Primavera, whose work illustrates the power of the individual in making a difference. She had, against all odds, worked in an aquaculture research centre and produced a powerful mangrove conservation agenda, which is quite hard to do as aquaculture has been responsible for the loss of most mangroves. She came up with the science to show that the 'mangrove to aquaculture pond ratio' needed to have *both* side by side. This recognised that aquaculture wasn't going to go away – you need food security – but mangrove protection was also needed.

Working with Jurgenne and a colleague of mine (Alison Debney) at ZSL we developed a project that confronted some big questions. How do you protect what mangroves are left and restore them when parts of the Philippines have lost over 70–80% and the country as a whole has lost over half? We looked at how to translate that scientific knowledge, the ecosystem services and multiple benefits that go with mangroves, to action on the ground with communities and strong synergies with our work on Project Seahorse. The loss of mangroves meant the coast had become very exposed to wave action, and communities were very concerned about this as land has a high value for farming. With the help of the communities, scientists and engineers we produced a series of manuals – how-to guides – on mangrove restoration (Figure 13.3). These covered how you approach communities in the right way, work with them, what training they need, how to do it, working with government, and how to do all this from an ecological perspective. In some cases we ended up building very simple breakwaters out of rocks to stabilise eroded coastlines enough to allow planting, in others restoring abandoned fish ponds back to mangroves. Through that project Jurgenne led the team and now ZSL-Philippines is a registered NGO locally, staffed by biologists, community organisers and operational team members, all of whom are Filipino.

One of our strategies was to encompass mangrove habitats in MPAs, and that has been incredibly effective with up to a tenfold increase in our MPA sizes, moving beyond coral-reef habitats. Communities recognise that mangroves are already protected but hadn't been aware of their value from a fisheries perspective, even though most reef fish

Figure 13.3 Planting mangroves in the Philippines. Source: Heather Koldewey, ZSL

species really need mangrove habitat as part of their life cycle. Such expansion of MPAs is needed to meet the global target under the UN Convention on Biological Diversity (UNCBD), which is to protect 10% of the ocean by 2020, and Philippines national legislation to protect 15% of coastal waters. Having been a fish person for so long I got very into mangroves, as trees don't move – so monitoring their recovery is much easier than with fish. Building these community groups to support the mangrove rehabilitation efforts is amazing, so-called people's organisations, which we train and support, include building partnerships with the local government. You see this transformation of energy, the people are so excited, realising how important the mangroves are, realising they're not just where the mosquitos and waste goes; this is a huge transition.

Disaster strikes – earthquake and typhoon Haiyan

We were on a really positive trajectory and then we had two disasters within three weeks at the end of 2013. Firstly, there was a 7.2 magnitude earthquake in Bohol, which was devastating and affected around half of our sites. We had dive surveys under way at the time, which was quite an extraordinary experience for the team in the water, though luckily everyone was safe. However, people died, and there was huge damage to property and infrastructure, and some of our team members' homes were affected; it was a very worrying and difficult time. Three weeks later, a bit further north at a number of our other sites, super typhoon Haiyan hit, the strongest typhoon to ever make landfall, and it devastated a vast swathe of the country.

We moved into disaster relief mode, partly because the communities reached out to the team because they knew them, and also because some of the team were the first people to reach those sites. This was a new experience for me, and I was in the UK talking to disaster relief agencies like Oxfam and ShelterBox, and trying to work out what to do to help. We worked with our local government partners, who were organising relief, and our community groups, who were set up to restore mangroves. We liaised with them about what they needed, and because they were very organised and used to working together they had systems in place like book-keeping that could be adapted to this post-disaster context. So even though our work with communities was initially around marine conservation, these people's organisations became the centre points for organising the aid relief. They worked out who needed what, an equitable system that got the right things to the right people, and documented the distribution and ongoing needs (Figure 3.2). At the same time, we also had to negotiate supplies, because things became scarce very quickly. One of our finance team, who was very heavily pregnant at the time, was negotiating materials and arranging for the rental of a big truck to help with delivery.

It really was incredible to see the team in operation. One of the team in the earthquake area had his wedding planned, so he got married in the morning and spent the afternoon with his new wife packing aid boxes. Getting tarpaulins to schools helped the children so they could have some sense of structure to their day, to be able to meet with their friends and have a sense of normality. Our Net-Works project, which helps convert used fishing nets into carpet, also came into its own, because the people needed money and collected more nets, and we then used our supply chain of trucks to take this out for recycling and bring in aid on the return journey.

In the aftermath there have been some really big benefits. It was clear that areas protected by mangrove forests survived much better than those without. This has prompted huge public support for MPAs and mangroves and a major government-led mangrove restoration programme. We continue working hard to ensure this enthusiasm doesn't ignore science, and unfortunately we do see a lot of mangroves planted in the wrong place that results in their mortality or impacts on other vulnerable marine habitats such as seagrass beds. Our scientific advice with our how-to guides and now our training courses has been invaluable.

THE POWER OF INDIVIDUALS AND COLLABORATION

The power of individuals

It is striking that you meet extraordinary people who are driven and have the extra spark to really make a difference. They stick their necks out to challenge the status quo. Qualities like determination, stubbornness, not always keeping everyone happy, not necessarily following the system or doing what you're supposed to do, often figure prominently. Recently I talked to a Filipina women whose community was the first place typhoon Haiyan hit, with 100% of homes completely devastated, and she had lost her husband and had a young child. We were in the process of setting up a substantial MPA and I asked why it was so important to her. She gave an answer that any marine conservationist would have been proud of, about how important this was to her. She was so committed, talking about how some of the people in the community were less than enthusiastic and how she felt it was her job to explain it, to engage and inspire them. She had no formal education, but I feel that she was an inspiration just as much as conventional leaders. Amanda and Jurgenne also show that passion and drive, which really engages people – it's motivational and it gives hope to see such remarkable individuals in action.

The power of collaboration

Easy to say and tough to do. At the end of the Project Ocean work which I did with Selfridges in 2011, we debriefed with the Selfridges team and set out what went well and what didn't. For me one success was having 23 UK-based NGOs all with the *same message* about ocean conservation. The Selfridges team were genuinely shocked by my view on this, because they said, 'You're all small, all under-resourced, all care passionately, and surely you should be saying the same thing all the time?' You come to realise that for all sorts of reasons the NGOs have different ways to communicate their approach and messages. When you work with your own press office they always look for the right angle for *your* approach, and although ZSL has a long history of collaboration it's frustrating for them to have a press release that has ten different organisations to consider – they would much prefer for it to just be a ZSL story. There is also the competition for funds between NGOs: you're pitching for a limited environmental pot which in the UK is tiny, and a marine conservation pot which is smaller still.

At the same time, the problems we face are way too big to be dealt with alone. Every time you see an organisation charging ahead, being single-minded and not collaborative, it seems to me they are heading down a conservation cul-de-sac. We need to get over the focus on 'egos and logos'. For collaboration there has to be a conscious decision to put your own logo aside and remember the bigger picture. The director at the zoo once went off with one of my ideas that he took credit for, and my first boss said, 'Well, it got the job done didn't it?' That experience has since been a real guiding principle for me, because I feel that it is more important to get the job done than to take the credit for it.

In exploring what collaborations can achieve the Calouste Gulbenkian Foundation has been both supportive and innovative; they are trying to understand how these projects work, they have a conceptual lab, and they also use excellent facilitators, which is a vital in this process. Their approach is to tackle collaboration head on, as an experiment, to get NGOs into a room, give them some money to develop something bigger together than you could do on your own. We don't articulate experiments often enough in conservation, because we are scared of failure, things are too serious and funders never want to hear about failure. Even if things go wrong you're always trying to write your donor a report in a way that highlights the positive points. The underpinning agenda is, how do we look at a wider range of values for the oceans, an ocean-friendly society, as opposed to the economic one which tends to dominate the narrative.

One project to emerge from this process was the One Less campaign, which is trying to eliminate single-use plastic water bottles from London – which is big, ambitious, yet tangible. We've got this great set of organisations which bring different skills and thinking – for example, Forum for the Future brought disruptive systems thinking, whilst Thames Estuary Partnership brought more practical data that we can get from beach cleans. The Gulbenkian funding gave us the time to build trust, support and recognition of each other's strengths to really make it work – and as a result it is phenomenally powerful.

Another effective approach to collaboration is to set a common goal, and we did this with the Chagos Environment Network. The aim was to set up the Chagos Archipelago MPA on a fixed time line, with a job to be done, a clear focus, and different groups from the Linnean Society to the Chagos Conservation Trust working together for the bigger picture. This network provided the evidence and momentum that resulted in the declaration of the Chagos Archipelago as a fully protected MPA of 640,000 square kilometres in 2010.

The cool collaborations are where you can bring together people with very different skills, although this can lead to a culture clash. I have had some amazing friendships through collaborations. Being united in a common cause is incredibly reinforcing: everyone has highs and lows and gets stuck, but in a collaborative process someone will always push through with new energy or crazy ideas. I don't see any other way forward than collaborations, and it's encouraging to see funders really pushing harder for collaborative approaches; done well, they are hugely empowering. There is a growing number of coalitions, alliances among the marine conservation sector which are more or less effective. Sometimes points arise in coalitions where there is no agreement, and this can cause difficult situations. The trick is to recognise this and in some cases concede the point for the greater good and move on. I am fascinated about what makes a good collaboration; how we can replicate that is a major area of work for the future.

OCEAN OPTIMISM

The background to this idea arose from a combination of motivations. As well as seeing amazing individuals around the world, you see lots of different ideas happening and really working, but then you see the same mistakes again and again. If you were in Bangladesh, how would you ever find out about something that was happening in the Philippines that might be relevant to you? I'd been thinking about that for a while. I went to a talk by Ellen Kelsey, who has worked extensively on the psychology of communicating science and how to address the dominant narrative of doom and gloom. This is a phenomenon well known in the behavioural sciences which recognises that a constantly depressing narrative is a turn-off and that we need to redress the balance. Even if things are bad it is important to give people hope and share and celebrate the successes instead of always focusing on disaster. At the same time, Nancy Knowlton from the Smithsonian added a 'Beyond the Obituaries' day to the International Marine Conservation Congress to share stories of success.

In 2013 we all got together in Ellen's house in Monterey Bay to talk about this. We then convened a workshop with a small, eclectic group of individuals. We asked, 'How can we come up with some kind of movement and one hashtag?' with no egos or logos accompanying it. We had no budget, so for World Ocean Day in 2014 we came up with the idea of ocean optimism to define the moment, with the hashtag #OceanOptimism that we could use to share success stories. So if you had something that worked, you shared it on social media. We launched and it just captured the public imagination, reaching over 60 million people through social media since then. It now has a website, and if you search for it on social media you can see that's it used by a wide range of individuals and organisations, with no attempt to own it. It is really exciting to see such a positive response. What we're looking at now is moving beyond a hashtag, using ocean optimism as a platform for change together with the complementary Earth and Conservation Optimism movement that launched in 2016. The internet is amazing, so now Bangladeshis, Filipinos, whoever, can actually connect and share conservation solutions. It has been a great way to collate and start sharing – but how do we then target change? Being optimistic keeps me going. You can feel like such a tiny speck. You ask 'What can I do? What can my team do? What can we do through collaborations?' If you don't start with 'I can do something', you would give up.

THE FUTURE

What are the future challenges? What are the most important threats to the marine environment?
Climate change and bleaching. The striking thing about climate change is that it's beyond your local control. We are looking increasingly towards building resilience, but the challenges are everywhere, and when you still see such basic debates at policy level such as 'does it exist?' then it makes you realise the scale of the problem – that the science is still being questioned, that big countries have varied under political leadership, for example, Canada was disengaged and now has really engaged, the USA was engaged but has now disengaged. This is something the world needs to unite behind, and it's difficult to see how that will happen when you see how vulnerable it is to political change.

The demand for resources, and the technology. Technology can be a double-edged sword. There are pluses, not least with science, but there are also destructive technologies around fishing, bigger, deeper, and longer – and then deep-sea mining becoming an emerging industry. Our project in Mozambique faces huge threats from natural gas exploitation – the country has more natural gas than Qatar but in an area where 50% of the households struggle with malnutrition and the average life expectancy is about fifty, and in one of the world's marine biodiversity hotspots, the Northern Mozambique Channel, a proposed World Heritage site. That landscape will change beyond recognition over the next twenty years, and that community and environment has little resilience to that kind of change. Those are the things you see on the front line, but it is part of the bigger picture of industry and extraction. It can be scary to contemplate.

What are the real, hidden problems and barriers that make progress harder than it should be?
One of the things that frustrates me most is that even though countries sign up to commitments they don't deliver. We have a new sustainable development goal specifically for the ocean and many other sustainable development goals that make our ocean relevant – it's all there. We know what to do, and we know how to do it, and mostly the world has signed up to do it – but still progress is very slow. For example, if you take the MPA target from the UNCBD, in 2010 only 1% was achieved when it should have been 10%. Instead of there being any accountability the deadline was pushed back by ten years, and no one said that's not ok. Where is the accountability when targets aren't met?

The other thing which has emerged in the thinking on marine collaborations is valuing the ocean. It is an interesting conversation to have with people, getting them to open up and tell you what aspect of the ocean they care about. It allows storytelling about experiences, being on holiday and watching an ocean sunset with a loved one, fishing with a grandparent, and holidays in Cornwall as a kid. It's how you capture that and use it as a way forward to engage people more in ocean conservation. I was talking about this to a fellow passenger on the train recently – how many people on the commute would have thought about the ocean this morning, this week, this year? As a society, we seem to have become more distant and disengaged from nature, and understanding values could really help us make progress.

Regarding policy and politics, political cycles and donor cycles don't necessarily match environmental cycles. So we're always trying to get things done in three years, job done, ocean saved! You're vulnerable to those political cycles that can set things back, change trajectories or just slow things down; it's important to do things at the right time. In the Philippines recently there was an election which meant that everything was on hold for five months until officials had been elected; for an environmental project that's a big chunk out of the working year gone.

When you retire, what would you most like to have achieved for marine conservation?
For me, it's creating a fantastic team of people working around you. If you do a good job as a boss then your legacy is continued and the work is done for you. The team do great things, coming up with new ideas, enabling *everyone* to play their part in marine conservation.

Are you optimistic or pessimistic about the future?
Optimistic.

CHAPTER 14

Sarah Fowler

Sarah works for several shark conservation organisations around the world. These include advising the Save Our Seas Foundation and being a board member for the UK Shark Trust. Her interest in marine biology stemmed from summers on the Isle of Wight, rock-pooling and mucking about in boats, but also a shell-collecting grandmother who stimulated her interest in natural history. She set her mind on becoming a marine biologist from an early age and went to Bangor University where she graduated with a degree in zoology and marine biology. At the suggestion of Mike Gash (Nature Conservancy Council, Wales), she applied successfully for a place on the University College London (UCL) conservation MSc course. This was by then an established route to working in the UK nature conservation sector.

Her first job was as a higher scientific officer for the Nature Conservancy Council (NCC) with Roger Mitchell, one of the few people working on marine conservation at that time. Her appointment coincided with the passing of the Wildlife and Countryside Act (1981), which included the first statutory provision for Marine Nature Reserves (MNRs) in the UK.

Eight years' work with mixed results on MNRs, and on a wide variety of marine conservation case work, also introduced her to issues surrounding the conservation and management of the basking shark in UK waters. Sarah left NCC to join colleagues Sue Everett and Paul Goriup in setting up and becoming a director (subsequently managing director) of Nature Bureau International. In 1991, a chance conversation led to an invitation to become the deputy chair of the newly established IUCN Shark Specialist Group, and shark and ray conservation became one of her main activities. This included the development of the European Elasmobranch Association and the UK Shark Trust.

Sarah has for over 25 years been at the heart of international shark conservation efforts, arguing for rational, science-based management of shark fisheries and conservation of threatened species. Her work in many international forums, including the Convention on International Trade in Endangered Species (CITES) and the Convention on Migratory Species (CMS), has taken her into difficult and challenging debates all over the world. Most recently, her contribution to the development of the Global Shark and Ray Conservation Strategy is another major accomplishment, and her work has also been recognised with an OBE.

MARINE CONSERVATION AND ITS DEVELOPMENT

What is marine conservation? How do you frame or describe marine conservation, and what are its key ideas?

My idea of marine conservation is about sustainable use, and improving the status of the marine environment. It is not about preservation, but rather about stewardship: looking after many elements, including commercially important resources, rare and threatened species and human community benefits. Getting to the stage where one is promoting preservation implies that there's been a major failure of management. My view is that marine conservation means making sure you don't lose that badly.

What are the differences between when you started and now?

I have always been interested in marine biology. By the time I finished my first degree, I knew I wanted to work in marine conservation. At that time nature conservation was a little bit out in the wilderness. The conservation MSc at UCL was producing ten to a dozen people a year to work in conservation; not a lot of trained people. Conservation is now mainstream. It has changed from being viewed by some as slightly ridiculous to being fundamentally important (even if in some cases people may say that but not believe it). Today, it has much greater public support, which means it has also got a lot more political support.

Attitudes have also changed. For example, the other day I heard a gentleman on the radio talking about the capture of the largest thresher shark on record. They were clearly determined that this fish would not stay on the boat longer than necessary, and it should go back in the water alive. You would not have heard that thirty or forty years ago. There are still shark kill fishing tournaments in the US every year, but many now only allow one or two landings per boat; others none, with the competition results based on lengths and photographs. Tag-and-release projects have also made a major contribution to scientific knowledge of shark and ray species.

When I first started working for the NCC, back in the early 1980s, I went to a meeting where somebody mentioned the concept of self-regulating fisheries. I asked what this was. Apparently, it is when fishermen take all the fish, then have to stop fishing, therefore the fishery ends and has regulated itself! If anyone came out with this today, people would be appalled. Today we call that boom and bust. Back then fisheries collapse was 'self-regulation', which I think is an example of how far we have come.

Another difference: in the early days there was often great tension between conservation and the management of fish species. They were not viewed as being the same thing. Making a broad sweeping generalisation, there was a tendency for conservationists to be viewed as the enemy of fisheries management. Fisheries management belonged to fisheries managers and regional fisheries management organisations (RFMOs); wildlife conservation had nothing to do with this and should keep well clear. The greatest change that I have seen over the last twenty years or so has been that division blurring. Instead of having two opposing camps, we have people from both fields working together on the conservation *and* management of sharks.

How is marine conservation different from terrestrial conservation?

It is far more difficult when you cannot own wildlife sites, to conserve them directly. You can't put up fences, you can't easily have visitor centres. It is also a lot more expensive to achieve effective site-based marine conservation. There are also differences for species conservation. When you are trying to assess a population, certain things are taken for granted in the terrestrial world. You generally know how long a mammal lives, how quickly it breeds, and what area of ground it needs to live in. Marine animals are so much harder to count and to track. You don't know how many there are, where they are, how far they travel. We don't even know how long some species live, or when they become mature. Every year we discover new, extraordinary and complicated things about the marine environment.

THEMES – MARINE CONSERVATION – YOUR APPROACH

Developing awareness of shark conservation – fisheries management, shark finning, China and culture

The world's first protected shark species was the grey nurse shark, in New South Wales, Australia, in 1984. What had happened was that, following the film *Jaws*, macho Australians 'discovered' grey nurse sharks. This species has a mouth full of snaggly teeth and looks very fierce, but is actually rather docile. Divers realised that it was quite easy to go out with a bang stick, swim right up and kill a shark, drag it onto a beach and take a photo. The population of grey nurse sharks in New South Wales nosedived and people noticed. This led to their protected status, and that was followed in due course by protection in South Africa for the great white shark, which had also suffered directly from revenge killings following its PR disaster of *Jaws*.

Whilst people had recognised that there were issues with a couple of big charismatic species in certain parts of the world, there was virtually no commercial fisheries management for sharks. I think the USA was one of the first, in the late 1980s, to start developing a shark fisheries management plan because stocks were becoming depleted. What altered attention to the issues of shark conservation was the sudden explosion of the middle class and the growing acceptability of conspicuous consumption associated with China's huge economic growth during the 1980s. Shark fin soup was rapidly on the menu at a huge scale and, to fulfil this demand, fisheries were starting up not just for shark meat, for which there had always been a few fisheries, but specifically for shark fins (Figure 14.1).

It was around that time that I started working in shark conservation, but completely by accident. At NCC, I had been trying to get the basking shark scheduled for protection under the Wildlife and Countryside Act, then I joined Nature Bureau in 1989. Around about 1990, I happened to have a rant during a coffee break with one of my colleagues, Paul Goriup. My theme was that shark populations were being depleted and nobody was noticing, or if they were noticing, nobody cared. There was a real crisis building with shark conservation and something ought to be done. An hour later Paul put a piece of paper

Figure 14.1 Sharks fins drying. Source: Stan Shea/Bloom

in front of me. It was the fax he had just sent to International Union for Conservation of Nature (IUCN) in Switzerland saying, 'It sounds as if IUCN needs a shark specialist group, and we have got someone here who will help you set one up.' Almost immediately my phone rang: it was Simon Stuart, then head of the IUCN Species Programme. Simon informed me that IUCN was setting up a Shark Specialist Group. They had a chairman, Samuel (Sunny) Gruber, and after my conversation with Paul it seemed that I would be deputy chair!

The people involved early on in the Shark Specialist Group were primarily researchers who had noticed that their study animals were becoming depleted, with very few fisheries managers or policy advocates. Our first meeting was held during the 1991 'Sharks Down Under' Conference in Sydney, Australia. That week a local radio host actually had people cheering for killing sharks: *'Let's kill all the sharks, shout if you want to kill all the sharks!'* That is where we started – and, thank goodness, that is very unlikely to happen today.

From a UK perspective the basking shark was important and it kicked off our awareness of conservation for large charismatic marine species that were not birds or mammals. It was also one of the pioneering species for adding sharks to the appendices of the international wildlife conventions: the Convention on International Trade in Endangered Species (CITES) and the Convention on Migratory Species (CMS).

Starting the work on shark conservation – IUCN, Shark Trust, EEA

There were only a few scientific bodies interested in shark conservation in the late 1980s and 1990s, including the Japanese Society for Elasmobranch Studies and the American Elasmobranch Society (AES). In 1990 and 1992, scientific meetings were organised in London by Bob Earll from the Marine Conservation Society (MCS), where scientists from Europe agreed that we needed a European equivalent of the AES. Eventually we got some funding and established the European Elasmobranch Association (EEA) network. By the late 1990s national shark or elasmobranch organisations were being set up all over Europe, which was very satisfactory. Initially these were mostly scientific bodies, because scientists first noticed the need for conservation management, but today there are also various pressure groups, which is a broader mixture.

CITES became engaged in shark conservation management because booming shark fin imports to China in the 1980s and 1990s were stimulating unsustainable global fisheries and international trade. CITES involvement was ground-breaking: when the Ninth Conference of Parties (CoP9, 1994) adopted a Resolution on Sharks, this was the first time ever that it became active regarding a taxonomic group when not one single species was yet listed in the appendices to the Convention. The basking shark and the whale shark became the first two elasmobranch species eventually to be listed in CITES Appendix II in 2002 (Speedie 2017). They were followed at CoP10 (2004) by the great white shark. All three were species of conservation concern, but were no longer being commercially exploited. Their listings still upset fisheries management interests! It took a long time for CITES to be able to move on to listing commercially exploited species, like the porbeagle shark at CoP16 in 2013. Pure species conservation was therefore at the vanguard of getting a broader suite of species listed, and stimulating action which related to both fisheries management and nature conservation.

Actually, my belief is that early in the CITES and sharks story, in the late 1990s, the UN Food and Agriculture Organization (FAO) International Plan of Action for the Conservation and Management of Sharks (IPOA Sharks) was adopted to prevent CITES from becoming more closely engaged in shark conservation, and to avoid the listing of shark species in its appendices. The argument went that CITES did not need to get involved because there was an FAO IPOA Sharks and so everything was going to be fine – 'Stand aside, we have this under control!' As the years went past but management did not improve, it became apparent that this was not true.

There were very few people from marine conservation bodies at the FAO Committee on Fisheries (FAO COFI) meetings in the mid-1990s, when I first started attending; it was sometimes quite a lonely experience. In those days, there was a lot of tension between conservation and management, and I certainly got the feeling sometimes at COFI meetings that the conservationists were not welcome. This was a time when you might have people from environment departments never discussing marine policy overlaps with their counterparts in fisheries departments. Never, ever. Even if they worked in the same ministry with offices in the same corridor! Nevertheless, both FAO and CITES had started to urge their members to ensure that a broader representation of specialists attended both bodies.

Eventually, and slowly, we did start to see changes. Instead of a delegate starting a one-person debate in an FAO meeting regarding whether CITES should even exist and suggesting that it was contrary to World Trade Organization (WTO) rules (although their own government had been a CITES Party for twenty years), there is now much better awareness of marine conservation needs. There is also a lot more communication inside governments. This may sound trivial now, but it was incredibly important. This shift is still ongoing. It started with a small number of conservation-orientated countries and a larger number of countries that we called 'friends of fisheries'. There are still a lot of countries in the middle, swaying one way or the other, but the conservation-aware group is growing and many more countries in the centre understand, or at least I like to think they understand, that marine conservation and marine management are the same thing. Serious tensions do remain, however, and I have no doubt that there are still countries that would prefer marine conservation interests not to be so actively involved in fisheries issues. They are finding it harder to win, though, because international biodiversity meetings take decisions based on a two-thirds majority. Ironically, this makes progress easier than in RFMOs, where a single member can block fisheries management proposals.

Maybe it would help to change perceptions if conservation was called management. CITES is mostly about sustainable management. Most CITES species are listed in Appendix II, which is purely and simply about sustainable trade. Commercial trade is only prohibited in a relatively small number of species listed in Appendix I. There are two main things that a Party needs to do to export products from these species: to confirm that it was legally obtained, and to demonstrate sustainability through 'a non-detriment finding' (NDF). This basically states that the export will not be detrimental to the survival of the populations. It is exactly the same as sustainable fisheries, which is what countries are supposed to be doing, yet implementing CITES is often seen as incredibly hard, scary and difficult. A key factor though is that there is a requirement to follow CITES rules, and deterrents for countries that do not.

The importance of capacity building

Capacity building is time-consuming but so important. To achieve marine conservation policy change, you need people who are interested in achieving the same things, who know how to work with decision-making bodies and with politicians. They need to understand how to set their objectives and recognise when they get there. Part of this is understanding what objectives are feasible and sensible, which is something that comes with experience.

One very interesting public capacity-building initiative in Europe was the establishment of the Shark Alliance. In 2005, I was asked by the Pew Charitable Trusts to help them scope out a three- to five-year campaign leading to a UN General Assembly ban on shark finning. This is the process of chopping the fins off sharks and dumping the body in the sea. My response was, well yes, sure you could do that. In fact, it would probably be quite easy to get such a resolution agreed by the UN General Assembly. But you know, so what? Implementation would be poor and it is unlikely to make a huge difference. My suggestion was that to achieve the world's biggest improvement for shark conservation, including addressing finning, the place to work was Europe. At that time, Europe not only had the largest fleets and was catching more sharks than anyone else (which is still the case), but also it did not exactly have a stellar reputation for fisheries management. This was agreed, and the Shark Alliance was formed, based in Brussels. It had two very simple, clear objectives. One was to get a decent shark finning ban adopted across Europe, as the existing ban had many flaws, not least being unenforceable and using the wrong mechanisms. The other was to get Europe to adopt an effective Community Shark Action Plan under the framework of the FAO IPOA Sharks. Achieving these aims should also lead the EU to promote these policies globally in the RFMOs.

The Shark Alliance was a brilliant campaign. It was time-limited but geared up for massive capacity building across Europe that would outlive it. The alliance comprised individuals and organisations interested in shark conservation from as many of the EU countries as possible. Alliance members were trained; how to go to Brussels, how to identify their permanent representatives, how to find their Members of the European Parliament, how to ask their representatives in Europe to do things that were sensible and achievable, and even how to dress for the occasion. Sorry, but they will not take you seriously if you turn up to meetings in a ripped T-shirt. You must have the right camouflage! The campaign was fantastic and achieved both its objectives as well as giving us a huge insight into the importance of capacity building. Since the alliance closed, there has been some loss of momentum, but there are now a lot of people in Europe who know how to take part in the democratic process that is the European Union. We should do a lot more capacity building on this scale.

Another form of capacity building takes place at government level. Now that we have a few commercially exploited sharks such as porbeagle and hammerheads (Figure 14.2) listed in CITES, some environment or wildlife departments are panicking about

Figure 14.2 A shoal of scalloped hammerhead sharks off the Cocos Islands, Costa Rica. Source: Jeremy Stafford-Deitsch

implementation. A few countries have a knee-jerk reaction: if a species goes onto CITES Appendix II, management becomes too difficult. They just ban trade. That is not a success story, nor is it what we want, which is sustainability. So we have been helping to develop the tools that countries need to implement CITES for marine species, which is essentially fisheries management. The problem is that for many years there has been almost an intentional strategy to separate wildlife management from fisheries management, and this has been successful in many people's minds. We are now trying to stitch them back together, including by providing countries with guidance, for example for developing a non-detriment finding (NDF), which is the basic certificate of sustainability needed by CITES.

A few years ago in Bangkok, at the sixteenth CITES CoP (2013), after all Appendix II shark listings had been adopted in the final plenary, many NGO delegates were rushing around high-fiving, hugging, taking selfies and cheering. I sat in my seat thinking, 'Oh my goodness, now we have got to implement this. If we don't, it will be an absolute disaster!' I was there as an advisor to the German delegation, one of whom turned to me and asked, 'Well, what do we do now?' This was just fantastic. While many conservationists were thinking 'we've got them listed, we can go home', my colleague was immediately looking forward to the next step. My response was that we needed capacity building, particularly guidance on how to put together NDFs. He said OK, and made sure that it happened. Since then, NDF guidance has been produced, there have been many workshops, and we are encouraging countries to share their NDFs (which is not a CITES requirement) to help others. In 2016, Germany commissioned shark experts Paddy Walker and François Poisson to develop 'example NDFs' for thresher sharks in the northeast Atlantic, and silky sharks in the Indian Ocean, because these were two of the shark species proposed for listing at CoP 17 that year. This was in anticipation of arguments at CITES that these species shouldn't be listed because it is too difficult to do NDFs, and an inability to issue NDFs would drive international trade underground, and/or result in trade bans. Our aim was to go to the conference and say, actually NDFs are quite simple. This is how you do it, and here are some we prepared earlier. Implementation workshops for governments have continued throughout 2017. Capacity building for countries, and demystifying the fisheries management process for wildlife departments, is vital. If we can't make this work, we risk sliding back to where we were over a decade ago.

Capacity building is also important in other forums, including regional bodies. In the northeast Atlantic, OSPAR has been working on shark and ray conservation for a long time. In the Mediterranean, the Barcelona Convention has also been plugging away at elasmobranch biodiversity conservation for years. OSPAR and the Barcelona Convention are not really supposed to 'do' fisheries, but their efforts are starting to cross over into fisheries management now. The General Fisheries Commission for the Mediterranean (GFCM, an RFMO) has adopted the Barcelona Convention annex listing the threatened sharks and rays that should be non-target, prohibited species. Unfortunately, many GFCM members are struggling with these measures, and we have to remind countries that they signed up to do this. They will also need to be helped with capacity building for implementation, and reporting back on compliance. You do need the regulations in place first, though, because until then you can't do anything about implementation. We still have a lot of 'paper management' for species, and turning those pieces of paper into conservation management on the ground is a big step.

The challenges of working in the main conservation and fisheries forums – CITES, IUCN, ICCAT, CMS

The Convention on Migratory Species (CMS, 1983) has been around for nearly as long as CITES (1975). CMS has a similar list of elasmobranch species to CITES, but agrees almost every proposal by consensus, rather than by vote. However, CMS does not have the same 'teeth' as CITES, which can as a last resort stop the incredibly valuable wildlife trade of a country (with severe economic consequences) if it doesn't regulate trade properly. What CMS has, that CITES does not, is a memorandum of understanding (MoU) and action plan for migratory sharks. Anyone can sign up to the MoU and action plan, including non-Party countries and NGOs: for example, the Shark Trust is a cooperating partner to the CMS Shark Action Plan. These measures are also blurring the distinction between fisheries management and species conservation, which is great.

The International Commission for the Conservation of Atlantic Tunas (ICCAT) and other RFMOs are also becoming engaged in international shark management. It is good to see that several conservation NGOs now attend RFMO meetings, where they issue statements and talk to member countries, and fisheries bodies have been listening. For example, ICCAT had adopted a prohibition on the retention of big-eyed thresher shark before thresher sharks were successfully proposed for listing in CITES Appendix II (the sustainability appendix) in 2016. It is debatable, I suppose, whether ICCAT was just giving itself a little bit of a greenwash. Saying, 'Look, we are doing shark conservation; we have adopted a prohibition on this species that is hardly seen any more.' But, you know, things *are* changing.

The Global Sharks and Rays Initiative

The Global Sharks and Rays Initiative (GSRI) partnership has set global priorities for conserving sharks and rays, under a ten-year strategy for 2015–2025 (Bräutigam *et al.* 2015). It originated from the Wildlife Conservation Society in partnership with several other organisations that have been working on sharks for a long time: the IUCN Shark Specialist Group, TRAFFIC, WWF, Shark Trust and Shark Advocates International. It is amazing that someone came up with such a bold idea, and it certainly could not have happened twenty years ago. We identified four main planks or strategies, which are equally important and mutually supportive, to get shark *and ray* conservation management onto a firm footing. One is saving species. Sustainable exploitation is not really possible for the most seriously threatened species, so the aim is to protect them. The second plank is the sustainable management of species that are not yet endangered, by improving fisheries management. The third is responsible trade. That is making sure that the volumes of shark fins and other shark and ray products in trade does not undermine fisheries management and threaten species. The fourth plank is encouraging responsible public consumption of these products. That's not saying 'don't eat shark fin soup' but rather ensuring that consumers make sustainable choices. The previous 25 years' efforts have, in effect, established the framework and the footing for this big push. The other amazing thing is that there are now several Foundations engaged in a Global Partnership for Sharks and Rays that are prepared to put millions of dollars into this, every year for ten years. That is remarkable.

The tensions between science-based management and direct action

We now have a really broad spectrum of organisations, of techniques, and of policies working towards the conservation of sharks and rays. At one end is pure science, grading into fisheries management and biodiversity conservation strategies. At the other is direct action, sink the fishing boats and ban everything. The direct-action groups are very good at getting attention, but I have major concerns about legality and safety, as well as other issues. I think they are engaged in some very dangerous activities, but direct action is undeniably really good at raising public awareness. In the early days, I used to be quite wary of what was going on at that end of the spectrum because, while I wouldn't personally want to pull wings off flies, I saw myself as a scientist and more interested in species conservation and stock management than welfare issues. But a friend explained to me that these organisations are very important for our cause, sitting on the fence in the middle, because they shift the centre of debate. The more extreme the organisations and the individuals at the far end of the spectrum, the more that moves people working in species conservation and management towards the centre. Remember, there was a time when we were right out on a limb, working on marine conservation and shark conservation. We were once the weirdos, but now we occupy the middle ground.

Another advantage of direct action is that a black and white, simple message is so easy to convey. Something is good or bad. Coming from an old-school conservation background, I see all shades of grey and focus my efforts on trying to wipe out some of the darker shades and move towards the lighter shades. It's a complicated global issue, involving conservation, fisheries management, trade regulation, and not eating too many sharks. If one is trying to engage the public and trying to raise money for our work, grey doesn't really cut it, although the reality is that life is shades of grey.

As an example, there are some really well-managed shark fisheries and an effective shark finning ban in the USA. They have addressed overfishing, and they have some stocks that have recovered, which are now producing delicious shark steaks and sustainable sources of shark fin in well-managed fisheries. This is good for the environment, for fishermen who need to earn an income, and for people who like eating sharks or shark fin soup. The shark management plans that have been developed since the 1980s have got real traction now. We have moved well away from a situation where some fishers catching sharks were keeping the fins and chucking away the carcasses, to sustainable shark fisheries with minimal waste (Shiffman & Hueter 2017). However, there are now campaigns in the USA seeking to prohibit trade in any shark fins, to prevent possible imports of fins that might come from poorly managed fisheries elsewhere. That does not really make sense to me, in the context of the USA's domestic shark fisheries management and conservation. Proponents may not be thinking about the shades of grey. They are just thinking about a single issue that needs to be addressed outside the USA, and I'm not convinced it is helpful.

THE FUTURE

What are the biggest threats?

I think the biggest threat is still unregulated fishing. For example, some deep-water fisheries are horribly unsustainable, akin to mining out the resource. In extreme cases, fisheries had to be shut down after the damage had been done. It would be good to have had sustainable management that anticipated the potential for damage. I believe that managing the big oceanic shark fisheries is possible, because these are primarily carried out by developed nations. The really difficult challenges are managing subsistence, mixed-species catches in shallow coastal waters and estuaries. They are so important for food security, for income, and for basic human survival – and very hard to manage because of their complexity and lack of data. Although not an issue in the UK, this certainly is a problem where developing countries overlap with major centres of fish biodiversity, for example in the Indo-Pacific triangle.

Another threat is failure to implement what we've been fighting so hard to achieve. Working very hard to achieve a listing in CITES or CMS appendices, then sticking it on the shelf and forgetting about implementation. It is critical not continually to move on to new objectives, but also to keep looking over your shoulder and making sure the earlier achievements are not being neglected. The Shark Alliance achieved its objectives, including the adoption of a Community Plan of Action (CPOA) for sharks, and was disbanded, but where is the EU CPOA now? We must continually keep moving these things back up to the top of the agenda and reviewing progress. The CPOA had time-limited objectives; can we now check whether these were delivered, and review the new objectives for the next ten years? With CITES, the challenge is implementing the Appendix II listings and convincing the countries who must now make them work that it is not that hard, and they have got to do it. It may seem boring, going back over old ground, but it is essential.

The scary big future challenge is ocean acidification, but that's for someone younger, I think! Dealing with that will present many of the same challenges, but more so. It is just so big and all-encompassing that I tend to pretend it isn't there. That is one I can't handle.

What are the real, hidden problems and barriers that make progress harder than it should be?

To start with, resources, funding, and attention-deficit disorder. There is a tendency for people, including funders, to support something for just a few years and then get bored and go to do something different. That is why we set up the GSRI. The fact that a global partnership of foundations say they are going to work on this for ten years is really unusual. Next there are still a few countries in the world whose policies appear to be dominated by the wish to exploit fisheries to as great an extent as possible. Short-term economics, influenced by wanting to make your investment pay back in five years, or the focus on a five-year timetable by decision makers who don't look beyond the next election. That does to some extent provide an opportunity, in that newly appointed decision makers are quite keen on photo-opportunities and making grand statements, but then you may have to hold their feet to the fire for the next twenty years, and of course they definitely won't be there for twenty years, so you have soon got to start again.

Another barrier is the need to have more people working on the issue, because I do sometimes get grumpy and think, why am I still the only person doing x, y or z? Then I look around at a CITES conference, and it is amazingly full of people working on shark conservation, with more shark conservation organisations present than you can shake a stick at. OK, I think, it took me 25 years to get to where I am now, and those coming along behind have got to put in a bit more time before they know what I know, so it's important to be patient. I didn't have a clue what I was doing when I started – and if I had, I probably wouldn't have gone there!

What are the most interesting and promising new approaches?

Seeing the marine conservation wildlife sector in governments working closely with the fisheries management sector, and with industry. While not new, this is something that has mostly developed over the past ten years or so, and is really promising. It's so good to see different parts of government working together, instead of spending far too much time defending their patch of turf from each other. Shark tourism is not without its issues, but it is fast growing and various estimates put its worth at $200 million a year – and it may well come to play an increasing role in conservation efforts.

We have got much better science. Some really good stock assessment people, and the development of novel approaches for data-poor fisheries. There has been a tendency in the past for people to say 'We can't do this yet because we need more data,' or 'We can't really advise how to manage this stock because we don't have enough data.' I believe we are getting over that inertia, and I think that among the most important assets are new approaches to data-poor management and rapid assessments, because we're never going to have all the information we need. We cannot afford to wait until we know exactly how everything works, because by then there will be more bits missing from the ecosystem. Don't let the best be the enemy of the good.

When you retire, what would you most like to have achieved for marine conservation?

Retiring itself would be a personal achievement! However, I would like to see more people doing what I've been working on over the last few decades, to the point where I was completely superfluous and unnecessary. That would be a major achievement.

Are you optimistic or pessimistic about the future?

To cope with the sheer negativity of the situation, one focuses on the little tiny steps and on what is possible. It is amazing how much can be achieved like that. Last year, I was trying to identify a suitable fishery for François Poisson to test a CITES NDF for silky sharks. My initial idea was that we would do this for the French purse seine fleet in the Indian Ocean, because they set their nets around floating fish aggregating devices (FADs) and get a huge bycatch of silky sharks. François said, well no, because the entire huge purse seine fleet will no longer keep a shark on board. They are not necessarily bleeding hearts about shark conservation, but aware that sharks are a real pain politically, and that landing them could result in all sorts of trouble. This means that in the last thirty years, we have gone from a situation where people thought you were insane if you mentioned the words 'shark' and 'conservation' together, to one where the entire French purse seine fleet have decided that they will not keep any sharks in their boats.

I have learnt so much about how difficult and complicated shark conservation is, but we have made huge strides – and there is no reason to imagine that things can't continue to improve. I am not expecting massive changes, and I recognise that it takes time. Overall, I think there will continue to be more gains than losses, which means I am generally optimistic for the future.

CHAPTER 15

Euan Dunn

Euan is currently Principal Policy Officer for the RSPB's marine work and is based in Sandy, Bedfordshire. His career has focused on seabirds and fisheries themes which stretch back to influences from his childhood. His grandfather was a fisherman based in Gorleston (Norfolk) working out on the Dogger Bank, later moving to Aberdeen when steam trawling took off there. Euan's father, having been out on a fishing boat once, decided it was not for him but did become the fisheries correspondent and Aberdeen editor for the *Scotsman* newspaper. What prompted Euan to pursue a lifelong interest in birdwatching from the age of five when he got his first bird book and binoculars he finds rather less easy to explain.

Euan studied zoology at Aberdeen University, where his interest in the marine environment developed on field courses to the likes of Millport in the Firth of Clyde, and an honours project on the bivalve *Macoma balthica* in the Ythan estuary. The two New Naturalist classics on fish and fisheries and plankton by Sir Alister Hardy were also a memorable inspiration. He gained first-class honours at Aberdeen and went on to do a PhD at Durham on the feeding ecology of terns, spending the summer working on Coquet Island (Northumberland) and plying back and forth between there and Amble in his small clinker-built boat *Rissa*.

Six years at the Edward Grey Institute of Field Ornithology in Oxford with more work on seabirds, two years in New Zealand, including a stint on Campbell Island working with albatrosses, and then thirteen years as one of the author-editors of *The Birds of the Western Palearctic* led him eventually to the door of the RSPB. As a freelancer he produced a report for RSPB looking at the research needs on interactions between fisheries and seabirds, with longlines and gill-nets a particular concern. When Nancy Harrison, who had commissioned this report, left the RSPB Euan applied for and got her marine policy officer job in 1994. So started his work with the RSPB and BirdLife International, dealing with the impacts of fisheries on seabirds in the UK, Europe and internationally. In 2007, Euan was awarded an MBE for contributions to marine conservation.

The last twenty years have seen the evolution and recognition of NGOs working with the fishing industry to develop effective solutions that have helped resolve many harmful seabird–fishing interactions. Euan's work has also involved looking at the wider issues of how fisheries interact with the marine environment and the developing thinking on the implementation of the ecosystem-based and precautionary approaches to fisheries management.

MARINE CONSERVATION AND ITS DEVELOPMENT

How do you frame and think about marine conservation?
The themes I've worked on for most of my career, and most recently, have focused on the protection of seabirds, but that has also included considerable effort to improve the management of fisheries for wider environmental benefits. This has also implicitly recognised, as many marine conservationists do, that you have to work on the larger 'landscape', whole-seas scale. 'Birds and fish have no passports', boundaries are less evident, and effects of climate change on the marine environment illustrate the broad canvas we have to work on. Much of my work has also involved applying ideas like the ecosystem approach and the precautionary principle to the issues of bird, fish and marine ecosystem protection and management. To many people these might seem like rarefied concepts, but my work with fishermen at the regional sea scale and with gear changes to save seabirds at a more local level are practical manifestations of this.

What are the differences between now and when you started?
Although I started out as a seabird ecologist and explored other areas of natural history in between, I came back to marine conservation when I started working for the RSPB in 1994 on offshore fisheries issues. The context was very different then in many ways, the seas were considered a factory floor for the fishing industry with chronic overfishing the going rate, and fish stocks – white fish in particular – were in an appalling state. In the worst years (2002 and 2003) the International Council for the Exploration of the Sea (ICES) recommended a zero North Sea cod quota to the ministers at their December Common Fisheries Policy (CFP) council meeting and they routinely overturned this recommendation; overfishing was rife and on an industrial scale.

Whilst from an NGO perspective the dreadful state of the stocks was a clear story and helpful in one sense because it was possible to mount cogent arguments, the issues were very raw and often seen in a very black and white, binary way. The reality is that it's hard to do conservation in a crisis. When I started, the fishing sector and the NGOs were effectively in silos, everything was highly confrontational, shouty, and 'dialogue' was often conducted through the media (Dunn 2005). This was also a time when the quality broadsheets had bold and effective environmental correspondents, most notably Charles Clover in the *Daily Telegraph*, Paul Brown in the *Guardian* and Mike McCarthy in the *Independent*, who were the mouthpieces for the stuttering dialogue between the NGOs and the fishermen.

So at this time the NGOs essentially had a whistle-blowing role and there were very few opportunities to sit down together with fishermen to discuss issues of mutual interest. The fishing industry's fire was directed at the European Commission and the CFP, whose focus was firmly fixed on trying to regulate exploitation of the stocks, not on wider conservation issues. My mantra was that fishing had to adapt to the environment, not the other way round. The challenge was to get the emerging thinking on the ecosystem approach into fisheries policy and recognised as one of the legitimate costs of fisheries doing business. It's been a struggle, but it's improving.

Another major change has been the development of the conservation movement into a more powerful and cohesive force. When I started at RSPB, rather few NGOs were working on marine issues, but now marine conservation has really come of age. A good example is the representation of NGOs on the North Sea Advisory Council (NSAC). When this began over ten years ago there were just three NGOs on board, namely RSPB, WWF and Seas at Risk. Now it also includes, among others, Oceana, Pew, ClientEarth, and the Environmental Defense Fund; they make a big difference and really strengthen the NGO voice.

How do you think marine conservation is different from terrestrial nature conservation?
This goes back to my initial remarks. I see marine conservation working at a series of spatial scales – a bit like Russian dolls – so whilst there are sites of special importance that require protection it is also essential to work at a wider spatial scale. Climate change provides a particularly good example of this because its effects in the sea are everywhere and we have to work with this. This is different from the traditional focus on land, where there has been greater emphasis on sites, and species tend not to move around quite as much. There is a real buzz currently with conservationists on land promoting 'landscape-scale' conservation, but I think marine conservationists have been working on this scale for a very long time and it somewhat sets them apart from their terrestrial colleagues. Whilst we recognise that there are clearly hotspots, special places, that need protection, marine conservationists have seen the need to work at the level of the entire North Sea, and even larger regions such as the North Atlantic and Arctic. Overall, we also face greater data deficiency in the marine environment, and this places constant and growing demands on research, monitoring and the application of a precautionary approach to management.

The development of an ecosystem-based approach has become an important driver applicable at a range of spatial scales and has provided a framework for ICES work and the OSPAR regions. Early on, the fishermen often complained that, there being no accepted definition of an ecosystem approach, no one knew how to implement it, and for too long this gave them a let-out clause. However, the EU Marine Strategy Framework Directive (MSFD) effectively provided everyone with the operational programme and standards for applying an ecosystem approach; it is now a well-understood and tractable baseline for the wider community and marine conservationists to conduct their work.

THEMES

Effective working relationships between the NGOs and the fishermen
The North Sea Ministerial (NSM) meetings provide a huge impetus towards building effective working relationships between the NGOs and the fishermen (see Box 6, page 170). They started in Germany (1983), followed by London (1987) and the process of the ministerial meetings had the effect of pulling together a wide range of stakeholders from

Figure 15.1 Danish sandeel trawler. Source: Chris Gomersall, RSPB Images

industry and the NGOs, often for the first time. The backdrop to this was that marine pollution was still a major issue and the UK was regarded as the 'dirty man of Europe' (Rose 1990). These meetings were important because both the precautionary principle and the ecosystem approach were key elements in the discussion. This was really the start of a dialogue between NGOs and fishermen in an international forum, and by 1995 NGOs had a seat at the table when fisheries were on the agenda.

I was involved in the preparatory Intermediate Ministerial Meeting (IMM) in Copenhagen (1993) for the North Sea Conference in Esbjerg (1995). Put simply, North Sea fish stocks had been trashed and there was still a major, totally unregulated industrial fishery for sandeels (Figure 15.1). There was an animated exchange when the British environment minister at the time, John Gummer (now Lord Deben), challenged his Danish counterpart, Svend Auken, over the issue of sandeel oil being used as part of the fuel mix for Danish power stations. Auken snapped back that this practice had now ceased and he'd 'cut off the head' of anyone found doing it! Some conference delegates were a bit sniffy about the participation of Greenpeace, who were particularly active on ocean issues during this period. However, the recognition of Greenpeace and the other NGOs as legitimate stakeholders in the NSM conferences was wonderfully summed up by Svend Auken in his opening address: 'There was a time when people wondered what such organisations were doing in a serious debate; on the contrary, it is these organisations that make the debate serious.'

I have come to reflect that perhaps the most important meeting I ever attended was the so-called Intermediate Ministerial Meeting (IMM) on the Integration of Fisheries and Environmental Issues held in Bergen in 1997. It was exceptional because for the first time gathered round the table, and I can't recall it being repeated since, were both the fisheries and environment ministers from the participating countries – you could almost smell the sulphur between them. The NGOs and stakeholders were there too, but this forum was very powerful in breaking down the barriers between the ministers on many of the issues, which set the scene for much that was to follow. On the Danish-led industrial fishery for sandeels there was recognition of the need to restrict fishing in sensitive areas, and Svend Auken went further in calling for suspension of the fishery in such areas. That was a watershed. Shortly after this meeting, Unilever announced that it would be suspending sourcing fish meal and oil from industrial fisheries in European waters to protect its future supply of fish dependent on eating sandeels. Interestingly, it was also the first meeting to call for an investigation into the economic and ecological effects of applying a discard ban. That ban only came to pass in the CFP sixteen years later in 2013; marine conservation can take a long time, and you have to be prepared to play the long game! You could see that people in Bergen saw the bigger picture and that things had to change. Although the NSM declarations were not legally binding, everyone realised that they had to be taken seriously as powerful statements of political intent, and that ministers would need to take action when they returned home.

Do the changes made then still hold true? Broadly I think so: for example, a seat for NGOs at the table was a key point, not least the way the Brussels institutional framework operates in relation to stakeholder involvement in the CFP. A game changer in this regard was the then EU Fisheries Commissioner Emma Bonino, who, as a bit of a maverick, wanted to open up access to the European Commission's Advisory Committee on Fisheries and Aquaculture (ACFA), and she oversaw NGOs being admitted to this forum. Although there was some grumbling about this from the fishing industry, it was another step towards a more integrated approach to fisheries management.

A second major development was the establishment of the Regional Advisory Councils (RACs), now known as just Advisory Councils (ACs), to cover all the major European seas. This stemmed from a reform of the CFP in 2004, and I have sat on the North Sea RAC ever since and chair its Ecosystem Working Group. I well remember attending the very first meeting of the 'shadow' North Sea RAC, which felt rather like joining a private men's club. The big guns of the fishing industry were there, with people like Ben Daalder, the then king of the Dutch beam trawler fleet; it did take a while for the NGOs to find their way. Although the clear purpose of the ACs is to advise the European Commission and Parliament on specific measures under the CFP, for me an equally tangible outcome was to develop a working relationship and regular dialogue between the NGOs and the fishing industry. I think at least some industry members were surprised by the technical knowledge the NGOs could provide, and over the years there has been a great deal of learning on both sides. So this was an important legacy of the NSM conferences, cementing a fundamental shift in relationships between the NGOs and industry which has led to other progressive initiatives.

Following on from this we did begin to see various collaborative projects, with WWF working with the Scottish industry on conservation credits, and RSPB with the Danish sandeel scientists and fishermen's leaders. Currently there are many more NGOs involved, and I think there is a new dynamic emerging in which the industry is retrenching to an extent. Perhaps this is putting it too strongly but there are two reasons for this: firstly, the arrival on the scene of new, powerful and well-funded NGOs, and secondly, NGOs working on the 2013 CFP reform had a lot more influence than the industry anticipated. The NGOs worked incredibly successfully together on the reform, helped by appointing a full-time coordinator who helped marshal the Brussels-based NGO coalition. The NGOs picked apart the CFP proposal, article by article and clause by clause, and worked hard to align Members of the European Parliament (who for the first time had co-decision with Council over the bulk of the CFP reform) with desired changes.

We also worked on the European Maritime and Fisheries Fund (EMFF), which was being reviewed in parallel. I think the industry were taken aback by how effective the NGOs were, particularly in their influence on the European Parliament, and weren't fully aware of what was going on until it was too late. As a result of this the fishing industry subsequently came up with their own advocacy platform – Blue Fish – in part to counter the NGOs. There was no doubt that the arrival on the EU scene of Pew and Oceana, along with the legal clout of ClientEarth, meant that the fishing industry clearly saw that it had a fight on its hands with an increasingly effective NGO voice. This is in effect rather like an arms race between the so-called opposing parties. Whilst they share the goal of sustainable fisheries, in reality there are differing priorities: industry prioritises socioeconomic gains over environmental impacts and pushes against the ecosystem approach adding to their costs.

In conclusion, I think you can say that what we have now is a much more grown-up fisheries debate, more professional, better informed, with well-established channels of communication. However, we are still playing catch-up with best practice elsewhere in the world, and the EU's record is appalling in comparison with, say, Australasia and especially the Commission for the Conservation of Antarctic Marine Living Resources (CCAMLR) (1982) convention area, where management of ecosystem impacts was at the heart of operational fisheries from the outset. The way the EU historically divorced fish from the rest of the environment, including biodiversity, was very unhelpful and created a spurious division that led to all sorts of management problems. This division lies at the heart of trying to develop a more sustainable approach to management. Since the 1990s CCAMLR has required, for example, that all longline vessels carry observers, paid for by the industry, and that they deploy agreed best-practice measures to mitigate seabird bycatch. Unless they do this they don't get their licence to enter the fishery. All that was established many years ago and provides a valuable example that we've highlighted in our arguments to the European Commission about where the bar should be set. Everyone is becoming far more aware of evidence on a global scale, and this again has made European policies look short-sighted and introspective, especially where tried-and-tested solutions exist elsewhere.

Industrial fisheries and sandeels

Ever since the 1980s when Mark Avery was the RSPB's head of conservation we've had an interest in sandeels because they comprise a vitally important element of the diet of many of our seabirds. In some species, sandeels comprise 90% or more of the diet of their chicks. The RSPB had raised issues over the local Shetland sandeel fishery and its association with steeply declining numbers of arctic terns and other seabirds (Dunn 1998b). Our position was that there was so much uncertainty in the sandeel stock assessment that it was not being managed in a sufficiently precautionary way. This set the scene for the conflict that erupted in the early 1990s around the Wee Bankie, a Danish sandeel fishing ground just off the Firth of Forth where nearby kittiwake colonies, notably on the Isle of May and St Abbs Head, were in chronic decline. By 1993 the Danish catch from the Wee Bankie peaked at over 100,000 tonnes (Dunn 1998a), destined for conversion to fish meal and oil used nowadays principally to make feed for rearing farmed salmon. Things came to a head when Greenpeace took direct action by allegedly throwing concrete blocks into the Danes' sandeel trawl nets. Inevitably this stymied dialogue between Greenpeace and the Danes, but with the door kicked open it enabled the RSPB to enter and play a bigger role in helping resolve the issue. The Wee Bankie was in effect a microcosm of a much larger issue afflicting the whole North Sea sandeel fishery, where over a million tonnes of sandeels were being taken annually by 1997/98.

With scientific assessment lacking on the impact of the North Sea fishery on regional sub-stocks of sandeels critical to the needs of seabirds, ICES asked two Danish fisheries scientists, Henrik Gislason and Eskild Kirkegaard, to evaluate the management of the sandeel fishery. They concluded that the fishery failed on nearly every count to fulfil a precautionary approach and recommended closing the fishery around sensitive areas until more was known about the stock structure and interactions with seabirds (Gislason & Kirkegaard 1997, 1998). It was hugely influential to have this scientific endorsement of the NGOs' campaign for much tighter regulation of the fishery.

Set against the background of massive media coverage caused by the Greenpeace action and the scientific criticism, the Danes were under pressure to accept a compromise. The RSPB met with the Danish fishermen's leaders and Stuart Barlow, then Director General of the International Fishmeal and Oil Manufacturers Association (IFOMA) to discuss a potential solution. Behind the scenes, Mark Tasker from JNCC was pivotal in pressing the conservation case. This led to the European Commission establishing an area of 20,000 square kilometres closed to sandeel fishing, from Rattray Head in northeast Scotland south to Northumberland. The driver for this closure was the breeding failure of the sandeel-dependent kittiwake population on the Isle of May, the first time such a proxy had served for regulation of a commercial fishery in EU waters, and a pioneering case study of the ecosystem approach to fisheries management in action. The closure was set up in 2000 and is still in place now, although it comes under regular review.

Meanwhile in Shetland we were able to sit down with the local sandeel fishermen and Scottish Natural Heritage (SNH) and work out a management programme that included catch limits, fixed landing ports and a closed season for the fishery to protect the critical breeding months of the seabirds. Again this provided a precedent for how an ecosystem approach could be applied in practice, and it proved to be a seminal moment in the conservation of seabirds in this area (Dunn 1998b). The Shetland sandeel fishery has since collapsed – but not because of overfishing; rather the evidence is compelling that climate change has led to chronically poor sandeel recruitment for fishermen and seabirds alike.

We now have a much better understanding of the ecology of sandeel stocks in the North Sea. We know that, once settled as adults, sandeels stick to a given sandbank, forming a patchwork of aggregations, with each patch capable of being overfished. To take account of this, the North Sea management regime treats these sub-stocks separately in seven management units. Moreover, there used to be a blanket single total allowable catch (TAC) for the whole of the North Sea, but now fishing effort varies spatially with the scientific monitoring of recruitment into each of the seven areas. On this basis, in 2005 – the nadir of sandeel abundance in the North Sea – the fishery was shut down altogether mid-season. So the fishery is better managed now, and is even Marine Stewardship Council (MSC) certified. However, in the ICES assessment model, the needs of seabirds and other predators are not adequately factored in to setting harvest levels, so this fishery still has a way to go to be precautionary and environmentally friendly.

In the late 1990s the Labour government's fisheries minister Elliot Morley contemplated an outright ban on sandeel fishing, but politically this was problematic. The relationship between countries in the EU subject to the CFP is a trade-off not least because 'relative stability', by which fishing quotas back in 1983 at the start of the CFP were not parcelled out equally among countries. The UK, for example, was allotted relatively more of the white fish while Denmark was allocated the sandeels. Given the relationship with the Danes under relative stability it was clear that any change in the status quo sought by pressing for a sandeel fishing ban would have been nigh impossible to agree and would potentially have had significant repercussions for the UK's own fisheries. However, Brexit presents the UK with an opportunity to revisit the management of the sandeel fishery in its waters and ensure that this vital forage fish layer of the food chain continues to sustain the seabirds and the other apex predators, including cod and other fish, that depend so much on it.

Seabirds and fisheries bycatch – working with the fishing industry to develop technical solutions

In 1993, before I worked for the RSPB, I'd written a report for them that included areas of research that they might undertake on conflicts between fisheries and seabirds, including seabird bycatch in gears like gill-nets and longlines. It was Nigel Brothers working in the Southern Ocean in the 1980s as an observer on a Japanese tuna longliner who first highlighted the scale of the bycatch problem, when he estimated that something like 40,000 albatrosses were being killed every year on longlines. This began to shed light on the population declines that had been recorded from albatross breeding colonies, which until then had been unexplained.

What turned out to be a real milestone was the FAO International Plan of Action for reducing the incidental bycatch of seabirds in longline fisheries (IPOA Seabirds: FAO 1999). The FAO plan is not legally binding but guidance to fishing nations, yet it did provide valuable background on how this problem might be tackled in Europe. It was clear to me that while the focus had been on the extinction risk to albatrosses in the Southern Ocean, there was also a serious, hitherto overlooked problem in European waters, and an opportunity to start work on this arose at the Intermediate Ministerial Meeting in 1997 in Bergen. Over a coffee with my Norwegian NGO counterpart Christian Steel we hatched a plan to put observers on Norwegian longliners where there was a known problem with the bycatch of fulmars. Remarkably, the Norwegian Ornithological Society (NOF) managed to put its own observers on a couple of longliners in the Norwegian Sea and we got fresh insights into bycatch rates. We reckoned that the Norwegian fleet was catching 20,000 fulmars a year, and if you added in the Faroese and Icelandic fishing effort using the same sorts of gear you're talking about a potential annual toll of 50,000–100,000 fulmars (Dunn & Steel 2001). Discussions with the Norwegian government followed and helped inform new fleet guidance on mitigating seabird bycatch.

In 2001 I attended the FAO Committee on Fisheries (COFI) meeting in Rome, which began with the customary ritual of all the world's fishing nations making opening statements. Our interest lay in the FAO's IPOA Seabirds, which called on all States to develop their own national plans of action. The EU submitted their 'preliminary draft' plan, by a long way the most appalling official Commission document I have ever seen. There then followed years of us pressuring the European Commission with endless meetings, seeking support from the European Parliament, and drafting a shadow plan to help shape the Commission's thinking. Finally, in November 2012, over ten years after their first FAO toe-dip, the Commission produced a European seabird action plan – which is a pretty good document with some very strong measures. The thing that pleased me most was the objective to not just *reduce* seabird bycatch but to *minimise and where possible eliminate* it. The EU seabird action plan also tackles bycatch of seabirds in *all* fishing gears, not just longlines, and gill-nets emerge as the most ubiquitous threat. The problem is that monofilament nets are cheap and easy to deploy but, as work nearly thirty years ago by Simon Northridge and others since has shown, they can pose a serious bycatch risk to seabirds and cetaceans.

Although the EU seabird action plan is only voluntary we are seeing some real progress on the ground, thanks to the legislative teeth of two revised framework regulations under the CFP. The first of these, which we had to fight for but is yet to be adopted, is the legal obligation to take measures to combat seabird bycatch in the EU Technical Measures regulation. The second one we pressed for is the onus on vessels to collect and report data on seabird bycatch in the new EU Data Collection regulation. This was not before time; the

Figure 15.2 Black-browed albatross caught on a squid-baited hook. Source: Fabiano Peppes (Projeto Albatroz)

EU has trailed abysmally behind the pace of global advances in tackling seabird bycatch. It was actually UK ministers, first Huw Irranca-Davies and then Richard Benyon, who championed the cause in Council and kept the pressure up for the measures we have today.

The BirdLife Albatross Task Force – harnessing collaboration and innovation with fishermen

Led by the RSPB, BirdLife International's Albatross Task Force (ATF) was launched in 2005 with the backing of Prince Charles, who from the outset has lent his voice to our 'Save the Albatross' campaign (Figure 15.2). With many albatross species facing extinction we desperately needed to engage fleets directly. The South African deep-water trawl fishery for hake provides a good example. In 2004/05 it was taking 18,000 seabirds a year, mainly endangered albatrosses, when it applied for Marine Stewardship Council (MSC) certification. Such were the stringent MSC conditions adopted, notably the requirement to deploy streamer lines to scare the birds away from the trawl, that bycatch has since been reduced by a staggering 99%, and ATF played a pivotal role. It was a win–win, not least for the fishermen, who no longer had to waste time disentangling birds and also gained a market edge with MSC certification.

The ATF puts BirdLife instructors on board vessels with fishermen to develop and tailor the bycatch solution to their particular vessel and fishing conditions (Dunn 2011). It takes a lot of port-time to prepare and set up because you first have to build a relationship with the fishermen, and the solutions need to be fine-tuned by trial and error. We have deliberately targeted seven developing countries including Argentina and Uruguay, which have seabird bycatch hotspots and where an injection of NGO energy and resources can bolster the lack of national infrastructure. The ATF had an ambitious target of an 80% reduction in bycatch by 2015 in its chosen fisheries, and that this has largely been achieved is a remarkable success. The aim now is to scale up from demonstration vessels to the whole fleet level and then get the measures hard-wired into national fisheries regulations; Namibia has recently done precisely that.

A really important point is that developing partnerships with fishermen to solve problems of mutual interest leaves a very strong legacy. It is often easier to do this with smaller communities, but this creates a powerful dynamic which we are now importing into Europe. The ATF is a really powerful model that holds huge promise and has led to some highly innovative gear modifications. One of the most exciting of these is the experimental 'Hookpod' being used with pelagic longline fisheries. Each baited hook is enclosed in a plastic, seabird-proof capsule till it sinks to a certain depth, whereupon a pressure valve springs the casing open like the petals of a flower, releasing the baited hook to start fishing.

Nearer home, working with fishermen in Filey Bay, Yorkshire, led to a modification of their nets to help stop guillemots from nearby Bempton cliffs getting entangled. It was a local fisherman, Rex Harrison, who came up with the idea of a high-visibility band of netting across the top of the net to make it more detectable to approaching seabirds. This is a really important point, because fishermen have an incredible wealth of untapped knowledge; they are practical, inventive people and given the right incentives know how to fix things. This is relevant to the EU Landing Obligation, better known as a discard ban, because fishermen often know how to fish more selectively and it just needs the right conditions to activate it. By tackling seabird bycatch from lots of different angles, challenging governments, developing and testing technical fixes and working collaboratively with fishing communities, I'm proud that the RSPB and BirdLife have made a really significant contribution.

THE FUTURE

Continuing challenges – important threats and impacts

The biggest threat we face is climate change. The North Sea for example is warming four times faster than the global average for marine waters, as highlighted by a paper that came out in late 2015 in which one of the authors called the northeast Atlantic a 'cauldron of climate change' (Rutterford *et al.* 2015). Sea warming really stepped up in the mid-1980s, and thanks to the Continuous Plankton Recorder (CPR) programme run by the Sir Alister Hardy Foundation for Ocean Science (SAHFOS) in Plymouth we have a long-term record of key changes that have been occurring before and since. This is a wonderful time series which shows that since the 1960s the biomass of zooplankton in the North Sea has declined by over 70%, a massive change which amounts to a 'regime shift' (Beaugrand *et al.* 2002, Beaugrand 2004), and ecologists don't use that term lightly. This is a similar scale of change to the flip that saw the collapse in the early 1990s of the Newfoundland cod stock and the rise of shrimps and snow crabs in its place. SAHFOS data show that the cold-water zooplankton *Calanus finmarchicus* has progressively been replaced by a warmer-water species, *Calanus helgolandicus*, but at a much lower biomass: it is a smaller, less nutritious species, with a different seasonal cycle. When you think that fish breeding cycles have evolved over eons to synchronise with *Calanus finmarchicus*, this is a seismic shock that must reverberate up the food chain.

One of the things we think might be happening is a mismatch between the timing and quality of plankton abundance and the breeding of sandeels in late December and January, such that emerging sandeel larvae encounter a poorer diet of the plankton they need to survive and grow, and it appears to be worse the further north you go in the Atlantic. Puffins are disappearing from Fair Isle and Shetland, and on the Faroes and southwest Iceland they haven't bred successfully for the past ten years or more, so much so that the locally cherished tradition of puffin hunting on Iceland's Westman Islands has voluntarily stopped. So you begin to see how the whole food web has been disrupted and transformed by the warming of our seas.

It is clear that on a political level we need to see concerted international action to arrest the rate of climate change following the Paris Conference of Parties (COP 2015). As for other practical steps we can take right now, we need to try and build resilience into our seabird populations to give them a fighting chance to meet this challenge. This is why we see the establishment of a coherent network of marine protected areas (MPAs) as being important for seabirds, and it is lamentable that, Scottish waters apart, we still haven't got any truly offshore SPAs (Liverpool Bay SPA and the Outer Thames SPA straddle the 12 nautical mile limit). In addition, the Westminster government has so far been resistant to including seabirds and other mobile species as qualifying features for Marine Conservation Zones (MCZs, i.e. national MPAs), and we see it as a major campaigning challenge to convince them otherwise. We have done lots of work, not least with highly sophisticated remote tracking, to identify the specific offshore areas that are critically important for seabirds year after year and ought to be designated as MPAs.

There are just two things that seabirds really need: the first is secure breeding sites on land, the second is an adequate and accessible food supply at sea, so much of our work has focused on these two elements. Invasive species, most notably rats, on the outer 'halo' of islands around the British Isles are the outstanding threat to seabirds in need of a safe breeding place. The RSPB has been using the incredible pioneering work and know-how of the New Zealanders to help eradicate rats on a number of priority islands, enabling long-lost seabird species to return, as on Lundy where we now have many more breeding puffins, and have seen the return of Manx shearwaters and storm petrels. We've had similar success following rat eradication on the Isles of Scilly and the Shiant Isles in Scotland. An outstanding concern, however, is recent news of stoats establishing a foothold on mainland Orkney, in terms of the havoc they could wreak on both seabirds and breeding waders. That's high on our list of island invasives to tackle next.

Another major worry is offshore wind energy, because although we fully understand and promote the importance of offshore wind in terms of renewable energy, there are some major potential conflicts. I've been involved with the consultations on a wind farm on the Dogger Bank, which when it comes on stream will be the largest in the world, and also the site at south Hornsea. The issue is really the siting of these farms and how these interact with major seabird flight lines. We appreciate why sandbanks are attractive to wind-farm developers in relation to pile-driving, and perhaps in deeper water floating turbines will eliminate seabed disruption in the future. There are two issues of concern for seabirds, collision risk and displacement. There is evidence that gannets, for example, will fly around the edges of a wind farm rather than through it, which means that they could have less access to critical foraging areas – and this could have knock-on effects on adult energy budgets and their ability to feed chicks.

What reflections do you have on the hidden barriers to getting things done?

The themes I've highlighted are full of examples of simple procrastination, often because ministers don't want to do something because of extra costs or regulatory burden. There was a debate a few years ago about trying to find MPA sites that were to protect both fish and other marine biodiversity, and whilst this is a nice idea it proved to be nigh impossible in practice, all because the minister put too much emphasis on cost-effectiveness.

Currently we are very concerned about the *hollowing out* of government in two ways, not least following Prime Minister Cameron's infamous phrase about '*getting rid of all the green crap*'. His administration had an agenda to do away with regulations that have often been put in place for good reason, not least to protect the marine environment, and we have had to invest heavily in countering this attack on useful 'red tape'. Similarly, government departments and their agencies are being starved of funds, death by a thousand cuts, and not fulfilling some of their basic commitments – and there are some really unhelpful symptoms appearing which affect the RSPB's core interests directly. For example, at this time of unparalleled climate change, it's unacceptable that we haven't had a major national census of seabird populations since 2000, and that JNCC was starved of the resources needed to mount a fit-for-purpose project in 2015. So we are having to plan around this to get a new census funded and off the ground. This is symptomatic of decay in a government infrastructure that has taken years to put in place.

NGOs' engagement with government – the art of being partly pregnant
A challenge for the NGOs, and I've described it as the 'condition of being partly pregnant' (Dunn 2005), arises when they become part of the governance arrangements. I've described my role in the North Sea RAC, but it occurs all the time with stakeholder engagement or being on the inside track with advising governments or their agencies. With governments increasingly charged to engage stakeholders and uphold transparency, combined with their desire to decentralise and delegate, the challenge for NGOs is to maintain a campaigning edge and avoid the risk of being co-opted into a consensus which delivers weaker outcomes. Working with other organisations can produce hybrid vigour, but it is a constant challenge to balance campaigning with stakeholder collaboration to achieve real change. Greenpeace's highly visible campaigning has sometimes served as grit in the oyster to create the opportunity for other NGOs to participate and assist beneficial policy outcomes, as happened with the challenge of radically improving the management of the North Sea sandeel fishery in the 1990s.

How do you view the actions of the newer big organisations like Pew and Oceana?
I saw their contribution first-hand, Pew in particular, with the reform of the CFP – and their influence was very impressive. They are highly committed to a very clear policy goal and are inclusive in the way they go about their work with the other NGOs, serving as a hub to enable the collective work and progress on CFP reform with resources that went way beyond what we have experienced in the past. The interesting thing about the CFP – and we also saw this with the UK Marine Bill – is that these topics are big enough for all the NGOs to work constructively together in ways they don't necessarily do on smaller issues, where individual NGO 'ownership' can become divisive. I'm positive about the involvement of these larger organisations in these big issues.

New technologies – seabird tracking
The development of seabird remote tracking technology, with the devices becoming ever smaller and lighter, has been transformative. Researchers have found from attaching geolocators, for example, that wintering kittiwakes go from UK and Norwegian colonies as far as Newfoundland and Labrador to feed on pteropods (Thecosomata) in the winter (Frederiksen *et al.* 2012), and that post-breeding puffins from the Skelligs on the west coast of Ireland went briefly to Newfoundland after the breeding season to feed, probably on a flush of sandeels and capelin, before coming back to the mid-Atlantic ridge, en route eventually to Ireland again (Jessopp *et al.* 2013). This is an astonishing addition to our knowledge. Of course it's helping us enormously with our understanding of the spatial distribution of seabirds and plays into our thinking on marine spatial planning, which of course in its own right is a relatively recent concept

developed by Bud Ehler and Fanny Douvere (2009). It is also helping us develop our evidence base for MPAs for seabirds, and with year-on-year records we can say that certain areas are systematically important and not just a flash in the pan. Another very important approach is our ability to use stable isotope analysis of feathers and faeces to determine what birds have been feeding on. This is enabling a much broader picture of bird dietary requirements to be made without the need to see a 'fish in the bill'.

When you retire – if you retire, bearing in mind you are still working for RSPB at 72 – what would like to be remembered for having achieved?
I would like to think my legacy has contributed to our seas being perceived not as a factory floor for commercial fisheries but as a complex environment that needs managing carefully for all its biodiversity. Notably the ecosystem approach is now recognised at the policy level as a key objective in the CFP. I do feel I have helped move this shift in perception up the curve over twenty years. Seabirds have been talismanic for doing this because they are good, visible indicators and they resonate with people. Although the puffin, for example, has recently been put on the IUCN Red List, which is a matter for real concern, such is its popularity amongst the public that this move will help us with campaigning not just for this species but for marine conservation more widely.

Are you optimistic or pessimistic about the future?
Optimistic. Why? I think the fishing industry, other marine users and their managers are gradually becoming more aware of their stewardship role, and that gives me hope. They are beginning to recognise their responsibilities to the wider marine environment, and that is encouraging. The worry is that climate change undoes the good work, and that is my biggest fear. The opening up of the Arctic as ice melts, and the risk of a gold rush to exploit its resources, is a particularly alarming prospect.

BOX 6. NORTH SEA MINISTERIAL MEETINGS

Bob Earll, Euan Dunn and Paul Horsman
The first North Sea Ministerial (NSM) meeting took place in Bremen in 1984. At the time, green politics were particularly strong in Germany and it developed a theme that the international organisations responsible for protecting the marine environment – the Oslo and Paris Commission (OSPAR) and the London Dumping Convention – were simply not making sufficient progress. Coincident with this was the birth of *Vorsorgeprinzip* or the precautionary principle and its application to marine pollution (Paul Horsman). There were six conferences in all (Bremen 1984, London 1987, The Hague 1990, Esbjerg 1995, Bergen 2002 and Gothenburg 2006) set up to enable the governments of Belgium, Denmark, France, Germany, the Netherlands, Norway, Sweden, Switzerland and the United Kingdom to work together in an effort to protect the North Sea. Each conference generated a ministerial declaration which, while not legally binding, was a strong statement of political intent. The universal recognition of this strong political intent made ministers very wary of over-committing, and so there was always a strong counter-current of NGO ambition against a race to the bottom rather than striving for the highest common denominator from the ministers. The political intent of the North Sea meetings was often transferred to OSPAR).

The UK government was less than enthusiastic about this process in the mid-1980s and thought that by bringing the meeting to London in 1987 and by marshalling the scientific evidence into what in effect was a status report it would be possible to put an end to the meetings. The reality turned out rather differently. It became quite clear, not least from the wide discrepancies in monitoring and measuring contaminants between laboratories and countries, that the science wasn't up to the task. In 1985 the then Department of the Environment agreed to support a collaboration of both environmental and industry groups to prepare evidence for the meeting. It proved to be a highly useful forum, bringing multiple sectors operating in the marine environment together to describe a range of issues, an initiative which continued into the late 1990s as the Marine Forum and established a pattern of multi-sectoral forums that continue to this day. For many environmentalists in the 1980s it was their first experience of working in the European context, and it spawned collaborations between European NGOs, notably the Seas at Risk coalition.

During this critical period of rising political awareness and engagement the UK was known as the 'dirty man of Europe' (Rose 1990). The UK's intransigence in wanting to maintain its policies of dumping industrial waste and sewage sludge at sea (the only country to be doing so) was singled out at the Hague NSM meeting (1990) and agreements there led to the UK ending the practice of dumping wastes at sea. The Esbjerg 1995 ministerial meeting coincided with Shell's decision, supported by the UK government, to tow the decommissioned Brent Spar oil storage buoy through the North Sea for dumping in the northeast Atlantic (Paul Horsman, Chris Rose) which prompted a huge public outcry across Europe that forced further fundamental reforms, eventually leading to ending the dumping of decommissioned offshore platforms at sea (see Timeline 5, page 93).

In 1996 the London Dumping Convention (now called the London Convention) held a special meeting that rewrote the 1972 convention in favour of a prevention and precautionary approach. In the 1990s the issues surrounding fisheries gradually came on to the agenda with a higher profile. Arising from the Convention on Biological Diversity (1992), the ecosystem approach emerged as set of guiding ideas. Euan Dunn describes how a range of meetings led to major changes in the way that North Sea fisheries and in particular sandeels were managed, and how this facilitated a much wider level of participation by NGOs in the EU's Common Fisheries Policy. By the time of the last NSM meeting in Gothenburg (2006), international bodies like OSPAR, the London Convention and the International Council for the Exploration of the Sea (ICES) had undergone comprehensive reforms and the European Union had taken a much higher profile in the sustainable management of the marine environment. The NSM process provided the platform and political impetus to generate fundamental changes and improvements in the way we now manage the marine environment, not just in the North Sea as a test bed for innovation but in the wider northeast Atlantic under OSPAR, and globally under the London Convention.

CHAPTER 16

Simon Brockington

All views expressed are the personal opinion of the author and do not reflect the position of any organisation mentioned in this chapter.

Simon is Deputy Director for Marine Science and Analysis at the Department for Environment, Food and Rural Affairs (Defra). However, at the time of interview he was the Executive Secretary of the International Whaling Commission (IWC), a post he held for seven years. In that role he was head of the permanent secretariat and responsible for coordinating the Commission's work programme and supporting its development. It required liaison with the 87 contracting governments and the many scientists, non-governmental organisations (NGOs) and commentators from around the world who are active in the Commission's work.

Simon grew up in Birmingham in the 1970s, and his interests in the environment and the oceans began with family holidays by the coast, especially at Staithes in Yorkshire and on the west coast of Wales. His interests at school and with early work experience helped him gain awareness of the environment and of various pollution issues which were topical in the early 1980s. This led to a decision to study aquatic biology at University College of Wales, Aberystwyth. Another, perhaps sharper, reality influencing his decision was the Athletic Union prospectus, which highlighted the active diving club! Fundamentally, Simon really enjoys being outside *in the environment*, and through the outdoor pursuits at Aberystwyth he developed those interests to the full. Diving opened up a new horizon, and as a student Simon recalls 'feeling a tremendous sense of discovery and excitement after seeing fields of brittle stars, vibrant cuttlefish and huge conger eels'.

One evening in 1992, his final year at Aberystwyth, a friend invited him to a talk by a scientist from the British Antarctic Survey (BAS). The talk was hugely motivating, and on hearing that BAS was recruiting he prepared an application the same evening. His diving experience played a big part in getting his first appointment, a two-and-a-half-year term living, working and diving in Antarctica as an assistant marine biologist. This increased his enthusiasm for the marine environment, and upon returning to the UK he started an MSc degree at the University of Plymouth in applied marine science. However, even before the course ended, he returned to BAS for a second term in Antarctica to conduct fieldwork for a PhD on the ecology of sea urchins and how they adapt to different latitudes. His time in Antarctica was formative and inspiring, and the diversity of marine wildlife – from killer whales through to albatross and wondrous planktonic beasties – was just amazing. It was a productive time as well, with his PhD awarded in 2001 and then subsequently the Polar Medal in 2008 in recognition of his studies.

Returning to the UK, Simon was struck by the differences in the conservation status of European waters compared to the polar regions, and he became progressively more interested in conservation. He worked for four years at Natural England, firstly as a marine monitoring officer and latterly as a policy specialist with the marine protected area programme. In 2007 he moved to the Marine Conservation Society (MCS) as head of conservation, and then to the IWC in 2010.

This chapter reflects on the different perspectives of marine conservation that emerge from working with organisations in different sectors: for a governmental non-departmental public body (Natural England); for an NGO (MCS) and for an intergovernmental organisation (IWC). Although the days of the 'Save the Whales' campaigns are long past, the chapter reflects the fact that today's IWC involves conservation and management on a global scale and the integration of many disciplines including science, policy, organisational governance and diplomacy.

MARINE CONSERVATION AND ITS DEVELOPMENT

How do you frame or describe marine conservation, and what are its key ideas?
My own definition of conservation is ensuring the capacity to enjoy the benefits which arise from the environment. These include extractive uses such as the option to harvest fish as well as the ability to benefit from ecosystem services, including climate regulation. As importantly, it also includes the ability to enjoy and gain happiness from the intrinsic beauty of the landscape (or seascape) and the status of the wildlife. In contrast, conservation is more usually defined at IWC in terms of the numbers of animals that can be removed from a population before there is a consequence for future viability. This is very much a utilitarian approach to stock conservation. Whilst this is the stated definition, in reality the emerging pattern of activities at IWC implies a broader approach to conservation which is increasingly based around fairness and the need to reflect different priorities in different parts of the world. By way of example, in some locations IWC is taking a leading role in supporting whale-watching tourism. In other regions it has set aside large areas of ocean as whale sanctuaries, and in other areas it sets sustainable catch limits for subsistence whaling by indigenous peoples. Given these activities, the overriding discourse is one of fairness and respect for different priorities in different geographical locations. This implies a diplomatic, situational and social approach to conservation, and one which recognises different human relationships and priorities with the environment in different geographical locations.

At IWC, concerns for welfare are increasingly compared and contrasted with conservation. The ability to hunt small whales – especially through coastal or subsistence whaling – has dramatically improved in recent years, and times to death are now close to instantaneous. This almost removes welfare concerns from whale hunting. However, in stark contrast, there have been many instances of whales becoming accidentally entangled in fishing gear or marine debris and sustaining terrible injuries which cause death over many months. These are a very serious welfare concern, and one where the IWC is leading the response. However, this is not a conservation issue in the traditional sense of the term because the species involved – frequently humpback whales – are numerous and often have increasing population sizes. Nonetheless I believe these are conservation issues, and it is one of the reasons why I define conservation in terms of human enjoyment of the environment: a whale which is mortally entangled in marine debris clearly detracts from any ability for humans to be satisfied with, and hence enjoy, the environment. A definition of conservation based on capacity for human enjoyment of environmental benefits necessarily includes concern for welfare.

What are the differences between when you started and now?
My first employment, in Antarctica as an assistant marine biologist, started in 1992, and my first role in European conservation started in 2003. My early impressions of working in both the polar and the European environment were the same, which is that we know far more about the seas and oceans than we often recognise. Even the remote waters around Antarctica have fairly comprehensive species lists, thanks in large part to the Challenger expeditions which took place over 140 years ago. Here in the UK we have an even richer history of marine science, and this, coupled to the Marine Nature Conservation Review from 1987 to 1998, provides a solid foundation for today's marine conservation work.

The parts that have changed since I started in conservation are improvements in the way we use the sea and its resources. The earliest issues in my career were about preventing direct harm to the environment, and these often focused on reduction of pollution. Those issues were tackled very successfully: ocean dumping was addressed in 1974, there are no longer direct untreated sewage discharges to the marine environment, and the UK as well as many other countries has enacted measures to control a wide range of other pollutants, from heavy metals to antifoulants to organic chemicals.

More recently, conservation initiatives have started to address first sustainability and then recovery of the ecosystem. For me, the publication of Charles Clover's 2004 book, *The End of the Line*, was a milestone in fisheries management because of the increased public attention, and in 2007 Callum Robert's book, *The Unnatural History of the Sea*, gave a similar rallying call for recovery measures. It was hugely reassuring to see that the 2009 Marine and Coastal Access Act and the 2012 reform of the Common Fisheries Policy gave fisheries sustainability much greater prominence than previously, and the 2009 Act also introduced the UK's first marine planning system as well as measures aimed at marine biodiversity recovery, including provision for a network of marine protected areas (MPAs). With these recovery measures in place it is wonderful to see the UK becoming progressively more involved with international marine conservation work. As is so often the case, this progress is led by science. UK research institutions are increasingly active in supporting blue economy development in a growing number of other countries.

How is marine conservation different from terrestrial conservation?
At its heart, I'm not sure that it is. There are superficial differences of course. No one actually lives in the marine environment and it does not need to be 'managed' in the

way that a terrestrial habitat may need grazing or draining or any other measure. Marine conservation is often considered to be playing 'catch-up' to terrestrial conservation initiatives because key elements of legislation, such as establishing a national series of nature reserves or protected areas, have often taken place forty to fifty years earlier for the terrestrial environment.

With the advent and increasing maturity of the marine protection and recovery measures mentioned above, the similarities between marine conservation and terrestrial conservation become more apparent. The ultimate goal for both is the enjoyment of the benefits provided by the environment without causing harm or detracting from productivity or intrinsic value.

THEMES AND STYLES OF MARINE CONSERVATION

Contrasting marine conservation at Natural England and the Marine Conservation Society

I worked for Natural England from 2003 to 2007, and for MCS from 2007 to 2010. I've changed between organisations quite often, and each move has brought an increased chance for learning. I moved to MCS because I wanted to have a chance to be more outspoken. In 2007 the marine environment was in increasingly urgent need of protection, and the freedom of a campaigning charitable organisation was a great attraction! At that time Natural England and MCS were both working on very similar issues. The main difference was the customer: Natural England is a non-departmental public body reporting to Defra, while MCS is an independent NGO accountable to its members. This difference shaped the culture and behaviours: Natural England invested resources in its statutory and advisory functions and MCS invested time in public awareness, campaigning and coordinating the many thousands of volunteers who participated in beach cleans, fundraising and even the occasional march on parliament. Meeting the MCS volunteers was always a real high point. They were always very motivated indeed, they cared hugely about the marine environment, and it meant that working for MCS became a privilege.

Leaving these differences aside, there were many similarities between the two organisations. Both relied heavily on science as an evidence base, both were held equally accountable (although to different groups of people), and staff in both organisations shared an almost identical desire for improved environmental stewardship. My role at MCS as head of conservation meant that I had started to meet a very wide range of people including volunteers and supporters, funders, civil servants and representatives from the many industry groups which have an interest in the seas and oceans. I started to become fascinated by the many different ways that different people relate to the sea, and then one day, while browsing a copy of *New Scientist* I saw an advertisement for Secretary to the International Whaling Commission. If ever there was a chance to understand different perspectives of how people relate to the sea and its wildlife, then this was it! A few weeks later I was hugely excited to be offered to attend interview in Hawaii. In June 2010 I took up post as IWC secretary.

The International Whaling Commission

The IWC was seventy years old in December 2016, and the International Convention for the Regulation of Whaling under which it was established is the oldest of the extant multilateral environmental/fisheries agreements. At the time of writing the Commission has a membership of 87 contracting governments. Around 250 civil servants and scientists representing 66 governments, plus 150 observers from NGOs, attended the last meeting of the parties in October 2016. The IWC's evolution is remarkable. It was established in 1946 at a time when post-war rationing meant that there was a huge need for dietary fats, and of course this could be readily supplied by the whale fisheries in the Southern Ocean. Seventy years later the IWC's operating environment could not be more different: there are concerns about too much fat in our diets, whale numbers for many populations remain greatly reduced following overexploitation in the 1950s and 1960s, and whale hunting per se has become socially unacceptable for many countries. Thus the IWC's story is one of continuous change and adaptation. The preamble to the IWC's Convention states that its purpose is to 'provide for the proper conservation of whale stocks and thus make possible the orderly development of the whaling industry'. However, in its early years its role was, in effect, as an economic regulator. If too many whales were caught in any one season it would depress the market price of whale products, leading the industry to become economically unsustainable, and hence a form of regulation was required.

By the mid-1950s concern was starting to grow that the whale populations were reducing and catch numbers were decreasing and hence the IWC's role naturally transitioned from a regulator concerned with economic stability to a regulator increasingly concerned with environmental sustainability and stock conservation. At around this time the Commission developed its Scientific Committee to provide independent advice on whale population numbers. Their advice was to reduce catches, but overexploitation continued. This led civil society organisations to become progressively more interested in the IWC's efforts to achieve sustainability, and in 1972 the Stockholm Conference on the Human Environment called for the contracting governments of the IWC to implement a moratorium on whale hunting to allow recovery to take place.

Scrutiny of commercial whaling continued to grow, and in 1982 the Commission decided to set zero catch limits for all commercial whale hunts effective from 1985 onwards. Today, with the moratorium now over thirty years old, there are clear signs of recovery for some populations, especially for humpback whales, which in some places are approaching their pre-exploitation levels of abundance. The picture for other species remains mixed, with blue whales (the largest of all whales, and the largest animals which have ever lived) at still only a fraction of their pre-exploitation levels.

The IWC makes an exception for whales caught by indigenous communities, often in remote parts of the world including in northern Alaska, Greenland and the far east of Russia. Termed aboriginal subsistence whaling (ASW), this is an area where the Commission remains active in setting non-zero catch limits. These catch limits are agreed by the contracting governments, often by vote, and are very closely scrutinised. The IWC's Scientific Committee plays a vital role in maintaining sustainability and it's a huge compliment to their work, as well as to the engagement of the indigenous communities themselves, that sustainability in these hunts has been carefully managed for many years. The bowhead hunt, in Alaska, is an excellent example of collaboration between local communities, scientists and national government which has resulted in a long-term increase in the whale population.

Nonetheless, not all members and observers to IWC support whale hunting by indigenous communities. Concerns often centre on sustainability issues, on questions of ownership of animals which are highly migratory, and sometimes simply on a principled objection to hunting of large mammals. In recent years this has led IWC to expand the

Figure 16.1 Whale watching boat and humpback whales, Cape Cod. Source: Bob Earll

discourse on subsistence whaling to take account of human rights issues as they relate to indigenous communities. It's a fascinating area, and leads the IWC into pioneering policy discussions.

Another important development for IWC has been the growth in whale watching. This represents a very different use of whales from that which was envisaged at the time of drafting of the 1946 Convention. Whale watching can trace its origins to the mid-1950s and has grown rapidly since that time (Figure 16.1). In 2006, for example, over 12 million people took a whale-watching trip, creating a total global expenditure in excess of $1.5 billion (Hoyt 2009). This is now a vital industry, often situated in areas where more traditional revenues from fishing may have decreased. This new industry has its own sustainability issues, which centre upon visitor experience and ensuring the whale populations themselves do not suffer undue disturbance, and the IWC is playing an increasing role in supporting contracting governments as they oversee development of the industry.

How the IWC works – the combined roles of strategy, governance, science and diplomacy

The IWC was established as a regulatory organisation. Its members, each of whom is a contracting government, work together to agree annual catch limits for whales. But the IWC has journeyed through a significant period of change since its foundation in 1946 and the regulatory model alone is now too simple to explain how the organisation does its job in today's more complex world. Instead, today the main product of the IWC is its discourse. That is to say, the organisation has become a place where contracting governments and non-governmental observers can assemble to discuss differing views on conservation and management of whales. Views differ greatly from country to country depending on national sentiments and differences in geography, ecology and economic status.

In 2008 Calestous Juma, a social scientist and professor at Harvard University, came to the IWC's annual meeting. He identified four basic interests amongst the contracting governments:

1. That whales as special creatures which should not be hunted in any circumstances. This represents an ideological philosophy, centred upon protection of large wild animals.

2.   That whales cannot be killed humanely and therefore should not be hunted. This is a welfare argument.

3.   That whales are natural resources like any other that can be exploited as long as it is done sustainably. This represents the traditional sustainable development argument.

4.   Finally that whale hunting is the only sustainable use of whales, given the previous history of overexploitation. This is a non-consumptive use and contributes to discussions on overall ecosystem benefits.

These contrasting views provide a basic framework for analysing discussions at IWC. But they are also opposing views which are not easily reconcilable, and this had led to a long-standing and very deep division amongst the membership. During my appointment as secretary it became increasingly clear that four disciplines, those of strategy, governance, science and diplomacy, were key to supporting the organisation and thus to the success of global efforts for whale conservation and management.

Strategy
I joined the IWC in 2010 at the end of what was called the 'future of the IWC' process. At that time the Commission was at an almost complete impasse and potentially close to collapse. The 'future' process was a significant period of negotiation which had lasted over three years and was designed to resolve the main issues. These issues were significant and numerous, and a document prepared by the secretariat categorised a total of 33 issues dealing with everything from the controversy over whaling to the frequency of IWC meetings. The method employed to resolve difficulties was compromise. The idea was that governments would meet in the middle such that whale hunting would take place at reduced numbers in response to the introduction of significant conservation measures. Unfortunately the initiative failed almost completely and was paused and then later abandoned. Governments were just not able to compromise on areas where they had principled objections. Soon after the process ended the chair and vice-chair both resigned, leaving a very real strategic problem for the IWC.

But in the pause which ensued after the 2010 meeting two interesting things started to happen. Firstly, far from leaving the organisation, individual contracting governments continued to work on the areas which interested them. Secondly, the absence of a chair meant that the agenda for the next meeting was based not around what an individual (or the secretariat) felt was important, but instead on what members wanted to report or wanted the Commission to address. In this way, the IWC transitioned from being led through top-down leadership (via the chair and secretary) to bottom-up development sustained by individual members. It meant that initiatives came forward because they were of significant importance to individual governments and they were able to advocate these reasons from a fresh perspective. It also restored the opportunity for the new chair to take a neutral role, and as such the pressure on other members to compromise was reduced. The development of initiatives by individual governments also allowed the secretariat to strengthen its communications function, not only to explain items under consideration at IWC to the outside world, but also to improve internal communication amongst the membership and so to start to rebuild the organisation.

Governance
At the same time as these developments, the Commission started working on its governance arrangements. One of the frustrations was that members perceived the

Commission met too regularly. It was expensive to attend meetings each year, and many other comparable organisations met every two or every three years and focused instead on making progress through agreed intersessional work programmes. After a year of discussions we put forwards a proposal to reduce meeting frequency, and crucially to introduce a bureau. The bureau would be a small group of commissioners who would oversee the organisation's progress in the extended period between meetings. This was a vital development for two reasons: firstly, it created a small group of commissioners who were close to the chair and who were tasked with ensuring the organisation was progressing against its objectives, and secondly, it caused the Commission to clarify what its objectives were. This was a significant step forward from the disagreements which had previously characterised debate. The bureau was able to look at other aspects of governance as well, including the formal process for agreeing the Commission's agenda and the arrangement of subcommittees, and it provided a mechanism to advise the chair and secretary on matters arising during the period between Commission meetings.

Science

One of the IWC's most important strengths is its scientific committee. It meets every year for two weeks in plenary session and is attended by 150–200 of the world's leading cetacean scientists, many of whom are appointed by the contracting governments of the Commission. Importantly it also maintains a very active intersessional research programme including original fieldwork as well as workshops and reviews on cetacean issues worldwide. The scientific committee's proceedings are held in high regard and published in its own in-house peer-reviewed journal. These scientific endeavours are essential in informing the Commission on the status of whale populations and the threats to their recovery. It also advises the Commission on sustainability levels for aboriginal subsistence whaling and gives opinion on whaling under special permit (often referred to as scientific whaling).

However, the scientific committee achieves more than its recommendations on conservation and management. Composed as it is of scientists from the Commission's member countries, it provides an alternative forum for discussions to that provided by the Commission, and because the scientific committee is not a decision-making body, those discussions can often be more open and relaxed than at the Commission. Disagreement at the scientific committee is not unusual, especially when addressing questions such as the sustainability of proposed commercial whale harvests, or when passing opinion on scientific whaling programmes. Since the committee is not required to come to decision it will often record differing views in the form of 'some and others' – recording that some members believed x while others believed y. This is a vital mechanism for ensuring the views of all contracting governments can be recorded, while at the same time identifying and distinguishing areas where there is genuine consensus.

Diplomacy

Diplomacy can mean many different things at different times, but for IWC it revolves around the art of bringing governments together. Above all that meant efforts to rebuild trust. Trust is a concept with which we are all very familiar, but it is rather hard to define precisely, especially when considering relations between different governments. At IWC two approaches were especially effective.

The first was as simple as listening. With divisions on whaling as deep as they were, the act of listening to all individual governments and ensuring their views were noted, often through the written record or by adjusting agendas and also just informally through conversations in the margins of meetings, was vitally important.

The second approach was to focus not on outcome, but instead on process. Clear processes are relatively easy to establish – after all, who would disagree with efforts to organise discussions – and they also mean that governments know their views will be heard and so invites their contribution to the debate and thus increases support for the organisation.

These two approaches together – of listening and providing clear process – also had the benefit of allowing areas of agreement between governments to naturally come to the surface. This in turn meant that agendas could be constructed for subsequent meetings which identified areas where successful agreement could be expected.

Personal learning in the IWC role and others

For me at IWC, the act of being a conservationist required the skills to run a small to medium-sized organisation, a constant awareness of strategy and governance, a remarkably small amount of knowledge about whales, and above all an ability to listen to people's views while remaining strictly neutral on all debates. None of these were skills that I had developed during my career prior to joining IWC, and in response to this new challenge in 2011 I embarked on an MBA course with the Open University. I found this training invaluable, and without doubt it helped me support the IWCs development.

One or two areas in particular were entirely new to me. The first, on strategy, examined in part differences between planned strategy as undertaken by a leadership team (similar to IWC during the early 2000s) versus emergent strategy as driven by developments within an organisation, and this was very similar to the IWC's development after the end of the 'future' process. Having been shown that this was a natural method for organisations to develop, it meant I started to foster this approach at IWC and invested increased time in strategy and governance – with very successful results. Another area which was new to me, but just as important to the development of an organisation, was an introduction to techniques and models for change management. All organisations need to change all the time, and this was especially true at IWC. Having knowledge of these frameworks meant it was possible to analyse events and support the processes.

Thus the MBA course was transformational for me by introducing me to techniques for helping the development of organisations which I had not been exposed to during my earlier scientific studies. I believe these skills are vital for conservationists, and it's wonderful to see courses now starting to emerge which link conservation studies with business development schools.

FUTURE CHALLENGES

What do you see as the threats to the marine environment?

I am struck by the inherent productivity of the marine ecosystem. Some areas of the world, including the waters around Antarctica and the upwelling regions of the west coasts of Africa and South America, have tremendous productivity. Given this exists, I find it disturbing that history has shown that it is possible to almost entirely remove populations of large animals from the sea. Right whales are extinct from the waters around Europe, blue whales across the globe were only just saved from the same fate, and the absence of recovery of cod from the waters of the Grand Banks where a major fishery once existed is hugely concerning. Hence one of the greatest threats is our own ability to understand these dynamics and to enact precautionary management. History shows we've not always been able to do this, and so for me it's still a major risk.

Climate change is very real, and warming waters are already leading to changed species distributions around the UK and elsewhere. But the threat posed by acidification is potentially greater. Only time will tell if calcareous plankton can respond to the speed at which seawater pH is changing. If this is not possible then the risks for ocean biodiversity seem very significant. Although we've made much progress on reducing pollution there's still significant threats. Jepson *et al.* (2016) reported that killer whales in European waters have body burdens of PCBs which are among the highest in the animal kingdom and are likely to cause infertility and the extinction of local populations. Although such chemicals were controlled many years ago, their longevity and bioaccumulation through the food chain is creating a severe legacy for higher predators in European and Mediterranean waters.

What are the real problems that make progress harder than it should be?

Some years ago, before the introduction of the Marine and Coastal Access Act in 2009 and its provisions for MPAs, marine planning and a strengthened conservation role for the Inshore Fisheries and Conservation Associations, I would have said the real problems were that it was hard to get government to listen. Nonetheless, the Act is a major step in the development of the UK's marine governance and it's clear that government can and does listen. A significant part of that MBA course was given over to study of organisational change and the techniques used achieve it. There is tremendous similarity between those techniques, which are well known throughout business schools, and the campaigning methods developed by NGOs in response to the need to ask governments to listen. It's wonderful to see conservation courses starting to link with business schools to embrace these techniques. As such, I don't believe getting traction behind ideas, and asking governments to listen, is the barrier that perhaps it once was.

In contrast, I do believe there are some very real barriers emerging in dealing with current marine environmental problems. The policy response to stop overexploitation of whales was very straightforward, even though it took too long to implement. However, the policy responses to dealing with plastic pollution, or with ocean acidification, are far more significant barriers. It's good to see increasing instances of NGOs, industry and governments collaborating to solve these issues, rather than resorting to campaigning in order to move single ideas forwards.

When you retire, what would you most like to have achieved for marine conservation?

I would like to see two achievements. Firstly, I would like to see humpback whale populations fully recovered across the globe and supporting an even more vibrant whale-watching industry. This is highly likely in our lifetimes and would be an amazing conservation success. The other achievement is the vision for UK seas which Charles Clover paints in Chapter 18 of his 2004 book *The End of the Line*. Not everything in his vision will appeal to everyone, and the idea of a heritage fishing industry will surely be unpopular. But the underlying premise of a highly productive fishery with abundant wildlife, improving water clarity and where humans have a deep connection to the marine environment is surely achievable. Progress since 2004 is already immense. The network of MPAs which Clover includes in his vision is in development, as is a marine planning system, and oceans are now much higher on the political agenda than they were ten years ago.

Are you optimistic or pessimistic about the future?

I feel hugely optimistic, but there is still a long way to go. It feels to me that we are discovering a hierarchy of 'environmental rights'. Similar in character to human rights, I suggest that primary environmental rights are a right to a clean environment, free of pollution and toxicity. Here we have made a great deal of progress in the marine environment, with the ending of dumping and sewage discharges, plus measures for a great range of other pollutants. Secondary rights are those which ensure sustainability and recovery so that we have thriving and productive ecosystems. This is the area where we find ourselves in 2017 with current policy measures aimed at recovery of whale stocks, recovery of fish stocks and recovery and protection of biodiversity.

Finally, I believe there is a developmental stage ahead, which could be considered tertiary environmental rights, where we start to measure development not in terms of gross domestic product and other measures of economic success, but instead in terms of social development. At the moment happiness indexes are in their infancy and the first 'World Happiness Report' was only published as recently as 2012, but they are produced annually (Helliwell *et al.* 2017). Such indexes place social progress, ideas like health and wellbeing, rather than economic progress, as central to the growth and development agenda. An important part of this shift in emphasis is greater attention to sustainable development and people's ability to connect with and enjoy benefits from the natural environment, over and above the ability to profit from it. I find this prospect hugely exciting, and so yes I'm very optimistic for the future.

CHAPTER 17

Sue Sayer

Sue is the founder and chair of the Cornwall Seal Group Research Trust (CSGRT) an evidence-based charity aimed at conserving seals. She has had an interest in seals from childhood holidays to the west coast of Scotland, but her love of seals, and her work on them, began much later in life well away from the Midlands where she grew up. Getting a decent pair of binoculars fuelled her awareness and interest. After graduating with a degree in geography from King's College, London and then doing a postgraduate certificate in education in 1985 she taught in London and found that she was very good at it, rising to a point where headships beckoned. In 1991 she moved to Cornwall with her partner Chris. Her teaching interests changed focus into vocational work on NVQs and GNVQs, which brought multiple disciplines together holistically to focus on practical projects of real-world relevance. For her, education was not about teaching but focused on learning. Her final teaching job was to set up and run the 'Learning Space', Cornwall's only Classroom of the Future. As director of learning, she developed creative cross-curricular, student-led programmes for all age groups offering lifelong as well as family learning.

After paying off the mortgage in 2008, Sue made the life-changing decision to leave the security of a salary to take a year out to write a book on seals, and this eventually led to her becoming self-employed and dedicating her life to seal research. In 2004 she set up the Cornwall Seal Group with people she had met during her seal surveys, and by 2015 this had morphed into CSGRT, a registered charity.

This chapter describes the development of Sue's interests and the growth of CSGRT, the observations and scientific work being done on seals through photographic recognition studies, the methodologies that have been developed and the successful development of volunteer citizen science networks, along with reflections on the difference that individuals can make through their passion and commitment to marine conservation.

MARINE CONSERVATION AND ITS DEVELOPMENT

How do you describe or frame your beliefs in marine conservation?

I believe that the marine environment is a shared resource that belongs to no one. It plays a critical role in regulating our planet, it is very precious and frankly incredible. Because it is out of sight and out of mind, it has been under-studied and under-appreciated and there is still so much for us to learn and discover. Cousteau inspired me when I was young and showed me that there was something amazing under water. I really enjoy the research we do on seals, and new discoveries drive my motivation. I am concerned about welfare and work closely with British Divers Marine Life Rescue (BDMLR) on seal entanglement because I don't like to see individuals suffer.

What are the differences between when you started and now?

I'm very aware that change is accelerating on many fronts, and even during my short time involved with seal research, technology has moved on from hand drawings and photographic slides to digital cameras. The ease with which photos and information can be handled, manipulated and stored on computers is gobsmacking. Changes in computing have enabled much more effective citizen science networking with easy sharing of data and the provision of feedback. I am able to communicate with volunteers and experts that I have never met. In the last decade there has been a growing public awareness about the marine environment, with huge networks of people now voluntarily engaged in marine science. It is obvious down here in Cornwall, where our environment underpins our financial economy (Cornwall Council 2017), that there are so many more people enjoying being out at sea. The proliferation of this activity can't go on unchecked so we need effective, proactive management.

How is marine conservation different from terrestrial conservation?

The impression I have is that the sea is much more complicated, in terms of both physical and natural processes, so that what you think is simple often doesn't work out that way. The interconnections within the marine environment also impress me: for example we routinely find lost fishing gear here that has come from America, and El Niño in the Pacific affects our shores too. You can't simply buy an area of sea and shut it off as you can on land. The dynamics of water movement make everything more connected globally, as well as actions on land affecting the sea.

THEMES AND STYLES OF MARINE CONSERVATION – WHAT WORKS?

The life-changing experience – setting up the Cornwall Seal Group Research Trust

I'd been interested in seals since seeing them as a child, and I was the only person to reply to Stephen Westcott's notice in our local newspaper to set up a seal group in Cornwall. During this early period I had the sense that seals were relative underdogs and even scapegoats in the marine world. Apart from three historical papers and Stephen's work, seals were not well studied in Cornwall or the southwest of England. Stephen had told me that individual seal fur patterns were unique, and that prompted me to ask whether the seals I routinely saw on my visits to the coast were the same ones. There was one particular site, 'the washing machine', where this first began. Whilst out watching seals I'd meet like-minded people, and Stephen suggested I try to set up a seal group. At first I was rather reluctant to do this, but in 2004 I had some seal friends round to my home for a meeting, we set up some group aims, and since then we have met every two months – and since 2009 we've met monthly – simply because there is so much to talk about. We are always looking for ways to streamline what we do. In 2015 we formalised the group and gained charitable status as the Cornwall Seal Group Research Trust (CSGRT). In 2008 James Barnett highlighted the need for us to move beyond just one site to the rest of Cornwall and beyond, and that sparked our engagement across the southwest.

The issue for me was that the seal work was involving more and more of my time and there simply weren't enough hours in each day. I loved teaching and had a wonderful job as an advanced skills teacher and director of the Learning Space, but in 2006 my partner Chris and I threw everything at paying off our mortgage. We knew we could live on one salary and we took the momentous decision for me to give up teaching to have a year off to write a book on seals (Sayer 2012). The book was completed within a year, but the key thing was that the year out provided me with headspace and time to think, so when people suggested ideas I had time to make them happen. One such activity was helping to set up a systematic seal survey on Looe Island in December 2008 with the support of Abby Crosby from Cornwall Wildlife Trust (CWT). We have been doing monthly day-long, low-water surveys there ever since, and the project became something of a trail blazer, because it made me think about how to engage more volunteers in seal research in other parts of the southwest.

Researching and developing methodologies

We have a developed a number of strands to our work, including a process for recruiting volunteers and protocols for seal photo identification systems at individual sites; a programme of systematic boat-based coastal transect surveys of marine life and human activity; as well as ways of communicating our research findings to other networks of volunteers, statutory agencies and non-governmental organisations (NGOs) for developing marine policy, planning and management.

The identification of individual seals – 'spot the difference' pattern matching – was at the heart of my initial interest and approach to research on grey seals. As with lots of methodologies you become wiser with time, but the system I devised at the outset in 2000 still works today. For each site visit, a discrete survey 'album' of information and photos is created. Seal album photos are compared to a cumulative 'catalogue' of seals previously identified at that site, as well as catalogues compiled from other sites. CSGRT now has over fifty different site-based seal identification catalogues, which I effectively manage as one. The whole system has grown organically, and although we have found lots of ways of improving it, you have start somewhere. The system enables records of individual seals to be retrospectively, independently verified by anyone. Photo identification is a very powerful research tool enabling a huge range of life-history parameters to be studied (Macleod et al. 2010). For example, we can create individual seal calendars so that we can track our observations for that animal over time and space. We have developed a strict and rigorous protocol for confirming ID matches, and all IDs are moderated by at least two people, one of whom is always me. To my amazement, in the last two years, other people have shown an interest and willingness to undertake this challenging and time-consuming but highly rewarding role for their local site, and this has revolutionised our organisational capacity. It used to be just me, and now we have over fifty long-term volunteers undertaking administrative roles in addition to those conducting field research.

One serendipitous offer from BDMLR to use their boat for surveys in 2008 led us to the development of a programme of long-term systematic boat-based surveys along a 100-kilometre stretch of the north Cornwall coast. Our first boat survey was memorable and in horrible conditions of horizontal driving rain, but it clearly highlighted that we could get access to a greater variety of sites that were otherwise inaccessible to our land-based observers. Now our programme is quarterly and paid for by participating volunteers – making them self-sustaining in between the grants we are awarded from ethical companies and charities.

We survey three roughly similar lengths of coast from Boscastle to Trevose, Trevose to St Agnes and St Agnes to St Ives. Whilst our main focus is seals, we also record all other megafauna, human activity and more recently lost fishing gear. Undertaking boat survey transects has validated our site-based land surveys, confirming that seals are not evenly spaced along the coast but are observed in hotspots, and these inform us about where land-based point surveys need to be focused. These trips achieve other outcomes beyond direct surveying, as they give us all a different perspective of our coast, we see what is happening and changing out there (the massive cliff falls and increased human activity are obvious things), and we can observe what is going on, being eyes and ears 'policing' our coast in a very positive way. Boats also became an obvious survey platform for lost fishing gear and marine litter projects. Each boat survey has a report summarising the detail of observations made, and this process has been embedded within the local marine group covering that stretch of coast.

Special reports on other species have also been written. For example for the Joint Nature Conservation Committee (JNCC) consultation on a harbour porpoise Special Area of Conservation, we were able to collate and report on all our sightings, only to discover that the wider Camel estuary is an important site for harbour porpoise activity during the winter months (Sayer & Millward 2016), and our bottlenose dolphin fin photo identification work has been submitted for Cornwall Wildlife Trust's imminent report on the use this small pod makes of our inshore waters (Dudley et al. 2017). We will soon start to publish the cumulative results of these surveys, which have been going for almost ten years now.

What are we learning?

Over the last seventeen years, there has been a wide range of discoveries made about our mainly grey seal populations in the southwest. One of our first posters for the European Cetacean Society conference showed the huge numbers of seals identified at our west Cornwall site. This had been considered a static colony of around thirty seals. Our observations challenged this conclusion. We now understand that in any given year, many hundreds of different seals can pass through this site. Our discovery curve for the west Cornwall site is still growing and hasn't yet plateaued, so perhaps every Celtic Sea seal passes through this site at some point in its life? Dave Boyle from the Wildlife Trust of South and West Wales shared Skomer seal catalogues, so we now know that at least forty individual seals have used both these two sites. I have reached out to people working in Brittany and Dublin and found that some Celtic Sea seals are using the French coast as well. Seals from Cornwall have also been identified in Wales, France, north Devon, south Devon and Dorset.

One of the other things we notice is that although some seals use the same areas for part of the year, year after year, sometimes they suddenly move on. We have a couple of ideas about why this might be, relating to disturbance or them simply practising the equivalent of 'crop-rotation' feeding in the sea. We now have the data to explore these questions and are discovering that the visit patterns to particular sites, whilst seasonally repeated, can change with life stage and maturity. The analysis of this data is extremely complicated, though, and we are working with Dr Matthew Witt at Exeter University to make sense of this. We are also able to look at pupping sites, and there are some mothers who pup at the same site textbook style, but other mums pup at different sites – and unusually one even pupped at her home site where she spends the rest of the year. So, if there is one thing I've learned, it is that is there is no such thing as an average seal. This increasing body of knowledge enables us to help other seal-related organisations countrywide who see our website and contact us to ask for advice.

Publications
I was told very early on by Professor Brendan Godley that if we didn't publish, then we wouldn't be taken seriously – and so we do, using a variety of research reports and scientific papers. For example:

- Each boat survey is written up using a standard template – mainly to thank the volunteers taking part, but this disciplines us into analysing the data in detail.

- Reports are written for key land-based sites as and when this becomes timely, and we now have a number of these covering multiple years.

- Commissioned contract reports have been produced for a series of different clients: for example on lost fishing gear for World Animal Protection, assessments for Wave Hub, and focused reviews for the Area of Outstanding Natural Beauty in the Isles of Scilly, the Marine Renewables in Far Island Communities Interreg Project and Natural England.

- A number of scientific publications have included collaborations with colleagues, and more recently I have begun writing my own.

- Numerous poster presentations have been made at international conferences in Europe and North America.

- Various popular materials (deliberately written for and aimed at outreach) communicate key messages, as we know that the effective conservation of seals requires wider appeal to hearts and minds.

Then there is of course the book that started me on this new career (Sayer 2012). More and more people are getting involved in this process, and as a network manager I spend time mentoring, training, encouraging and supporting our network. It is not that far removed from what I was doing in my previous teaching career.

How do you see your conservation aims?
The CSGRT aim is to protect grey seals, a globally rare species, for which the UK has a special legal responsibility. To achieve this, we take a holistic approach to research, recording and reporting on a range of other marine and bird life, as well as investigating

impacts arising from human activity, including disturbance, pollution (lost fishing gear), climate change and habitat loss. Outcomes from research are used to engage, enthuse and inform statutory agencies, NGOs, related charities as well as the wider public in the conservation of our precious marine environment. CSGRT is one of very few charities globally working to put grey seals on the agenda of planning and policy-making institutions, encouraging all to 'respect and protect'.

The main threats CSGRT witnesses include unintentional disturbance by the public from land and sea, and accidental entanglement in gill and trawl nets. Clearly there is the Conservation of Seals Act (1970), but really important protection also comes from the Sites of Special Scientific Interest (SSSI) legislation. If seals are listed on an SSSI citation then this confers considerably more effective protection, including from disturbance. CSGRT will be pushing to get seals more protection with every review of SSSIs or Marine Conservation Zones (MCZs) to ensure seals are part of their future management. We were concerned about Brexit, but this has been partially addressed by the realisation that the UK is a signatory to the Bern Convention (1982), which includes seals on the species listed. The UK will still need to meet its obligations under this convention outside the EU. Paul St Pierre of the RSPB has helped CSGRT a great deal, because there are close parallels between seabirds and seals as they both use terrestrial and marine habitat, exposing them to similar issues.

I have a couple of scientific papers in the pipeline. One is about seal movements across the Celtic Sea and another will be on haul-out sites. Making the location of these sites widely known is not without its risks, and that is why we publicly refer to sites in generic terms (like 'west Cornwall'). Research information about sensitive sites critically informs marine spatial planning, and once the public and the powers that be have this information, then management can be enacted – which, alongside greater public awareness and pressure, will help to protect seals better. The longer I have been doing this, the more I realise that what I do is more about people than seals – I need to engage people to protect seals.

Entanglement

CSGRT has been recording entanglement data from the outset because it is very obvious when seals have been caught up in nets (Figure 17.1). I have collaborated with Rebecca 'Bex' Allen at Cornwall College since 2003 and we had our first paper published on this issue in the *Marine Pollution Bulletin* (Allen *et al.* 2012). The clever thing Bex did was to compare the survival of entangled seals with similar 'matched' counterparts, and this gave us evidence of the implications of entanglement showing statistically significant reductions in survivorship for seals with deep constrictions and trailing gear.

The paper attracted the attention of World Animal Protection, who were running a 'Sea Change' campaign, and they invited us and paid our expenses to attend a conference they were running in Miami. It was amazing being able to spend a week sharing expertise on this issue, and we realised what good data we had collected over an extended period of time. World Animal Protection then funded a report update on the analysis of our work up to 2014 and in passing asked us whether we would be interested in doing lost fishing ghost gear surveys. In the end, they funded a whole year's worth of boat surveys, boat charters, coordinators and report writing. Land-based surveys were conducted voluntarily, and these are beginning to show us the differences between 'historical accumulations' on the shore and the new gear being washed up in subsequent years, as well as enabling us to assess the effectiveness and impact of clean-up operations. CSGRT is now a member of the Global Ghost Gear Initiative with a particular focus on the

Figure 17.1 Grey seal entangled in a net. Source: Sue Sayer

Building Evidence working party, and the global Pinniped Entanglement Group. More recently we have relied on other grant funding, and the clothing company Patagonia has generously funded a further two years of our boat surveys, culminating in them nominating us for membership of the prestigious 'One Percent for the Planet' network.

How do you reconcile the large number of dead seal strandings in the southwest?

Seal populations are a big puzzle, the dynamics of which continue to elude us. In a recent bycatch report, even Simon Northridge at the Sea Mammal Research Unit (SMRU) acknowledges that it is difficult to reconcile what we know about the size of the southwest's grey seal population with the numbers of bycaught seals. Dead seals have been recorded systematically in the southwest since 2000 by the Cornwall Wildlife Trust's Marine Strandings Network. Very roughly about a hundred seals are washed ashore dead each year. You can learn a lot from these animals, and we are getting a much clearer picture of their biology and causes of death. In addition to this, the annual report on the implementation of Council Regulation (EC) No 812/2004 during 2015 (Northridge *et al.* 2016) reported an estimated 580 seals were bycaught in 2015, including 310 in the ICES areas around the southwest coast, mostly trapped in tangle and trammel nets. If you add this to the seals that we see entangled, we have a high level of mortality with potentially population-level effects. There is clearly something we don't understand about the size, scale and dynamics of the grey seal population in this region, as pupping is healthy, but numbers appear stable and mortality high – but to prove any of this, more research is needed!

How do you rationalise conservation with rescue and welfare?

From my early observations locally it was clear that there were seals entangled with nets, and an adult female called Lywans was my very first entangled seal, recorded on 19 August 2000. Initially I didn't know what to do or who best to contact, and once again Stephen Westcott and Paul Semmens came out to help. We now work very closely with Dave and Dan Jarvis of BDMLR, Tamara Cooper of the Cornish Seal Sanctuary at Gweek and the RSPCA wildlife hospital at West Hatch in Somerset on rescues, then seals enter their rehabilitation programmes and we monitor them once they are released back into the wild.

My views on welfare have developed in part because I routinely work with identifying individual seals using photo ID, and so for me the individual animal matters! The other element of this is that although conservationists often say you should allow 'nature to take its course' there really isn't much that is natural about entanglement on the scale we see it, nor for that matter other issues such as pup disturbance by humans. I've worked with Dan Jarvis a lot on this; he is a long-standing trustee of our charity and now employed by BDMLR. I've done the BDMLR Marine Mammal Medic course, for which I have occasionally done the seal lecture. I attend incidents as a volunteer using my abseiling/climbing skills and those of my partner Chris Howell, but there is now a very comprehensive BDMLR protocol for seal rescue which not only focuses greatly on human health and safety but also has very clear guidance on best practice, based on hard-won experience.

The issue of how long released rehabilitated seals survive is one of great importance, and Dan Jarvis, who also worked at Gweek, is collating this material. We await a detailed analysis, but our observations based on seals that have been released with flipper tags is encouraging. One individual female, Rabbit F, has been around for seventeen years post release and has bred successfully on a number of occasions. The numbering on tags lasts about four years, and by this time we hope to have their fur patterns fully documented. Staff at the rehab centres now take photos of fur patterns prior to release and we provide feedback from our field observations on their animals back in the wild, giving a much-needed boost to their hardworking animal care teams, who love to see how their charges are faring. Recently Gweek changed a protocol, lowering the release weight, and we are following this up to see what effect it might have.

What is your approach to citizen science?

Volunteer networks are very popular at the moment, and trained, long-term volunteer networks have been recently branded as 'citizen science', but this has been a well-known feature of birdwatching for decades. It is particularly popular and important now as statutory agencies that are pressed for cash see this as a way to collect the data that they need to do their jobs. With CSGRT, this approach evolved to enable local people to make long-term, sustainable recordings of seals, as this is important for winning the public's hearts and minds and empowering and enabling locals to become advocates for the seals on their patch. I do lots of talks, and usually at least one person approaches me afterwards asking how they can help us with our seal research – in effect they 'self-refer'. I send them an information pack, ideas on where they might look, and some simple guidance on looking at low tide, counting and photographing the seals. After some initial solo surveys, we sometimes do a shared survey or training and recording together and they end up coming to our monthly CSGRT meetings, which are open and free. CSGRT have over 200 volunteers routinely submitting surveys and photos. Getting public liability insurance for our unusual range of work was a challenge for most insurers, but Dive Master generously provide us with subsidised cover because of personal connections through BDMLR.

For citizen science there are some important lessons. People are always way more important than data, and it is essential to provide feedback to volunteers otherwise they'll not continue with their submissions. You need to aim to give more back than you take from each volunteer. I find that remaining active in the field myself as a role model (alongside my long-term seal-surveying partner Kate Hockley) helps a great deal to maintain my integrity and credibility. All my work running CSGRT is done voluntarily. I only earn money when we secure the occasional research contract. I start most days at 6.00 and am still doing CSGRT business last thing at night. Communication is the key – if you take the time to listen and work with people, even those you might not instantly take to, you can learn a great deal in return. I often learn more from people with different outlooks to my own. The plethora of skills that each volunteer – from all walks of life – brings to bear is frankly astonishing. For example, one of our volunteers, Annabelle Lowe, is an ex-midwife, and the insights I've learnt from discussing seal pupping with her have been fantastic. CSGRT manages and analyses our data in-house, enabling us to be independent and self-reliant. Our model for public engagement is easily transferrable and has been adopted by other seal-related groups all around the UK for whom we aim to provide back-up, advice, guidance and moral support wherever we can.

Communication – crystallising key messages

This is where my experience as a teacher comes to the fore – because I used to teach travel and tourism, which included modules such as marketing, health and safety, and running a business. To be an effective teacher you need to be able to crystallise what you're trying to say, distilling it into a simple message using as few words as possible. Before our interview I was working on how to succinctly and positively phrase our 'take home' messages on seal disturbance. I love the idea of ocean optimism (Heather Koldewey, Chapter 13) and aim to use this in all that we do.

Getting the words right is hard and takes time, the reality is that people's brains simply cannot process negative messages. Some of the key phrases we use include 'admire from a distance', 'respect and protect' and 'leave as you find'. For signs of disturbance we have our 1, 2, 3: (1) seal looking at you – too close; (2) seal moving to the sea – back off; and (3) seal in the sea – too late.

Summarising our organisational aims was a lot harder, and the key terms we've come up with are research, communicate, engage and conserve. We've also got several straplines depending on who we are communicating with. These include: 'identifying and monitoring local seals', 'passionately protecting Cornwall's precious marine environment', 'putting seals on everyone's agenda', 'giving seals a voice' and – depending on who is listening – the end point is 'in a world where money talks'. Using fewer words takes a lot more time and thought.

Learning who you need to know

From a researcher's point of view, trying to figure out how conservation works is tricky, complicated and hierarchical, and getting your head around the big picture of how the human system works is very difficult; you tend it pick it up in snippets. Understanding who the key players and cultural architects are who make a difference regionally, nationally or internationally, and engaging with them, has also taken a lot of time. Making meaningful connections can be serendipitous, but others we have developed by design. For example, George Eustice, our local MP, is currently a minister at Defra and he only lives 6 kilometres from me. He'd never been to his local site to see the seals. After one 'no show' we were able to fix up a two-hour meeting with him in the field. We prepared thoroughly and were able to talk about a host of issues using the visual prompts in front of him such as entangled seals, climate change, cliff falls and underweight pups, so it all worked brilliantly. At the local café, which benefits financially from seal visitors, we followed up our field session with reports and a framed photo of a seal, which he promised to put on his office wall at Westminster. We got lucky too, as it turned out to be his birthday. You always need a bit of luck. I routinely email him research reports from us and others including 'Brexit: The marine governance horrendogram just got more horrendous!' (Boyes & Elliott 2016), and we have an ongoing dialogue on all things marine.

FUTURE CHALLENGES

What are the most important threats to the marine environment and seals, and what needs to be done about these?

Disturbance for all marine creatures, particularly seals and birds, is ongoing and ubiquitous in Cornwall. The traditional holiday summer season has extended significantly. Disturbance is increasing and our data show some worrying trends. We work with key landowning organisations like Cornwall Council, the National Trust and RSPB to try and reduce unintentional disturbance by raising awareness. The public feeding of seals in harbours encourages seals to hang around what are very dangerous environments, and every year seals are reported hooked by fishing lines or hit by propellers, and boat fuel can be a killer for them.

We are getting a clearer idea of the impacts of extreme weather events, in particular rock falls from the cliffs. Even the death of one seal can have a huge effect. For example, we recently lost a beach master, the dominant male, to one of these rock falls. This created a power vacuum, resulting in a very unsettled haul-out throughout the rest of the pupping season. At sites where this has happened it can take several years for a successful beach master to re-establish harmony and equilibrium, and in the interim pupping numbers decline. St Ives Bay is one of the most rapidly eroding coasts in Europe, and one rock fall blocked off two really important pupping caves after the start of the season, most likely trapping all the seals inside. This cave system was out of action for at least five years and the knock-on effects could be seen at all surrounding sites. Another impact of climate change (sea-level rise) may well be slow enough for seals to adapt.

Shooting seals continues in a clandestine fashion in Cornwall, and in 2010 CSGRT worked with the newly formed Marine Management Organisation to get a voluntary 'no shoot' agreement in the Isles of Scilly, a designated Special Area of Conservation for seals. The majority of residents realised that this could be a huge PR disaster for the islands, even though fishermen admitted, remarkably in a public meeting, that they shot seals every winter. We are lucky that the shooting of seals is thought to be a small problem in the southwest, but we involve the police whenever we hear of incidents. A report entitled *A Seal's Fate* documents the nearly impossible task of killing seals humanely as moving targets from the moving platform of a boat (Advocates for Animals 2009).

Noise is something we need to learn more about after initial research we did for the Atlantic Array consultation. Pile driving can cause death if seals are close enough, and at the least can cause permanent deafness and abandonment of an area. We know from our recent work on fishing gear that we have the highest level of entanglement for any phocid seal species anywhere in the world, arising from our local fishing efforts as well as from lost gear drifting across the Atlantic (Allen *et al.* 2012).

We are also concerned about chemical pollution in our seal populations from various sources, including pharmaceutical runoff from prescription medications that we excrete down our loos; agricultural waste from riverine sources affecting marine water quality around the southwest; and recently proposed seabed mining and dredging activities risk disturbing heavy metals deposited from historic mining in our area. James Barnett, our local veterinary pathologist, routinely collects seal samples from post-mortems which we hope will reveal the contamination levels of seals – who, as top predators, are so critical for keeping our marine ecosystems in balance.

What are the hidden problems and barriers that make progress harder than it should be?

For me, not knowing the right people in the 'system' or the bigger picture is often an issue. Ocean optimism is important, as a negative approach often means your message doesn't get heard. Another is 'self-talk/belief': it's no good believing that as individuals we are powerless or thinking that because you haven't been doing conservation that long you can't make a difference, when you can!

Powerful attitudinal shifts have been achieved in society: wearing seatbelts and not smoking in public places are two great examples. A Dutch professor called Matthijs Schouten outlined a model of societal paradigm shifts in human attitudes towards our environment from one of exploitation, changing to one of stewardship, to one of seeing ourselves as part of the planetary ecosystem. However, he rightly predicted that when the economics gets tough we revert straight back to our model of environmental exploitation. He proposed that we needed to frame our messages according to the prevailing paradigm. I find it frustrating when communicating with the press, who invariably want to put me in the 'environmental/conservationist' silo and only want to hear my views from this singular perspective, when I know the social or economic arguments would be more effective in communicating my message.

Listening is a critical skill, and learning from the constructive feedback of others (even when painful) can teach you a great deal. Understanding where people are coming from in this context is critical. The recent approach by the Cornish Plastic Pollution Coalition has been very successful, achieved by joining a large number of organisations together as co-signatories on collaborative letters with multiple logos. This has achieved great things in a very short time. They recently persuaded Virgin to stop balloon releases and destroy their stocks because of the evidence that had been collected on local beaches; this was a quick and effective win.

When you retire, what would you most like to have achieved for marine conservation?

I'm not going to retire. I'm going to be doing this at ninety! I might not be out in the field as much, but I do this because I love it. Why would I stop? I want seals to be at least as prolific then as they are now, preferably even more abundant at the sites where they currently occur. I'd like there to be more volunteers watching their local patch, an army of people and children in Cornwall who appreciate the very special seals we have here. I'd like to get more people out and about appreciating the coast, marine environment and the seals, not least because it is good for their own health and wellbeing.

How do you know you've been successful? When other people quote you or reference your work in international papers – that is a huge compliment. It is the best feeling in the world when someone comes up to you and says that they hadn't realised they were interested in seals or that you have inspired them – that's a great buzz.

Are you optimistic or pessimistic about the future?

I'm an optimist, and I think wildlife is more resilient than we might imagine, but we need to keep on with our work. The comeback of seals since the low points of near extinction in the 1930s is encouraging, and watching children enjoy a wild experience means their generation will appreciate the value and importance of their natural environment way more than my generation ever did.

CHAPTER 18

Alan Knight

Alan is currently the Chief Executive of International Animal Rescue (IAR) and Chairman of British Divers Marine Life Rescue (BDMLR), both organisations that he helped set up in 1988. He was encouraged in his interest in wildlife and butterfly collecting by his father, but collecting butterflies made him aware that the animals needed to be killed, and this did not sit comfortably with him. A careers teacher at his school in Romford, east London, couldn't understand why Alan didn't want to work for the Ford Motor Company, suggesting that his ambition to work in a butterfly farm would not give him a 'proper job'.

Alan's interest in wildlife led to him reading biology at Sussex University (1975–1978), where he honed and developed his strong beliefs about ending cruelty and suffering in animals, a viewpoint that brought him into conflict with the course staff; eventually it was agreed that he did not have to experiment on animals. A formative debate on animal rights and welfare hosted in the Philosophy department saw Alan's argument win the day. Inspired by his partner Liz, he became a lifelong vegetarian, volunteering with a range of welfare and anti-vivisection organisations including joining a national committee for the Hunt Saboteurs Association. Despite volunteering for many conservation projects, paid employment eluded him, because he was having to compete with people who had PhDs who were also unable to get jobs. That experience prompted a change in career, and Alan started working for a scientific instrument company (Scientific Optics Limited), initially making microscopes and eventually becoming Chief Executive Officer. In 1999 another change of direction saw Alan redirect his energies into IAR, of which he is now CEO.

Having taken up diving in the 1970s, in 1988 Alan responded with other divers to the plight of common seals with phocine distemper virus (PDV). At that time there were no organisations capable of rescuing marine mammals at sea, so he and like-minded people set up BDMLR, a charity which has provided equipment and medical support to enable a very wide range of marine mammal rescue scenarios for seals and increasingly whales. Working for IAR now takes Alan to India and Indonesia to help rescue orangutans, slow lorises, sloth bears, macaques and many other species. In 2006 he was awarded an OBE for his outstanding work for animal welfare.

This chapter looks at Alan's work developing BDMLR and IAR; the challenges and interactions between the ethos of welfare and conservation; how specialising in rescue has borne fruit; how effective delegation to volunteers is a key element of success; the development of technology; and how whale rescue is developing worldwide.

MARINE CONSERVATION AND ITS DEVELOPMENT

How do you frame what you do, and what are its key ideas?
My beliefs are based absolutely around animal welfare, the idea that every animal counts and we should rescue individuals and stop their suffering. These beliefs can be traced back to my dad and my partner of 43 years, Liz, who is a vegetarian and guided me towards a whole set of other ideas about animal welfare and hunting. Up until I met her, I'd been a big meat eater and had even hunted, but I felt this was hypocritical. I changed and became vegan for eighteen years. Liz and I joined the hunt saboteurs because I didn't think chasing defenceless animals around with a pack of hounds was right, and this became a guiding influence. What I find is that once I commit, that stays with me, so my stance developed into anti-hunting, anti-vivisection, anti-captivity (animals in zoos and battery farms). Until 1999 I did this work as an unpaid volunteer in addition to my business work, so I've seen welfare and conservation from two very different perspectives.

What are the differences between when you started and now?
When we started there were no organisations that could either undertake rescue or look after the welfare at sea of larger animals such as seals and cetaceans. From the base of volunteers in 1988 BDMLR has grown and now we have over 9,000 qualified volunteers throughout the country, trained to be marine mammal medics who can respond to events. BDMLR training is now regarded as the gold standard for rescuing marine mammals worldwide, and I see lots of people who work in this area and see 'marine mammal medic' on their CVs, so it is something people take seriously. BDMLR is now widely recognised by the main players, the coastguard, the fire brigade, the police and RSPCA, who look to us when the need arises. Our work has evolved, and many innovations with the equipment we use have been developed. The different species protocols, and our close relationships with scientists and vets, have been the pillars of our approach.

How is marine conservation different from terrestrial conservation?
I work on marine and terrestrial animal rescue. The same welfare ethic applies, and there is no difference in our approach to animals in either environment. The biggest problem is human greed in exploiting the environment. We need to find ways to coexist with animals and the environment and educate people to leave space for animals. There is a happy medium, and IAR is finding, in Indonesia for example, that through our outreach programmes, palm oil companies are seeking advice on how to be more ethical. We are encouraging them to leave bigger corridors and undamaged habitats in palm oil plantations to enable the wildlife, including orangutans, to prosper; we are helping them to develop their strategies.

The phocine distemper virus and the start of British Divers Marine Life Rescue

In the summer of 1988, hundreds of mainly common seals were washing up dead on the beaches of the Wash and East Anglia, suffering from what we now know to be phocine distemper virus (PDV). There seemed nothing that people could do. One particular photograph in the *Daily Mail*, of a seal being shot with a captive-bolt pistol, shocked me beyond belief. I then saw an article in *Diver* magazine by one of their staff, Robin Eccles, who was rescuing seals – and this inspired me to go up to the Wash with other sports divers. Our diving background had given us the skills – navigation, boat handling, using radios – to go out to sandbanks safely and survey them for infected seals. We soon discovered that infected seals could be caught quite easily, which is not the case with healthy seals. None of the other organisations, the RSPCA or Greenpeace, had the capacity to get to the sandbanks. The RSPCA's only option was to shoot them on the beach when they got so ill they couldn't survive. As we did this survey work our views developed, and we suggested that if we could identify animals that were ill and then bring them back, something could be done to try and interrupt the infection before they became so ill that they washed up on the shore. That was the theory – but it posed a problem: what were we going to do with them?

To their credit both Greenpeace and the RSPCA got behind this idea, and they set up a 'halfway house' in an old bus station – a shed – at Docking. By this time there was quite a big group of divers involved who were builders, plumbers, electricians, in effect every type of profession we needed to renovate the shed. The RSPCA were keeping each seal in isolation; because they knew very little about the disease they just kept them in big round tubs. We all felt very bad about this because it seemed like sensory deprivation, but we assumed they knew better so we carried on.

Very quickly the RSPCA ran out of money, and it turned out that Greenpeace didn't put in any. We continued on our own, raising money through the newspapers, and put it all into Docking. Very little could be done and lots of those seals died. A few of them survived; they were probably immune within the natural population, and their survival probably didn't having anything to do with our intervention. If anything we probably shortened their suffering, so that a lot of those animals died before they would have done had they just washed up on the beach and died slowly. However, we didn't feel there was anything positive.

What we had achieved was the identification of a need – rescuing large marine animals – and discovered an unfilled niche, which was that people could go out and rescue animals at the coast and in the sea. In December 1988, following the end of the seal episode, we got together and agreed to start a charity. We met in a pub in Hunstanton, and tried to agree on a name. There were strong arguments for including all the key words, marine life rescue, divers, British – and so we came up with British Divers Marine Life Rescue (BDMLR). In hindsight that was probably the worst name we could have chosen, because to this day nobody gets it right! It is even more ironic that there is now very little diving in BDMLR. It has expanded its remit, and science is central to what we do. We had started a charity, and very quickly we got lots of people interested – but we didn't really know what to do next.

We were looking for other projects and soon got involved with the Born Free Foundation, with whom we formed an excellent relationship. They were tasked to transfer the last dolphins in captivity in the UK to the Turks and Caicos Islands in the Caribbean. It was an odd situation, as the dolphinarium trusted us but not the Born Free Foundation, which enabled us to go into the dolphinariums but not Born Free. We successfully removed three dolphins from the UK and flew them across to the Turks and Caicos, where we built a sea pen to enable them to acclimatise. We paid our own expenses, and received no payments.

In January 1993 the oil tanker *Braer* ran aground in Shetland, and we helped set up a rescue service for oiled birds, seals and dolphins. It was some of the worst weather I've ever dealt with, but we had three boats and the Shetland Islands Council subcontracted small craft to us which we managed through the whole of that *Braer* oil spill.

The marine mammal medic course and handbook – evolution, innovation and working with scientists

Our work on the *Braer* started us thinking that we needed to get more organised about what we were doing. We now had experience with seals and dolphins and knew about the practicalities of dealing with net entanglement. This led to us setting up an official training course, the first of which took place in 1997 in Thurso; it was memorable because it was the weekend Princess Diana died. Now we see these courses and our *Marine Mammal Medic Handbook* as the foundation of everything we do.

The courses have evolved and developed over the years to represent the highest standard of welfare for marine animals in the world. The first course had no precedents, and two of us made it up as we went along. We were showing people the best ways to pick up seals, how to transport them without getting bitten, and the best way to undertake assessments for seals and cetaceans. There are now courses all over the country, run by our area coordinators. The one-day course includes lectures on adaptations to the marine environment, the biology of whales, dolphins and seals. We then cover first-aid techniques and rescue techniques. Later we take people onto the water and use pontoons to inflate and re-float a full-size pilot whale, which weighs 2 tonnes when filled with water. We've also got life-size seals and dolphins. The training covers how to identify and assess an animal on the beach to see if it needs rescuing, and then how to actually rescue it. It is very much a baptism of fire. Valuably, the course fees have also been the mainstay of helping to finance the work of BDMLR. In 2017 we have 2.5 paid members of staff; everyone else is a volunteer.

Working with veterinary medicine practitioners and scientists

Our *Marine Mammal Medic Handbook* is our training bible and it's reprinted every eighteen months or so. We are now on our eighth edition, and update it with innovations in rescue techniques, assessment and diagnosis. For instance, if you get bitten by a seal, modern antibiotics won't fix it and you get septicaemia. So we have a special letter in our manual to tell doctors to make sure they give you original penicillin, not the broad-spectrum antibiotics. If you have a problem with being allergic to penicillin then we can warn our volunteers about this risk beforehand. Currently, we have a real problem with euthanising large animals like whales because the big-animal anaesthetic, Immobilon, often used on horses, has been banned. These techniques are evolving rapidly and we are working with scientists to come up with satisfactory methods. If we put an animal down we look at that as a success, as long as we've assessed it and it is not viable for release, on the basis that we have reduced its suffering.

Science is changing animal welfare, and we believe that all animals we are involved with that die should have a post-mortem to enable science to play its part in informing our activities. We are veterinary-based, and everything we do is agreed and prescribed in protocols. We work very closely with the national strandings programmes, and Rob

Deaville from ZSL is a good friend and supporter who works on the post-mortems, as does James Barnett who does post-mortems in Cornwall. We work hand in hand with them to help access and retrieve corpses, and I've attended more post-mortems than I care to remember. We're very practically minded: for example, I've used my Land Rover on the beach to pull blubber off a humpback whale so the vets can reach the internal organs for samples. We work closely with vets and scientists all the time and they are phenomenal; we really do learn a lot from them. This is where animal welfare, conservation and science are coming together. We don't agree on everything, and sometimes our emotions overtake us and we will release an animal because it might have the chance of survival even though they've said don't do it – but we normally agree.

One of our biggest rescues was what became known as the 'London whale' in 2006, when a northern bottlenose whale swam up the Thames. A team went out and conducted the rescue and we were in attendance on the boat going down the Thames keeping this animal alive. It was viewed on TV by 500 million people worldwide and was the biggest event on the Thames since Churchill was buried and went down the river on a barge; every bridge we went under had thousands of people on it. Unfortunately the animal died before it got to the open sea, so we took it out of the water and did a post-mortem. It was the most thorough post-mortem anyone has ever done on a northern bottlenose whale, and the scientists came up with all sorts of really interesting things which were later published.

Approaches to managing people and a welfare organisation

I came to managing BDMLR and IAR after a long career in business, which taught me a number of really important things. The ability to learn from what you're doing and respond to change is fundamental, and if I'd been talking to you five years ago my responses would have been very different; problem solving and innovation are very important. Another key lesson, one of my mantras, has always been to do a few things very well – because I've seen organisations which are spread too thin, and they end up doing lots of things very badly.

One of the hardest lessons we had to learn at BDMLR was the need for delegation. At one point we tried to control everything from the head office, but gave up because there is so much to do. If you delegate effectively you get a lot of people doing well, and if you can teach them to delegate well, then you get a very effective organisation doing all sorts of exceptional things, and massive change. We have delegated responsibility and set up area coordinators – we now have forty around the country – and within parameters we allow them to post rescues on the Facebook page. They don't advertise or do things that are against the ethics of the organisation, and this works very well and is self-regulating. We've become much more successful by enabling volunteers to make their own decisions, to operate their own rescue groups and undertake local rescues. It does take someone in the office to coordinate it, but generally local volunteers will get involved much more. By not being frightened to decentralise and delegate we've made the charity *its members*.

Developing technology can help both with the management of the organisation and with the rescue and welfare issues. Our recent investment in computing systems has enabled us to make enormous changes to our practice. Providing the positive feedback to volunteers used to be a major problem for us. A significant breakthrough came with investment in a database and an SMS text management system which has enabled us to make local volunteers instantly aware of rescues they could take part in. Now volunteers tell us that they are sorry they can't do more.

Another amazing technology is 'what3words'. It is three words that gives you every three metres square on the planet, so when we have a whale that's on a groyne or in the middle of a sandbank, you can't get a GPS mark. We can convert this geographic reference and email details to volunteers, which gets them to exact locations in areas where conventional methods don't work. Smart phones have been a real plus, as they have enabled us to get images to the office and our consultant vets. We have a triage process, and the accurate identification of the species determines the species protocol we use and our understanding of the likelihood of survival. This routinely makes the difference between rescue and release or putting the animal to sleep humanely.

Fundraising for any small charity is a continuous challenge. With BDMLR we are lucky in several respects, and our shared offices with IAR mean that our overheads are much lower than for other organisations. The membership and course fees are our bread and butter. We charge commercial concerns who have been happy to support our work. Recently we designed a lifting rig with netting to rescue seals trapped in the cooling-water culvert of a power station. There is no doubt that welfare organisations relate to a particular mindset of the British public, and we are lucky to get our fair share of donations. It is all based on people trusting us to do the right thing. We didn't use the Thames whale as a fundraiser – we should have had people running up and down asking for money; we would have made millions!

The development of whale rescue

Many of us became increasingly aware of the interaction between marine species and fishing gear and the problems that entanglement causes (Figure 18.1). Seals all around the UK can be seen with Elizabethan collars of netting. We decided to get a group together to learn more about how to disentangle whales and seals. In 2006, we took six people across to Cape Cod to the Center for Coastal Studies (CCS) in Provincetown to learn about disentanglement techniques. Much of our work is based on trust and strong, long-standing relationships. The trip was funded by IFAW (the International Fund for Animal Welfare) and Ian Robinson, executive vice president of IFAW, was the chief RSPCA vet we'd worked with on PDV and he trusted us to take this on.

The CCS has been going since 1984 and Charles 'Stormy' Mayo and Dave Mattila are two strong characters, both having seen the need to rescue marine animals. We went

Figure 18.1 Rescuing a whale in Iceland. Source: Megan Whittaker, Elding Whalewatching

out on the CCS boats and learnt how to disentangle whales, and we are currently the only group in Europe who can do this work. We have two sites with the kit we need, in Scotland and southern England, which means we can deploy in hours. We have been called out seven times in Scottish waters and once in Iceland, and we've been successful in disentangling most of those cases. One whale died, but overall it's a successful system, although very dangerous.

We have refined the course so it's more applicable to the situations we have here, and now we run a Europe-wide rescue service for entangled whales and have a tailored BDMLR course and protocols for this. We are now part of the World Alliance for Disentanglement, which is ironically controlled by the International Whaling Commission (IWC) and which is funding Dave Mattila from the CCS to go around the world training people. We had already done the course, but he came over here to give us a refresher and we were complimented on our state of readiness.

The funding of whale rescue is a challenge at the moment as it is very expensive. It's an area I'm worried about. It cost £30,000 to set up and we've just received a grant of 20,000 euros, and made up the rest ourselves through BDMLR. Just the one rescue in Iceland cost £10,000 to ship people and equipment across, to pay to replace equipment which was lost or damaged; we borrowed a boat and unfortunately it got damaged. It mounts up, and if this happens three or four times a year we're going to be in trouble.

In Cape Cod whale rescue started as a result of a conservationist approaching the government and then taking them to court. They argued that under their wildlife legislation there is an absolute need to rescue endangered species, in particular the northern right whale, which was down to 400 individuals and so critically endangered that every whale did count. The National Oceanic and Atmospheric Administration (NOAA) have since funded and regulated the CCS to provide a rescue service for northern right whales initially, but of course they go out for any species of whale or turtle and have a phenomenal success rate. They fully document their rescues and have an extensive database which has provided the scientific evidence that informs their work. The CCS is also active in preventing entanglement by working closely with local fishermen on fishing-gear modifications. For instance, the floating ropes have been replaced by sinking ropes, so the sea is less of a three-dimensional maze for the whales to navigate. They have also designed weak links in the fishing gear, which if the whale gets entangled break easily. It is by no means perfect, but by working with fishermen and talking their language progress is being made on preventive measures.

Welfare and conservation – ethos, challenges, interactions, the role of science and the blurring of the lines

BDMLR has come a very long way from those early days of 1988. When we first started there was a perception that we were cowboy divers racing around in fast boats enjoying ourselves. Many felt we had no interest in the welfare of the seals, and we did get a bad press. I realised very early on that there was conflict between conservationists, who were mainly scientists, and welfare people, who were seen as emotional, woolly thinkers. That made me concerned, because I'd trained as a scientist and have had a long career in scientific technology and I was determined that we should be taken seriously by scientists and conservationists.

A commitment to work to the highest professional standards has been central to the development of both BDMLR and IAR. In 1992 we set up a group called the Marine Animal Rescue Coalition (MARC). My aim for that organisation, as one of the

Figure 18.2 Pilot whales stranding, Durness, northern Scotland: on the beach with BDMLR members attending. Source: Jamie Dyer, Alan Knight, BDMLR

co-founders, was to bring together as many as possible of the groups in the country that had any interest at all in marine mammal rescue, to come up with a series of protocols for our methodologies. We got a positive response and lots of good people attended, which resulted in a set of protocols and we learnt a lot about PDV in seals. This is a morbillivirus, similar to measles, and immunity is not passed to offspring, so when you get an immune population of adults all the young are not immune. So at some stage when the offspring are mature enough and exposed to the virus, there could be another outbreak.

This is exactly what happened when in 2002 the PDV hit again. This time we were prepared. We convened a MARC meeting with scientists and the vets, and asked, what do you want us to do? The concerns were similar to 1988. What worried the scientists was that we would be displacing infected seals and that might increase disease transmission. They basically said, you cannot tell when an animal is infected, and my view, from our work in 1988, was that you could. I explained that PDV is an infection of the respiratory system and when the seals get compromised they don't breathe very well and they get exhausted quickly. Putting it simply, you can't catch a healthy seal. We took the RSPCA's chief vet Ian Robinson out to the sandbanks and demonstrated this. Our other point was that you could re-catch the infected seals and check them. So that's what we did, we picked up three seals, they were all infected, and in fact two of them died before we made it back to the hospital. After this we were taken more seriously. We then came up with a protocol that clarified how we could rescue seals, without spreading the disease from one sandbank to the next. Pretty quickly we found that the good we were doing was minimal and there was nothing we could do to treat the seals, and so we will not be launching a massive rescue operation if it happens again. More fundamentally, no one would invest in the development of a vaccine for wild seals. Morbillivirus is very closely related to dog distemper, so all the seals in Sealife centres and zoos were given dog distemper vaccine to try and protect them.

By working with vets and scientists to produce the *Marine Mammal Medic Handbook* and training courses, BDMLR became 'recognised' and taken seriously. Importantly, this recognition enabled us to obtain insurance cover for our medics. Defra suddenly started taking notice and listening to what we said, and this understanding meant that they gave us the licences to do rescue work. I think now that, whether we like it or not, we are considered to be a conservation organisation based on animal welfare principles. By taking seriously all of the threats that conservationists and scientists worry about, and trying to bring those into a protocol, we are saying that this is a group that can be trusted to go out and rescue a minke whale or a pod of pilots where other groups or individuals wouldn't normally be given a licence to do that (Figure 18.2).

The relationship between conservation and welfare

There are still major tensions between the conservation and welfare communities, though I think they are blurring and we are eroding them. This is vitally important, because we are working on highly endangered species to which both disciplines can contribute. A good example of this blurring is the work IAR has done with Indian dancing bears. We were focused on this cause because of the cruelty and welfare issues relating to the bears. Although Indian legislation banned this practice it was still rampant, and our programme invested in rescuing the bears and retraining the handlers to gain employment with new skills. This went down well simply because the handlers could then put down roots and stay at home rather than having to lead a nomadic life. Whilst we cannot release the captured bears back into their natural environment because they have suffered so much mental and physical damage, we have largely put a stop to young bears being taken from the wild and their mothers being killed in the process, and we have seen wild bear populations recover. This illustrates how a welfare programme can have a direct benefit for the conservation of an endangered species.

Conservationists often don't take animal welfare very seriously, and I believe that welfare organisations are taking the lead on many of the major programmes – with orangutans, for example, where welfare groups are leading on reintroduction programmes. We seem to have to work through the relationships between conservation and welfare interests with each species we work with, but at least now we have the background and understanding to explore the common ground. IAR is working in Borneo, but we have just lost a big grant ostensibly because they feel that we're an animal welfare organisation, and what we are doing is welfare when we're supposed to be doing conservation. So there is still a bias out there between conservationists and the animal welfare community.

Until my dying day I will fight to protect animal welfare groups, because these are the people that will have the animals that are capable of being released back into the wild. The welfare community is being charged with polluting and diluting the gene pool, allowing interbreeding and keeping subspecies together, so that animals being returned are not going to be good to restock the wild. I think this is plain wrong and too pedantic. IAR, for example, has the best blood-testing equipment in Borneo, better than any human hospital, and incidentally we donate some of the equipment to the human hospital to help them as well. We have the best PCR machine, analysing samples from animals to work out whether or not they have tuberculosis, because that is one of the main things we don't want to return to the wild. Funnily enough, that is a scientific approach that conservationists understand fully and respect us for, because we are absolutely determined not to introduce any viruses or bacteria back into the wild population. We are doing everything we can to break down the barriers between conservationists and welfare groups.

FUTURE CHALLENGES

What are the most important threats to the marine environment, and what needs to be done about these?

We routinely come into contact with human activities that are harmful to wildlife, and it is this that drives our efforts to mitigate the worst excesses. However, it is clear that action is needed on a variety of levels. The policy context, for example, is important to us. Decisions such as leaving the European Union, which has provided a robust framework for animal welfare, causes uncertainty and concern for our future work. Climate change is a major concern which people are more aware of; the fate of polar bears in the Arctic is clear. We see fires caused by delays to monsoon rains in Indonesia destroying forests and wildlife. Things are changing, and we support other organisations who campaign on climate change. Plastics, in particular fishing nets, their longevity in the marine environment and the entanglement of marine wildlife is an area where we are directly involved. Other problems such as pollution from persistent organic toxic chemicals and heavy metals are less directly obvious but no less worrying, as are declining funds for post-mortems.

What are the real, hidden problems and barriers that make progress harder than it should be?

Funding is probably the key issue which limits what we can do, and it seems to be becoming more complicated as public funding declines. If we can effect successful rescues then this saves public bodies, such as Local Authorities, the cost of disposing of the corpse. Recently it cost one local authority £18,000 to send a whale to landfill. Rescue and ongoing welfare is an expensive business, as we see with the seal sanctuaries – which are dependent on the generosity of the public. Welfare organisations are often very competitive. They didn't and often still don't work together enough, and we continue to spend a great deal of effort trying to work with other organisations. I find myself doing this more and more with conservation organisations, with whom we share common interests.

What are the most interesting and promising new approaches?

Innovation is fundamental to everything we do, whether it is the way technology has transformed our relationship with volunteers and members, or with the development of new protocols or new equipment that enables rescue. Every rescue is different and poses different challenges, sometimes we can use tried and tested techniques, but often we have to innovate and through trial and error develop equipment and techniques that help us. Our recent experiments with drones, GPS and tags for tracking orangutans in the forest is very promising and showing the way to what this work might look like in the future, since the traditional ways are very labour-intensive and expensive.

When you retire, what would you most like to have achieved for marine conservation?

I have no thoughts of retirement. I hope to leave BDMLR as a stable and self-sustaining organisation with systems that work and courses that work effectively to achieve successful outcomes for wildlife.

Are you optimistic or pessimistic about the future?

Always optimistic, but certain things do get to me. The Brexit vote was really depressing and we will need to fight to ensure that European animal protection protocols are in place.

CHAPTER 19

Paul Horsman

Paul is currently the acting Campaigns Director of the regional office for Greenpeace Andino covering Argentina, Chile and Colombia and is enjoying learning Spanish. Through his Catholic upbringing for a long time at school he'd anticipated becoming a priest, but after a good deal of thought opted for a degree in zoology at Newcastle University (1973–1976). This was because he was very interested in his best school subject, biology, but he'd also been inspired by Cousteau and others to have an interest in marine biology. He spent his final degree year working on his marine biology project on benthic diversity at the Dove Marine laboratory. The following year was spent doing an MSc at Portsmouth University working on the use of tributyl tin (TBT) biocides, raising questions of the cycle of how biocides were developed, licensed, used and their risk to the environment. In quick succession he moved from a teacher training course to working as a marine ecology tutor at the Field Studies Centre at Orielton, in Pembrokeshire. It was during his time there that he learnt to dive and took part in diving surveys for nature conservation. However, a chance return to the Dove Laboratory was to provide a very formative opportunity.

Frank Evans, one of the lecturers at the Dove, asked him if he'd be interested in a Leverhulme Fellowship working with merchant sailors, teaching them about marine biology as they travelled the world's oceans. He jumped at the chance, and the one-year contract was extended to three and then five. He toured the world, produced a 250-page textbook – *The Seafarer's Guide to Marine Life* (Horsman 1985) – and worked on a PhD collecting together observations of seafarers. During his last year at sea, 1984, he found himself in New Zealand, where he met some of the leading Greenpeace activists who were involved in the Nuclear Free Pacific campaign, and where a new government had just declared itself 'nuclear free'.

In 1985 he joined the Marine Conservation Society (MCS) in Ross on Wye as its first marine conservation officer, and got involved in a whole range of projects including achieving a ban on the use of TBT antifouling paints on small boats – back to his interest in biocides – and he also helped to set up the basking shark work that is still going today. He also worked on the North Sea Ministerial meetings, including the London meeting in 1987, and became chair of the newly formed European NGO group Seas at Risk.

At the end of 1988 he joined Greenpeace in London and continued work on marine pollution throughout the 1990s, including on the Brent Spar campaign. This chapter mainly reflects his work for Greenpeace through their campaigns in the North Sea to his current work on fish farming in Chile and climate and energy in Argentina. The chapter also covers the rationale and his views on direct action in the context of campaigning, the work of Greenpeace, overarching ideas like the precautionary principle and the different tools for campaigning.

MARINE CONSERVATION AND ITS DEVELOPMENT

How do you describe or frame your beliefs in marine conservation?
My beliefs have evolved over the years and matured in many ways. If I go back to the time at university in the 1970s and my work on ecology and the environment, measuring the problems in the field or the lab could be justified – it was valuable and needed protecting. My career quickly evolved beyond just looking at the science, and changed to trying to understand the reasons behind why environmental problems continue to arise. My main focus has been on the marine environment, not least because a lot of the stuff we put into the environment ends up in the sea. Over time I have come to think that our attitude to the environment and the way people treat it is fundamental. A current view of the environment is that it is a 'resource' that is un-costed. There are those who are now putting a value to this – if you cost it then you can work out the triple bottom line (social, environment and economic) – and believe that this will solve the problem. Well, often it doesn't – and indeed it is based on the very premise that created many of problems in the first place. So my view now is that the environment is not just something to be used. We must have regard for our integral part in the environment. This is where the arguments become much more philosophical and move away from direct campaigning to try and fix a problem to understanding the attitudes and power structures that created the problems in the first place.

What are the differences between when you started and now?
Being on merchant ships in the early 1980s I was appalled that rubbish was put into plastic bags and then thrown into the sea. The Marine Pollution Convention (MARPOL) Annex III was supposed to cover this issue, but it was only optional and unratified at the time. I wrote an article on garbage pollution from merchant shipping for the *Marine Pollution Bulletin* (Horsman 1982). I'd been travelling on many merchant ships but one of these was a BP tanker and when I was in London I was invited to meet someone from the BP shipping unit. He was pretty upset by the article and informed me that Annex III would never be ratified and he was very comfortable about that. We have recently come to fully realise the huge scale of plastic litter pollution in the ocean, illustrating that the situations we face now are of a different scale which are complex, global and existential in nature. Importantly, they are still about our attitude to the environment, and sadly, like the BP shipping man in the 1980s, attitudes haven't been changed sufficiently to effect the change needed.

How is marine conservation different from terrestrial conservation?

Clearly there is a difference in terms of the expertise people need, the science etc., similarly the physical boundaries and lines on a map which make a big difference on land. The sea and the air represent more fluid environments where many of the issues and the pollutants have few physical boundaries. Most people will know the sea from the coast, but working for Greenpeace, you can go to places where people would very seldom go and you can see at first hand what most people can't see or visit easily. Given that the sea is 70% of our planet, most people will seldom be able to see or experience what is really happening, and because so much of the ocean does not 'belong' to any nation – it is a global commons – it is also much more difficult to regulate activities at sea.

The North Sea conferences – the precautionary approach

My interest in the North Sea arose when we were working together in 1986–1987 on the London North Sea Ministerial (NSM) meeting (1987). In the 1980s scientists (particularly pioneered by Germany) developed the *Vorsorgeprinzip* or the precautionary principle, which from the point of view of the environment was a much better scientific rationale for turning the burden of proof of an issue on its head. When we were talking about stopping pollution or trying to protect the environment it was always the debate about 'well there's been no effect, you can't show any damage'. This harked back to my MSc work testing TBT biocides and bio-deterioration on wood. One of the conclusions I came to in the whole process of looking at biocides and pesticides was that a product would come onto the market and those who had invented and developed it would want

Figure 19.1 Ship dumping waste at sea. Source: Greenpeace

to get as many uses out of it as possible. If you look at the history of pesticides or biocides, in general there is development, then marketing, and then with use the problems of build-up in the environment. By the time problems (either for the environment or for human health) are noticed there is resistance to taking it off the market, which eventually happens, but by then the corporation has developed something else to take its place. So there's a cycle, which clearly wasn't doing any good for the environment, but the onus was always put on the environmentalists, or the public, to say this isn't good and prove that it was doing damage before anything was done about it.

What the precautionary principle did was to ensure that you demonstrate that a chemical would not have any damaging impact *before* using it in the environment. This was anathema to the Chemical Industries Association and the corporations because it put the burden of proof on them to show chemicals were going to be safe, not on environmentalists to demonstrate harm. Once you can demonstrate harm, then the question of banning use is a huge and expensive problem not least for the environment. The precautionary approach provided environmentalists with a very good and solid scientific basis for justifying a complete change in thinking, not just on pesticides and biocides but throughout the chemical industry, and for corporations in general. In the 1970s and 1980s the environment was seen as a cheap disposal site for waste, with the pipelines that came from chemical plants such as ICI and Ciba-Geigy in the UK, Bayer and BASF on the Rhine and others discharging into the rivers and the sea, which was viewed as a disposal right (Figure 19.1). The precautionary approach gave us the opportunity of 'pushing back up the pipe' into the corporation or the production, to put the onus on industry to make clear that what they were doing wasn't going to cause damage.

At the time there were a lot of problems in the North Sea, including the seal mortality incident in 1988, and massive algal blooms that were attributed to excess fertiliser runoff. Now whether or not the input of toxic chemicals had actually directly impacted the seals, whether or not it was the discharge of excess fertilisers that increased or exacerbated the algal blooms, the point was that those two impacts created massive public concern around Europe about how industry was treating the environment. This came on top of the discovery of the ozone hole by Joe Farman from British Antarctic Survey in 1985, alongside Margaret Thatcher's recognition of the problem of climate change and the French bombing of Greenpeace's *Rainbow Warrior* (1986) – so there was this massive awareness that something was wrong environmentally and, importantly, our elected leaders were doing very little about it. The public flocked to organisations like Greenpeace and MCS and at that time we grew quite substantially. Most news outlets began to have environment correspondents. It was a time when the environment was on the agenda, and it led through to 1992 and the Earth Summit in Rio.

The third NSM meeting was held in The Hague, Netherlands in 1990, and by that time I had moved from MCS to Greenpeace. The UK had a reputation as 'the dirty man of Europe' (Rose 1990), and we were the only European country still dumping industrial waste and sewage sludge at sea. I remember the final NSM press conference in 1990; the chair opened by saying that they had 'two problems at the meeting; one was nutrients and the other was the United Kingdom!' At that point the opprobrium levels heaped on the UK government resulted in a massive change in government policy: they declared an end to sewage sludge dumping in a set time frame and an end to industrial waste dumping within an even tighter time frame. There have been successes, and we now have a global ban on nuclear waste dumping and industrial waste dumping at sea.

Figure 19.2 The Brent Spar oil storage platform, during the Greenpeace occupation. Ships are hosing the activists. Source: Greenpeace

Brent Spar – attitudes to dumping

In the early 1990s I had been working on oil spills on a global scale: *Exxon Valdez* (1989), the Persian Gulf (1991), *Aegean Sea* in La Coruña, Spain (1992), *Braer* in Shetland (1993). I had some links within the offshore oil industry and I got a leaked report which showed that the UK was considering giving permission to Shell to dump the redundant Brent Spar in the North Atlantic. The Brent Spar was an oil storage platform, not a rig, and had been decommissioned. At the time Shell was the world's second largest company, and supported by the UK government they still held the attitude that they could dump their waste at sea. The issue was that Brent Spar would set a precedent, because the lifespan of a lot of North Sea oil installations was coming to an end, and if this was allowed there would be a whole gamut of offshore installations that would be dumped in the world's oceans. So we wanted to use this as an example.

The Brent Spar was being towed for dumping in the Atlantic shortly before the 1995 NSM meeting. The leaked report led the Greenpeace campaigners (from Germany, the UK, Netherlands and France) to get the North Sea ministers to declare that dumping redundant offshore installations was not going to be allowed in the North Sea or in the northeast Atlantic under OSPAR (Figure 19.2).

In the end Shell backed down and the Brent Spar was dismantled in Norway. In retrospect we know it was a monumental battle of ideas and of attitudes. Even today, 25 years later, the Harvard Business School use the Brent Spar as an example of how not to deal with civil society. Brent Spar captured the public's imagination in a way like nothing else had. It wasn't necessarily saying that this was the biggest environmental threat, but it represented an attitude. As you would not expect people to dump their old car into the village pond to get rid of it, so people said corporations shouldn't dump their waste in the oceans but should recycle it. Currently North Sea installations are being brought back to shore and their materials recycled, creating many jobs in the process and saving raw materials. Fundamentally, though, Shell believe that they 'lost the PR battle'. They still don't see it as something that was in principle wrong to do.

Direct action – personal and Greenpeace perspectives

Environmental NGOs often want large and rapid changes, rather than the incremental steps that are on offer. What Greenpeace and others have found is that direct action can be an effective tool to secure large-scale changes relatively quickly. I'd like to describe two strands to this, firstly my personal development with this and then the Greenpeace perspective.

My relationship with direct action goes back to the peace movement in the early 1980s and my awakening to some of the really big problems. My last voyage at sea was in 1984, and we travelled through the Pacific and I met a lot of Pacific islanders who talked about the impacts of nuclear testing. At this time Ronald Reagan and Margaret Thatcher were distributing nuclear missiles throughout Western Europe in the rather stupid belief that they would defend against a supposed Soviet attack. There was a lot of scientific and philosophical work by SANA (Scientists Against Nuclear Arms) and the International Physicians for Social Responsibility demonstrating that this was ludicrous, but it wasn't making any impact. I decided to do two things – to tell people about what was going on in the Pacific and to join the peace movement. My then wife was also taking part in the Greenham Common Women's Peace Camp. I did my non-violence induction and training through the peace movement. I recognised that direct action was much more philosophical; it wasn't just a question of sitting there and opposing the use and installation of nuclear weapons.

Direct action is a demonstration that we don't have to comply with rules and regulations that are wrong. Once you become aware of that you start looking at the Gandhi principles, at how the Suffragette movement changed the law, and you recognise that all laws are man-made and that they can be changed for the good. Sometimes the only way to change them is by breaking the law and raising the profile. So that became of interest to me, but as I gained experience of direct actions and getting arrested I also began to realise that in the confrontation between me and a policeman or a military person, there was communication happening not at a superficial level but at a much deeper level. I was taking my personal responsibility to stop a really bad law and the authorities had people who had also to take responsibility to arrest me to maintain the status quo. So direct action became for me, and for a lot of people I know, something that works on many levels – it's not just a stunt that gets pictures. It's a very personal commitment to doing what you can in order to effect change.

When I joined Greenpeace my perspective on direct action became clearer. Greenpeace was founded by some Quakers in Canada, and they have two fundamental principles. The first is that of bearing witness, showing what's going wrong and pointing to who or what is responsible – so if you know what's going wrong you become associated with it continuing or not. And the second is that of taking peaceful direct action, being non-violent and trying to stop what is going on (Figure 19.3). These are the two fundamental principles at the heart of Greenpeace. Their first action was to charter the fishing boat the *Phyllis Cormack* and to sail north in the Pacific to stop the US nuclear testing; they didn't get there, but the fact they tried resulted in so much public outrage at what was going on that it gave birth to Greenpeace. It was also part of that awareness of environmental problems that was coming up in the 1970s. On board the *Phyllis Cormack* was the now famous book *The Warriors of the Rainbow* (I have a copy), and the Quakers of early Greenpeace associated themselves with the story of the warriors of the rainbow who would return when the world was sick and rescue the earth. This became the narrative of early Greenpeacers, and it remains a strong narrative within the organisation.

Greenpeace captured global imagination because there were serious things happening in faraway places that people don't normally get to, and this became another

Figure 19.3 Greenpeace activists blocking an effluent pipe. Source Greenpeace

strand of Greenpeace's work because we could get to these places to send the pictures back of what was going on, what governments and corporations were actually doing.

The first *Rainbow Warrior* was a Scottish trawler, converted in London, using lots of volunteers and hard work. This was the ship that was used for various campaigns whether that be whaling, sealing, dumping nuclear waste at sea, testing nuclear weapons at sea, all of which we could expose because we could go there and bear witness. It was also the ship sunk by the French government's secret service by bombs and that killed a Greenpeace photographer, Fernando Pereira. In these early days a banner and direct action would get huge press coverage, but perhaps less so today – but we are discovering more creative ways of direct action and have greatly expanded our campaign tools.

Working for Greenpeace

Greenpeace works on global campaigns. Even with the current restructuring, all of our national offices are committed to 80% of their work being internationally relevant, 20% of it being more locally relevant, so it remains a global organisation with global strategies. Although responsibility for the coordination of campaigns has been dispersed to different national offices, those campaigns are still global. Greenpeace's management of its campaigns and staff can appear ruthless. I left in 2003 after one restructuring. It felt like I'd lost my right arm. In terms of the threats and how to get things done, there isn't a compromise – and that thread runs through the organisation.

Over the years that I have worked with Greenpeace, many people have come and gone, and some people come to Greenpeace and still believe it's just a bunch of hippies that are all very loving, but once you get into the organisation it is fairly ruthless, especially on deciding on strategies. It is democratic, we put the proposals together, we have very hard discussions about strategies and different tools and we make a decision. Once that decision is made then that's the strategy. There used to be an annual general meeting at which the global campaigns were planned and agreed, but now the nature of the planning is reflecting the nature of the problems. We can put together a proposal at virtually any time and put that to the organisation and there's an internal mechanism to get things done. I am absolutely in awe of many of my colleagues, who are amazing strategists, and I have learnt so much in this organisation in terms of being strategic and understanding the great difference between strategies and tactics, the narratives and stories that you need, the messaging, the selection of different audiences and the messengers needed. Campaigning is essentially about telling a story that is compelling and interesting enough to change behaviour.

South America – campaigning in the developing world – development pathways

Now I find myself as the interim Campaigns Director for Greenpeace Andino, which is the merger of Greenpeace Argentina, Colombia and Chile into one regional office. They haven't had a campaigns director for around three years, so a lot of my work there is starting again, building strategies, developing more effective campaigns, capacity building, restructuring the office to make it more efficient. I'm thoroughly enjoying it because it's a very different culture, noisy and vibrant, a very different way of campaigning, but it's still Greenpeace.

Being based in South America, we are working in countries that are adopting the same development pathways that Western nations are increasingly abandoning because of the damage they cause, and there is increasing recognition of the economic, social and environmental value of truly sustainable development. In Chile they are massively expanding salmon farming – it is the world's second-largest exporter of salmon – but they are growing Atlantic salmon in the South Pacific where you don't even get Pacific salmon. The industry has to use massive amounts of biocides just to make the salmon survive, and often this doesn't work. The salmon farms in Chile use 400 times the amounts of biocides as salmon farms in Norway, to combat lice. Ironically the biggest company operating in Chile is the Norwegian firm Marine Harvest. At Isla Chiloe there was a massive disaster early in 2016 where there was a huge marine wildlife kill that affected seals, otters, seabirds, fish and shellfish. It turns out that this was caused by a 'red tide', which is a naturally occurring phenomenon, but which had been exacerbated this time by the dumping of dead salmon in the affected area.

Fish farming has had a huge impact on the local communities and damaged livelihoods, and we have been working with them to try and find more equitable solutions. It's not just a problem that's been created by fish farming, but a problem of the chosen development path and an attitude toward a route to economic development. In the Greenpeace fish-farming campaigns in South America, and it's not just looking at fish farmers, and we are not going around saying ban fish farms. What we are doing is highlighting the problems that this development path is causing. Similarly, in the same area, which is a migration route and a breeding and feeding ground for blue whales, the Chilean government has given permission for an opencast coal mine.

So in Chile we have some iconic confrontations and issues around which we can hang a story highlighting that this is not the right way to go. These are particular battles on which to hang this story of a choice in development paths – one that is economically, environmentally and socially beneficial or one that destroys environmental and human health and welfare with stagnating economies. Peabody, the world's biggest coal company, went into bankruptcy earlier this year, and yet Colombia has hooked itself onto an economic pathway using coal that is rooted in the nineteenth century. Greenpeace's narrative is really a story about choosing a more effective development pathway and trajectory, and so we are not just going to sail up in the Greenpeace ship and stop something. What we are doing is forming allies and relationships with local communities whose artisanal fisheries and tourism industry are suffering, where the working conditions of people on fish farms is appalling, so you actually begin to look at all of the issues around which you can have a discussion.

Getting to the root causes of environmental problems – change – disturbance – opportunity – the continuing challenge

There is a thread that runs through to where we are now, and we can acknowledge we have had some major successes. Our role in the 1970s and 1980s was to jump up and down and say 'there's a big problem and we need to do something about it' and to confront resistance to change. The 1992 Rio Earth Summit was when the term 'greenwash' surfaced. This was essentially an attempt by governments, as well as industry, to say 'Yes, we know there's a problem, but look we're fixing it.' In effect, business wanted to continue as usual but needed to address the concerns. One of the memorable quotes from that Rio summit was George Bush Senior saying, 'the American way of life is not up for negotiation', ignoring the fact that sacrificing the environment ultimately destroys economic development.

The Earth Summit did result in three major UN global conventions: the UN Framework Convention on Climate Change (UNFCCC), the UN Convention on Biological Diversity (UNCBD), and the UN Convention to Combat Desertification (UNCCD). So we have enshrined in international conventions an acknowledgment of the problems and therefore some kind of commitment to do something about it. But throughout, it comes back to attitude. It's not about spinning your story better, it's about going inside, understanding what the fundamental problems are. There is a prevailing attitude that regards the environment as cheap. The Earth Summit in 2012 was in many ways a huge disappointment with pre-scripted speeches and no real decisions made. So the problems are *recognised* but as there is little action they are getting worse and more difficult. Greenpeace looks at power structures – analysing where the power actually lies and then making use of disruptions to get the required change

Greenpeace is also looking at the driving force for action, which is the fact that mankind is punching through the nine planetary boundaries defined by the Stockholm Resilience Centre (see Figure 25.1). This became for Greenpeace the clarion call to action. Whether it is problems of climate change or deforestation we try to address the mindset that has created the problem and look at the power structures behind this. This is why when I refer to the campaigning I'm developing in Latin America it's looking at the development pathways and where the power lies. Clearly when you sit back, the power lies with government and corporations – because it is economic power, and that's where it lies classically – but underneath that there are different power structures. Which is why this change of strategy, and therefore change of tactics, will I think create bigger change. We continue to confront corporations and governments directly, but to we also go beyond this and at a deeper level tackle the bigger problems, of a greater scale and complexity. The ways of tackling these problems are different but we have a much bigger toolbox and range of options for ways that these are tried and tested. When you hang a banner the first time, it's news, when you hang it a second time, it's still news – but the twentieth time people are bored. The tools change, they develop, and we use them appropriately. We are now not just saying there's a problem – people know this – but we are pointing to what we can all do about.

FUTURE CHALLENGES

What are the most important threats to the marine environment, and what needs to be done about these?

We've had a lot of successes. We've stopped some of the obvious pollution – industrial waste dumping, toxic discharges, nuclear weapons testing – in the developed world but these are still big problems in developing countries and we have many battles to fight. Overlying all of this are large and complex global problems, whether it be overfishing, ocean protection, deforestation or climate change, because of the politics involved.

Climate change is the biggest challenge we face. Here we have two main problems. One is that historically it is an issue that was defined by politics and science, not by a narrative from environmentalists – so we end up following an agenda, not forging it. The other big problem is that the issue has been closely tied to economic development. Historically, carbon emissions are a measure of economic success. This is changing: for example, China has disconnected growth in carbon emissions from economic growth. In the early 1990s the science was in place and Margaret Thatcher commented that 'We have been given a warning the like of which we ignore at our peril.' In 1992 the science said we needed a 60% cut in global emissions in order to stabilise atmospheric carbon dioxide concentrations. In 1997 we had the Kyoto Protocol that agreed a 5.2% cut in carbon emissions from industrialised countries alone. In 2015 we finally had a climate agreement on the need to keep emissions below a level that would maintain the global average temperature increase above pre-industrial levels to 'well below 2 degrees Celsius and close to 1.5' – a level agreed by science as that above which we enter very dangerous climate change – dangerous not just to ecosystems and communities but also to economies.

But turning that ship around is a monumental task, especially when you're talking about developing countries like the ones I'm working in, which are desperate to get out of poverty and one of the ways they can do that is by having cheap fossil-fuel energy – although increasingly renewable energy is proving to be cheaper than fossil-fuel energy and is certainly more sustainable. But it is in the area of fossil fuel where challenging the power structure becomes an issue, because it is the fossil-fuel energy industry that has developed a political power structure in order to maintain its business. For climate change we know the technological solutions are there, we know there's an enormous growth in alternative energy – it's unparalleled – whether you're talking wind or solar energy, both are growing massively and they present an investment opportunity for the future. But what is required is the political will to scale that up, against which you have incumbent corporations. From a Greenpeace point of view, we have an understanding of the problems, we're working with communities and providing the solutions. In the 1990s in London we doubled the number of houses with solar panels simply by putting these on one house, and now they are commonplace.

What are the barriers that make progress harder than it should be?

Deeply entrenched attitudes and corporate thinking providing a barrier to viable solutions is a recurring theme in this chapter. Getting the solutions and using the technology – but I'm not sure how much this is a 'change'. We focus a lot on solutions, as the environmental movement and not just as Greenpeace, but the fact is that they are real and it *has* created the change. What strikes me is that what you can read coming out of the Paris climate change agreement (COP 21) could have come out of a Greenpeace leaflet in the 1980s. To a large extent it's the same with the solutions: we've talked about these for years and now they're there. We jokingly talk about the T-shirt that says 'I told you so', but what is it that can create that change? Why do we have to go through all these battles, from the resistance, going back to the interview I had for my MSc, why do we have to keep going through these time and time and again?

What are the most interesting and promising new approaches?

A recurring theme in the chapter has been how Greenpeace has been developing a variety of different tools – same principles but different tools – demonstrating that our uncompromising engagement with corporations doesn't change but the tools we use have. A couple of examples illustrate how we have used YouTube videos to achieve important and successful outcomes. In our forestry campaign against palm oil the YouTube video had someone working in an office on their break, they opened the KitKat and they bit an orangutan finger, blood spatter. It was horrific, but the outcome was that Unilever couldn't move fast enough to move away from palm oil. Similarly with Lego and Shell – the London office decided that breaking the link between Shell and Lego would be a monumental shift, so produced a YouTube video called *Everything is NOT Awesome* linked to the Lego movie. Within a couple of months of that video coming out Lego broke away from their relationship with Shell. Those kinds of shifts are examples of having the capacity and vision to seek to greater change.

When you retire, what would you most like to have achieved for marine conservation?

Yes, I will retire, but what does retirement mean? I'm really fortunate, in that I really enjoy what I do and it is fulfilling on a very personal level. I have been able to travel the world and work with wonderful people. But on the other hand I do increasingly value my own time. I have a 120-year-old Dutch sailing barge which has taken me twenty years to get back to sail again and a wonderful partner, and she and I want to settle somewhere to go and watch the sunset. What I can envisage is that I will keep on doing what I do, but in two or three years I will do less jumping in front of dump ships and more writing.

Are you optimistic or pessimistic about the future?

I am hugely optimistic, much to the annoyance of a lot of people. I'm optimistic that we're going to get through, though I know that a lot of bad things will happen. The book that most closely comes to the way I've been thinking is Paul Gilding's *The Great Disruption* (2011). We are locked in to hundreds of thousands if not millions of people dying because of climate change, so to say I'm optimistic with that scenario sounds quite arrogant – but rather like the Second World War we will survive. But the scale of the cost will depend on how quickly we get our act together.

CHAPTER 20

Chris Rose

Chris currently runs a communications consultancy, Communications Strategy Ltd, with his partner Sarah Wise. They work for a very wide range of organisations including UNICEF, the Home Office, 38 Degrees, the Baltic Environmental Forum, the police, NGOs and businesses. Since 2001 they have also developed the Fairyland Trust, which engages families and children with the magic of nature, using methods that are now being widely copied. This was prompted by a detailed analysis of what was required to help the three- to eight-year age group engage with nature. Chris grew up in the suburbs of London and was inspired by David Attenborough and Peter Scott. From an early age he was very interested in watching and drawing birds, an interest encouraged by his father and a family friend – which saw holidays engineered to places like Christchurch Harbour. A move to Ruislip in north London prompted him to get involved with the site management work at a Local Nature Reserve and this led on to his first campaigning, which successfully stopped development on flower-rich meadows in the Colne valley.

He had a very clear desire to do the University College London (UCL) MSc course in nature conservation, and so at A-level he did sciences despite his school wanting him to try for Oxford to read English. He went to Aberystwyth University, where he read botany and zoology. Whilst on the UCL MSc course he and others set up the British Association of Nature Conservationists (BANC), including the journal *ECOS* so they could publish articles which challenged the way nature conservation was being handled at the time. The NCC Director was so annoyed by one of his articles that he declared that they would never employ him; ironically he'd already been one of their summer wardens. He did PhD research on lichens at Chelsea College and started part-time work for the Herts and Middlesex Wildlife Trust, where his conviction about popularising conservation led him to start the London Wildlife Group, which became the London Wildlife Trust. His first media event was with the Greater London Council's Ken Livingstone, who put up a bird box at County Hall. He has worked for other NGOs including Friends of the Earth, WWF International and Greenpeace, and started and ran a media charity – Media Natura – helping NGOs on communications projects.

From the very earliest days and throughout his career he has identified clear needs and helped create organisations and programmes to better deliver effective conservation. His particular expertise has been in effective campaigning and communication, and he has written books such as *The Dirty Man of Europe* (Rose 1990), *The Turning of the Brent Spar* (Rose 1998), *What Makes People Tick* (Rose 2011) and *How to Win Campaigns* (Rose 2012). This chapter describes Chris's work on campaigning and communications, essentially applied psychology and key lessons that have emerged from this.

CONSERVATION

How do you describe or frame your beliefs in conservation?

My reference points to conservation were developed when I was young and have evolved a bit over the years. From the outset, growing up in the suburbs of London, I had a love of looking for wildness and was excited by seeing golden plovers flying over in the winter. That was magical – for me they were emissaries from real, wild nature. I worked on issues such as acid rain, which was huge but regional; but since the late 1980s we have been grappling with human-induced climate change, and nature can't escape its global impact. I find the Bill McKibben view that 'everything is anthropogenic now' (McKibben 1989) a very depressing thought.

My approach is that when you re-engage people with nature it makes their lives better, and it is more likely that they will want to protect it. That can be profoundly encouraging. The big challenge is that people find it very difficult to read nature. By analogy, if one compared this to asking people to conserve books without being able to read, you get a sense of the problem. There is a challenge to develop people's nature literacy. Finding ways of re-engaging people whose lives are very divorced from nature, and equipping them with ways to be 'good at' nature, is a huge task. But it's essential if they are to have the confidence and ability to take part in nature conservation activities. I've often thought that one of people's main touch points for 'marine' conservation is their bathrooms: it's where they take their marine souvenirs and art, because of the connection with water. It is the equivalent of the domestic garden and offers great potential to marine groups.

There are lots of ways of defining marine conservation, conventional ways including biodiversity conservation, protecting the natural functioning of ecosystems, their natural biogeochemical cycles and functions of the seas and oceans. The way we understand the effect of pollutants is also very important, with, for example, nutrient cycles. One thing that wasn't realised at the outset of systematic work on marine conservation was the massive modifications to marine habitats and species that had already taken place over hundreds of years; this was described in Callum Roberts' book (2007). For example, from here on the north Norfolk coast wild oyster beds extended to Denmark until steam-powered trawlers dredged them away. I think the last oysters were landed here in Wells in 1906.

What are the differences between now and when you started?

It now seems inconceivable, but in the 1980s when I told a friend I was going to work as a campaigner for Friends of the Earth (FoE) he told me that I was consigning any hope of a [scientific] career to the graveyard. That doesn't happen now. One of the main reasons I wanted to go on the UCL conservation course was that it was one of the few recognised ladders into conservation – now there are hundreds of them, through courses covering a huge range of topics, not just natural sciences. The United Nations Conference on Environment and Development (UNCED, the Rio Earth Summit) in 1992 led to a huge proliferation of new institutions, programmes and projects – for instance Agenda 21 – and businesses getting seriously involved in 'sustainable development' (although that concept grew out of the 1980 World Conservation Strategy). Conservation principles have been mainstreamed, in particular in business. This wasn't the case in the 1970s and 1980s. It is interesting that because a great deal of this pre-dated the internet the younger generation simply take it for granted. Most of the story isn't on Google.

How is marine conservation different from terrestrial conservation?

I'm not an expert, but I got some insights into this in 2007–2008 when we undertook a major study with Natural England (NE) and Karen Mitchell for their Marine Conservation Zone (MCZ) programme (Rose *et al.* 2008). NE wanted to know how best to engage the public with marine conservation. The sea around the UK is perceived to be cold, murky and rough. Culturally, it is heavily associated with danger in people's minds, something to be overcome and survived. We found there was a huge difference with the land in that very few people have a familiar or liking-based relationship with the sea. This is a massive problem. There is a cut-off – an event horizon. Because the sea looks like it's empty, that is the default expectation. People say, 'well it looks horrible' – and then the sun comes out and the wind dies, and then they say it looks beautiful. Their reference points don't include there being much in it, and it is often muddy and they can't see things. People have a relationship with the beach and the coast, not really with the sea, and they have a mental relationship with the sea through proxies like art, literature and imagination. Birdwatchers have a different relationship, and divers and fishermen do have some understanding of marine life – from good to rudimentary – because they see it directly and realise there are 'places' and habitats under the sea. I often describe this to people as trying to do nature conservation on land as if they are in a permanent fog where they can't visit or see the sites. So in our seas it is not easy to tell a story and there is little emotional connection.

In the warmer countries, where the water is clear and warm and where you can see things, there is a very different relationship. In these countries people are much more attracted to the sea, and they can go snorkelling and see the things living and it is much more like terrestrial conservation. Our work on MCZs highlighted a huge experiential difference between these warmer and sunny seas and our own, and this has a profound effect on the way we do conservation in practical terms. Not least of these differences is that marine conservation in northern seas is still essentially a preserve of professionals and marine scientists rather than a popular movement. There is a huge enthusiasm by the public for proxies of the marine environment like aquaria and TV – but this is a mixed blessing, because these are like an unsorted museum collection or alphabet soup where relationships and context are often missing. We found in this research that people couldn't necessarily put a size to the things they were seeing, which is simply not generally the case on land. Another problem is that many of the species – fish, birds, cetaceans – move.

For the NE research we used the Values Modes model (Rose 2012) to assess how people's psychology affected their understanding of the sea and how it might be protected, and this was one of the best examples of demonstrating the power of a Values Modes approach. We had sufficient funds and we were able to ask some open-ended questions, for example 'What do you imagine being at the bottom of the sea?' and 'Can you name things under the sea?' Divers, fishermen and others with experience were excluded, but the reality was that the vast majority of people had no idea of what was under the sea. But when you asked them whether they thought there was something 'worth saving' you got massively different answers based on the psychology of the respondents, not knowledge. We had segmented the responders using the Values Modes methodologies. The so-called 'inner-directed' pioneers segment responded much more positively because they thought things could be done, while the 'outer-directed' group, essentially the middle 'sway group', were responsive once you showed them images, but they wanted more, more proof, in particular to experience it. We concluded from this that if you wanted to engage them then you'd need an experiential device, and we started working on this but it was never used. The settlers, the 'security-driven' group, were most naturally pessimistic about there being anything worth saving, but if you could show that their region had something distinctive, they instantly wanted to protect it as part of their identity. We anticipated that this project's outcomes might cause nervousness with government and the agencies, and so we spent a good deal of effort writing this up (Rose et al. 2008). Shortly afterwards NE was told to stop attempting to influence or 'educate' the public about nature, which is tragic.

THEMES AND STYLES OF MARINE CONSERVATION – WHAT WORKS?

Why did you move from a classical nature conservation training into the communications and media side of the work?

The honest answer is lots of accidents. These take you in a different direction, but also my personality and my father, who was an English teacher and always encouraged me to question things, which for example used to get me into trouble at school for writing *Private Eye* style articles on school business. At UCL some friends and I started the nature conservation magazine *ECOS* so we could publish views which questioned mainstream conservation thinking. Early on I got to meet and talked many times with Max Nicholson, who actively encouraged us. He was one of the founders of environmental thinking between the 1940s and 1970s, and he and others founded the World Wildlife Fund (WWF) and set out their *Morges Manifesto* in 1961. He had a great influence on me and he encouraged me to 'shake things up'. In his book *The System* (1967), Nicholson outlined that one way of solving problems was to create new mechanisms. From *ECOS* to this day that is an approach I have taken. Pollution was a big issue in the 1970s and 1980s and the UK was known as the 'dirty man of Europe' (Rose 1990), and my early work as a campaigner at FoE took me down this path and away from the conventional nature conservation route.

Communication

I initially got interested in communications because my dad gave me two books about marketing, *The Shocking History of Advertising* (Turner 1953) and *The Hidden Persuaders* (Packard 1957). These explained how advertisers were using language and various techniques to communicate their messages, effectively deconstructing the meanings of the way words were used. Today the public deconstruct the meaning behind things in a sophisticated way all the time and they are not the idiots just 'accepting of the narratives' as some in the commentariat seem to think they are. Effective communication is very much more complicated than people think. It is not just a question of 'framing messages' verbally; most of it is, for a start, non-verbal. It's not about 'changing people's minds' but about equipping them to reach their own conclusions that lead to the action you want.

I learnt a lot from marketing people in different organisations I worked with. At WWF International they were excellent at public relations and came from high-flying advertising agencies. Greenpeace were massively good about using the power of myth and brand, and in the German tradition of turning this into action. One of the first such actions in Germany was led by the German Greens and Joseph Beuyss. To highlight our questionable dependency on the car, he and his art students dug up VW's headquarters car park and planted trees. Action-art using symbolism through doing, to highlight things that were wrong. In Greenpeace, all of those traditions have been mixed up with the 'bearing witness' principles of the Quakers. This was a religious tradition of seeing something was wrong and saying so. These things get mixed together, and although people don't know where they have come from, they can get good at using them (Rose 2012).

We started Media Natura in 1988, and it evolved by accident. I wanted to make a film about WWF's 25th anniversary but the company we asked to do this said that my budget wasn't big enough. They suggested that we ask media companies to donate their time and expertise. We got the film made and then these companies asked 'what's the next project?' and this developed into Media Natura. So it was an iterative and empirical process rather than having any grand design. I'd learnt from some earlier work how little NGOs knew about the different areas of media skills, and working with a wide variety of NGOs we were able to explain to them the wide range of different technical skills that were on offer and match these effectively to the NGOs' needs.

I have found the 'framing' work of George Lakoff, an American cognitive psychologist, very helpful. His view is that an individual's experience of and attitude towards sociopolitical issues are influenced by being framed in linguistic constructions. His book, *Don't Think of an Elephant!* (2004) explains this. Perhaps the best example of this is of using the word 'investment' as opposed to 'tax' in economic discussions. 'Investing' is framed as positive and good: so the more of it, the better. 'Tax' has been framed as bad (although this might now be changing), so the less of it the better. So he or she who first frames the debate, tends to win.

Language often works subconsciously because of the way it is framed – a tax burden or tax relief as if it is a thing that can be cured. If you can create a narrative around a frame, say red tape or regulation, and politicians do this all the time, then it is a powerful way of communicating. There are lots of ways of identifying unconscious thinking processes like heuristics and values which are used by professional communicators. Lakoff argues for a new enlightenment which puts forward the view that we should train people when they are at school in how people think, like literacy or numeracy, because otherwise there is a big divide between the people that understand this and use it and the rest of the population. I agree with him, and it is particularly important for people in conservation who are communicating campaigns or policies aimed at behaviour change to understand framing. My experience of over thirty years is that the commonest reason that organisations fail with their communication, is not that there is a lack of expertise, but there is a lack of understanding about communication amongst the people who call the shots, controlling budgets and working at higher levels. So whilst many organisations accept as normal a high level of professionalism with staff management, finance and governance, communication is often dealt with in a less than professional way: nice to have, but not essential.

Interestingly, when Max Nicholson was looking to take nature conservation forward in the 1950s he framed it as very much a 'scientific' exercise, because in post-war Britain there was a strong belief in science, thanks to recently invented vaccines, inventions like penicillin and radar. The scientific approach provided a constituency of people to take this forward. Perhaps the worst example of a poor decision on framing was over climate change when at the Toronto conference the Canadian hosts wanted a Law of the Atmosphere, analogous to the Law of the Sea. It was decided, almost by accident, to get the World Meteorological Organization to take it forward, which put the communication in the hands of climate scientists. This led to the International Panel on Climate Change (IPCC), which set the whole debate around the science of climate change, making it a 'scientific question' and thus open to endless doubt and the ever-present need for 'more research'. If this had gone to the lawyers then their legal frame would have set the debate about emissions in a completely different way, and it would also have been easier to engage the energy sector. I reckon that single decision on climate change set us back by twenty years.

Given the influence of psychology on communication, do you find it strange that it plays such a small role in helping deliver conservation?
That's one of the points I'm making, but there are lots of different types of psychologist. My sense of this is that environmentalists don't need the academic psychology so much as the applied psychology which is used in marketing and advertising. Interestingly you often find this expertise in the fundraising side of NGOs rather than in the campaign teams. A good example of this is the Values Modes model I've already mentioned, where lots of empirical data on huge samples show that this works. Academic psychologists looked at this model years ago on the basis of thinking that it was based on Maslow's hierarchy of needs; this is a theory in psychology concerning motivation and comprising a five-tier model of human needs. This research was claimed to show that Maslow's model didn't work, and so it was discredited – and yet empirical studies on the massive samples that companies can run for their advertising show that it does work in practice. If something works, then companies don't really care whether they know why it works. By analogy one might compare this to the recent discovery of the Higgs boson and running a nuclear reactor; we didn't need to understand the Higgs boson before we built and ran nuclear reactors. Fundraisers tend to be more pragmatic than campaigners: for example in Greenpeace it has been the fundraisers more than the campaigners who have used Values Modes successfully.

Interestingly, newer NGO start-ups, like for example Global Action Plan, have based their actions around psychology, and there are some particularly good examples in the corporate social responsibility (CSR) field where governments or companies actively want to get behaviour change amongst employees because it saves colossal sums of money, for example on energy conservation.

Campaigning
Campaigns are rather like John Mortimer's comment on the evolution of the horse: 'a long and tedious process'. My advice to people would be make sure they try everything else first, because campaigning is one of the worst way of trying to get things done.

Always try more direct routes first, asking for what you want, buying or persuading or educating if that is possible (and quick enough): you simply couldn't live in a society where everything had to be achieved by campaigning.

Having said that, I have written a couple of books on campaigning (Rose 2011, 2012). When I started out there were a small number of NGOs who were known for their campaigning, but more recently many people have been using campaign techniques in other sectors, including businesses. Similarly there are lots of people who study and research campaigning, and a variety of university courses. Once you have been doing campaigning for some time your transferrable skills are mainly about communication and trying to persuade people that this is the right thing to do and how change can be applied. My work with the police on the perception of crime, or with the Home Office on their drug policies, demonstrates that the transferrable skills extend well beyond conservation NGOs. For these public organisations, campaigning is often about social change, and there is a whole industry on 'social marketing'. For NGOs, campaigning is often about power relationships, changing corporate approaches to the way they operate, using a variety of approaches from affecting consumer demand to regulation and product substitutions.

One campaign that certainly made a difference to the way Greenpeace thought about how it went about its work concerned the Greenfreeze fridge. This was a German campaign which focused on the issue of the replacement of the ozone-depleting CFCs with highly potent greenhouse gases (HCFCs) in a wide range of refrigeration equipment. The industrial groups such as ICI, Dow and Dupont had in effect a hegemonic position where they controlled the supply of chemicals, the supply chain, the training and equipment providers and had a huge influence over 'technical' regulations. Their view was that there was no affordable substitute for HCFCs, which they just happened to make, and fridge manufacturers such as Bosch went along with it. Then a couple of refrigeration engineers contacted Greenpeace to tell them that there was an old factory in what had been East Germany, that produced a type of fridge that could be used with non-greenhouse hydrocarbons (this was shortly after the Berlin Wall had fallen). Greenpeace Germany pulled together funding from a reunification investment agency, who collaborated with technicians to build a fridge – the Greenfreeze – which demonstrated that this could be done, and it completely transformed that situation. Greenpeace also organised the market demand, working with a precursor of today's online companies, in the shape of a large catalogue-based firm selling household goods, to create pre-orders. The campaign went way beyond advocacy or protest, creating a new product, demonstrating a new technology and building a market. For Greenpeace it showed us that we could focus more on promoting solutions. Hydrocarbons are now standard in domestic fridges throughout most of the world.

Atlantic Frontier campaign

One of the campaigns I loved most at Greenpeace was the 'Atlantic Frontier' campaign in the late 1990s. We had spent three years developing a brand model, which included things like giving hope by doing things, transforming industrial society, the need for renewable energy, the defence of nature and our positioning: our supporters saw and wanted us to be 'out on the edge', which came from our marine campaigns. After the Brent Spar we started to develop the Atlantic Frontier campaign, incorporating our brand values and also the momentum from the Brent Spar, not least with oil industry. I was visiting the Hebrides on holiday with Sarah and walking along the spectacular shell-sand beach at

Figure 20.1 Greenpeace activists occupy Rockall in 1997 as a part of the Atlantic Frontier campaign – which resulted in the UK recognising the Habitats Directive out to its EEZ. Source: Greenpeace

Luskentyre, thinking about the fact that out to the west oil development was planned (known as 'west of Shetland'). Out there lay places like St Kilda and Rockall. The natural beauty and sanctity of the area – and the film *Local Hero* was still resonant – seemed to make it the perfect setting for a campaign to try and turn back oil development.

The campaign included a variety of ideas, including marine parks for places like Rockall and St Kilda, carbon logic – if you can't burn it why take it out? – and renewables, ocean governance and wildlife conservation. One of our actions occupied Rockall, and we staked our claim for the island alongside the competing countries who claimed it (Figure 20.1). They wanted to win rights to oil and fish. Rockall is an amazing place: when you stand on it you have a 360-degree view with no land in sight, water on every horizon. At this time a deep-water coral community (*Lophelia*) was discovered on the Darwin Mounds and we pressed the government to protect this, and as they wouldn't we and others took the UK government to court to ensure that the Habitats Directive was applied out to the extent of the UK EEZ, and in the end they lost. This was probably the most significant outcome of the campaign because the UK government has ever since actively supported this and many designations have followed. By coincidence, Adam Markham and I had made the first call for a Habitats Directive in *ECOS* in 1985 (Rose & Markham 1985), following a meeting where the issue of directives for species other than birds was aired by Nigel Haigh. Nigel later told me that he saw me as the father of the Directive – which made him the grandfather. I got a footnote in one of his books on EU environmental law, which is probably about as far as I will get in history!

Brent Spar

After the Rio Earth Summit (UNCED) in 1992 there was a wave of awareness about the environment, and a lot of political nods to sustainability being important. There was a spate of government activity, and things like giving departments new names. The NGOs were also riding this wave. There was widespread assumption that we'd somehow 'won', but in fact nothing essential had changed. I had made myself rather unpopular by openly doubting the value of the UNCED conference and refusing to send any of our Greenpeace UK staff to it. At the time I was programme director at Greenpeace UK and at a discussion of our annual plan with the board I put forward the view that we should be looking for actions that would puncture this bubble. I asked if they were up for it and one responded, 'If you mean do you want us to feel the government's collar, then the answer is yes.'

Then one morning Paul Horsman came up to me showing me a diagram of this monster thing – the 14,000 tonne oil storage vessel called Brent Spar – and said that Shell had a licence to dump it offshore and that it would be towed through the North Sea when the NSM meeting was taking place in Esbjerg, Denmark, in 1995. Although it didn't fit neatly into any of the Greenpeace work silos it did chime with the overall aim of protecting the sea, and it soon became clear that the Dutch and German Greenpeace offices were very interested in this for all sorts of different reasons, and our toxics team jumped at the opportunity – which is why that issue arose in the narrative. We saw this quite simply as a way of challenging the existing OSPAR regulations which supported dumping as being the 'Best Practicable Environmental Option' (BPEO). My domestic UK motivation was to show that despite its claims, the government was still doing 'business as usual'.

The campaign took over six months to plan, but we only came to understand all the ramifications afterwards. The UK Treasury and Shell saw it as a way of minimising a loss to the exchequer. The Ministry of Defence saw it as a test case in relation to dumping large parts of nuclear submarines. There was massive tension between the Department of Trade and Industry oil people in Aberdeen and MAFF over whether this should be allowed. So we ran our action using lots of Greenpeace resources and several ships, and gradually it took off and became all-consuming to such a degree that virtually nothing else was being done; to all intents and purposes it became like a war. Other things started to happen and we discovered that a German church group was boycotting Shell garages and the German police had decided not to buy their fuel from Shell. Soon it became clear that whilst we were the central actor, what was happening was outside our control and it seemed to capture the public's imagination. One German politician neatly captured the reason for the public reaction (it was breaking a norm) by pointing out 'This is why we have laws to stop people dumping cars in the village pond – Shell should not be allowed to dump its rubbish in the sea.'

When Shell backed down many of my colleagues were overjoyed – we'd won – often we didn't, and it was a huge achievement. But I was reluctant to join the celebrations, as I thought it was going to have huge repercussions. One of these focused on whether it was right that an NGO should be able to thwart the will of the government and the fifth-largest company in the world; and we did see an organised kick-back from government until the next election.

Looking back, it did change the way the sea was viewed in relation to dumping, and the oil companies did change their decommissioning plans. It also changed the way large corporations viewed their social licence to operate, and at least one marketing expert said it originated the term 'corporate social responsibility' (CSR). Many company execs thought 'There but for grace of God', and the oil companies in particular were glad they weren't in Shell's position. Corporate leaders saw that events of this sort could directly threaten their positions, and Greenpeace was added to many company risk registers.

Many people, scientists and regulators in particular, were deeply unhappy with the outcome because it shook the consensus on what was reasonable to dump in the sea and their world view, and in effect undermined their positions. It fundamentally undermined the consensus on BPEO, which was (somewhat ridiculously, seeing as BPEO was inherently unscientific and a science–economic fudge) portrayed as undermining science itself. Subsequently a senior government scientist pointed out that BPEO wasn't a scientific construct but a pragmatic value judgement based on many things including politics, a view that many politicians in the UK seemed to ignore. It was very clear that European politicians didn't take this view. We all got to resent the huge amounts of time it took after the event, but Greenpeace was emboldened by the result to try more ambitious marine actions. I wrote the book on this because I knew that once it was over there would be lots of people coming back to us to ask what had gone on and I would have forgotten (Rose 1998). The ramifications are still playing out in 2017, as the debate about how best to decommission oil rigs is still going on.

THE FUTURE

What are the most important threats to the marine environment, and what needs to be done about these?

One of my main concerns is lost biodiversity, because it's effectively irreversible as far as we know, or at any event it's impractical to do anything to bring back lost biodiversity, as there is just too much being lost. Earth has been an evolving system accumulating information content ever since life developed, and the energy requirement alone needed to re-evolve what is being lost would simply be impossibly large, even if we had the technology. We are de-evolving the planet.

There is also the loss of naturalness and continuity of connection and experience. So even if you could reconstruct it, and even if you couldn't tell if you walked through it if it *had* been lost, because of shifting baselines, you'd have lost the meaning and cultural connection and part of your soul, and I don't want our children to lose this. It is what Mike McCarthy wrote about in his book about the loss of abundance of birds, *Say Goodbye to the Cuckoo* (McCarthy 2009). I remember in the 1960s seeing apparently endless streams of starlings flying to their roosts in central London on winter evenings: a river of millions of birds. That doesn't exist any more, and the baseline has shifted. Now we think that a murmuration of a thousand starlings is noteworthy and is worth putting on Twitter. I have spent lots of my time on various aspects of climate change, and for me it's the saddest and most heart-breaking thing. I don't want to look at disappearing glaciers and melting tundra: it's just too awful. I find seeing it happen profoundly depressing. Instead I focus on things where you can make a difference, in particular engaging children in nature.

What are the real, hidden problems, the barriers that make progress harder than it should be?

One of the things that limits change is the lack of political will to regulate, to rule things out. Now we know that there all sorts of arguments, from economic to technical, that make solutions perfectly possible in practice – but they are not perceived to be politically possible. If you start digging into why this is, you often find it is based on outdated perceptions and assumptions that are not founded on very much and certainly no rational answer. Another area is the lack of scientific literacy among politicians, especially in Britain. Many of them really don't understand even the basics, and many of them don't understand statistics, even ideas like the mean and mode; neither do journalists. This continuously leads to a systematic failure to act on evidence simply because they don't understand it. In NGOs there is often resistance to trying new things, approaches that work, because they like doing the same old things, which they would like to work but don't.

With nature conservation the big unexploited area is the lack of realising the potential of linking nature to the great diversity of cultural connections, for example with wildlife gardening. The work that Richard Louv is doing, re-engaging kids with nature in the USA, is not really happening here; his book *The Nature Principle* (2011) covers this. There is an almost endless list of these barriers.

What do you think about innovation in the conservation world?

It was Max Nicholson who taught me about creating new mechanisms to solve problems. Whether you call this innovation or capacity building it is a thread that has run through my career and has included BANC and their magazine *ECOS*, the London Wildlife Trust, Media Natura and most recently the Fairyland Trust. These have arisen from a mixture of opportunity, timing and working with people to identify and meet the needs of the time. It seems to me that this is fundamental to the evolution of conservation, and that you need constant development of new entities to maintain momentum and to force and cope with change.

A report I read early on by Tom Cairns, which he did for his MSc on the environmental organisations in Bath, explained for me how the different organisations filled particular niches and appealed to different values groups. The local Campaign to Protect Rural England (CPRE) was one of the most outspoken and interesting because it was composed of high-ranking civil servants and judges who were much less deferential to authority. The notion that conservation organisations occupy space or territory that is inviolate is a crazy idea, and it would be like one supermarket saying there should be no other brands of supermarket. Organisations like Plantlife, Buglife, Froglife introduce a new dynamic, and it's not just slicing the same cake. It is also easy for conservation organisations to get left behind. There is a field at Glastonbury called the Green Futures field, and it is charming and idiosyncratic, but a few years ago many of the organisations there were still promoting the need for renewable energy, when that revolution has been well and truly under way for years, and it is now mainstream. The NGO community is always at risk of pursuing its favourite cause and not noticing that others are actually acting on it.

Innovation in the way we think and deliver conservation has often come from outside the NGOs and especially from new start-ups. For instance the inventor of the wetsuit, Jack O'Neill, who died recently, was a businessman and an environmentalist, and his invention and ongoing financial support enabled many good things to happen in marine conservation. For long-standing NGOs there is a problem of the existing mind space, often occupied by the larger organisations that set the wider agenda, for example through having nature reserves. I think one of the big missing pieces in the way the NGOs work is in developing stronger links between cultural activities, lifestyles and conservation. One example is the vast number of people who enjoy gardening. People like Miriam Rothschild and Chris Baines pioneered the ideas of nature-friendly gardening and growing wild flowers, and this approach is now widespread. There are, however, many other ways people live and spend their leisure time and money which are opportunities for developing new ways of engaging with nature through travel, leisure activities, food and clothing. By taking this approach, NGOs could give people more agency in their everyday lives to support nature conservation.

Real innovation, doing things really differently, is a much rarer process and often involves completely new actors. Often individuals are instrumental in this – for example Greenpeace grew as an international organisation because of David McTaggart's business-based approach and because he wanted, above all else, to win. Without him, and some would say without his hunger for power and influence, it would still be a 'hippy' network. Likewise the influence of the commercial people in WWF enabled it to grow and develop like a business.

Recently the aggregation entities like Avaaz and 38 Degrees have developed the power of social media and the internet to reach across populations to gather new constituencies across the world. They are a major innovation – in effect 'campaign service organisations' rather than NGOs. They are online brands that are organised to achieve change: that's the service bit. It is also clear that if you want to get the scale of change needed to solve many issues, then mainstream organisations, often both business and government, need to get behind the change. Getting issues mainstreamed has been a major part of what NGOs have achieved over the last fifty years.

I think the most important person in the world at the moment is Elon Musk, the CEO of Tesla, and I hope he realises the challenge of electric vehicles and survives. I suspect that one the greatest areas for change will be in Asia, driven by highly educated young people. Social media is like old media, only organised differently, with the producers and consumers all mixing up different roles and functions in a new ecosystem – but it is settling into patterns of use and people are coming to terms with it.

When you retire, what would you most like to have achieved for conservation?
I'd quite like to retire from being a campaign consultant, but I'd need a large piece of land that I could spend time looking at and doing things on, which some people might not see as retirement. Being a campaigning consultant most of the time means running workshops and looking at the computer, and doesn't involve going into the field looking at nature. I've not really got any thoughts on legacy. When I go to London I see the London Wildlife Trust projects I was involved with which are still there, and that is satisfying, but by comparison lots of the campaign stuff is ephemeral and I don't remember what was achieved.

Are you optimistic or pessimistic about the future?
Optimistic that the right things about conservation will become totally mainstream in a generation, or at most two. Pessimistic that there won't be that much left to save by then.

CHAPTER 21

Peter Barham

Peter is currently an environment consultant who has his own company, and amongst other clients he is engaged by the Crown Estate in his role as facilitator of the Seabed User and Developer Group (SUDG). He also chairs the Welland River Trust (WRT) and the Solent Forum. Peter grew up in a village near Cambridge and as a boy he had a very keen interest in natural history and freshwater life. His interests led him to study animal behaviour, biology and psychology at the University of Keele. A key point in developing his thinking about conservation was making the links between animal behaviour at the population scale and the importance of the environment in influencing this.

He started to work with Anglian Water in 1978 as a technician doing routine river sampling for macroinvertebrates and all the standard ecology that went with this. He'd been working on the rivers of Norfolk and Suffolk for three years, producing lots of reports on the impacts of water quality on freshwater invertebrates, when another penny dropped. Although he spent a great deal of time producing reports on biological water quality, engineering projects were canalising the rivers to enhance their flows and severely destroying rivers' ecology with little or no recourse to the impacts on the wildlife of rivers. This realisation took him away from simple water quality monitoring to talking and working with engineers to see if there were other ways to achieve their goals.

After a further four years as biologist and then fisheries scientist in the Welland and Nene catchments, he chose to join the newly formed National Rivers Authority (1988) at the time of the privatisation of the water industry. With increasing recognition of the importance of conservation of habitat as well as water quality, he was made Conservation and Fisheries Manager of the Great Ouse, where for the first time the importance of ecology and conservation became a real consideration in planning any river engineering work. In 1991, he was made responsible for managing conservation for the whole Anglian region, which included the largest flood defence and water resources capital programmes in the country, including the whole of the low-lying Anglian coastline. During this time he worked on capital engineering on coastal flood defences, Broadland, the Ouse Washes and water abstraction proposals, and, through working closely with external conservation bodies, he pioneered environmental appraisal and valuation procedures to ensure that engineering projects took effective account of environmental protection needs.

Throughout this time new approaches were constantly being sought for addressing flood defence work, which would also protect or even improve the environment, and with the increasing importance of conservation legislation there was an increasing need to make sure that any work would be developed and supported by conservation bodies. In 1996, when the Environment Agency (EA) was formed, he was made responsible for managing all the EA's activities in the Humber estuary with the aim of producing long-term flood defence and environmental strategies for the whole estuary across three regions (Anglian, Yorkshire and Midlands).

In 2001 he joined Associated British Ports (ABP) as their Environment Manager, a role which developed into Sustainable Development Manager, where he was responsible for seeking environmental approvals for major port developments, developing environmental management systems for the company's 21 ports and helping to develop corporate social responsibility (CSR) reporting. Following purchase of ABP by private equity, he left the company in 2009 and set up his own consultancy, mainly working with SUDG and, in particular, working on the Marine Bill (which became the Marine and Coastal Access Act in 2009). Since that time he has spent much of his time working with marine industries, regulatory bodies and conservation organisations to try and find better ways of delivering legislation and improving the environment while supporting economic growth.

This chapter reflects on Peter's perspectives of environmental management and protection from his time working for a government agency and as a developer when looking at major developments and, in particular, how negotiation and consultation is critical to deal making. Consequently, it also covers how throughout his career he has sought to work with stakeholders on projects, and his more recent work with bigger partnerships to help deliver sustainable development.

CONSERVATION

How do you frame your ideas around marine conservation?

My view of what marine conservation is has changed over the years, and my current feeling is moulded by recent work with the Seabed User and Developer Group (SUDG), which actively promotes sustainable development of the marine environment – increasingly referred to as Blue Growth – where the importance of the environment is fully recognised. In this context marine conservation is something that should happen as a result of sustainable development and management of the marine environment. There are some people who suggest development shouldn't take place in the sea; in my view this is not an option, but it does means that development must be done in a sustainable way that protects the environment using the wide array of methods at our disposal to mitigate or compensate for any environmental impact.

The interaction of science, conservation and environmental protection

It became apparent to me in my early work that just measuring things – doing science – simply wasn't enough to achieve the changes needed. About the time of the Wildlife and

Countryside Act (1981) I'd been working on the rivers of Norfolk and Suffolk for a few years, producing lots of reports on river invertebrates to assess water quality, but at the same time river engineers were trashing the ecology through canalisation work in the name of land drainage and flood protection. There were a number of us who questioned the point of measuring water quality if you're not looking after the habitats of the rivers. All we were doing was measuring how engineering was disrupting the ecology and making things worse. There was another aspect of this in the early days of the National Rivers Authority, which was known as 'hobbying', where staff would be perfectly happy just to do some form of science and leave it at that with no thought to how it might be applied.

There are many variants of just doing the science where little or no thought is given to the application of the data or the derivation of information that may be gained from the work. I can think of various examples, but a particularly expensive one was a project called LOIS – the Land Ocean Interface Studies – on the Humber and its environs in the 1990s. Twenty million pounds was spent on collecting a huge amount of data on the Humber as blue skies research, but when we tried to see how this might actually help us with the management of the estuary, the response was 'well here are the data.' In fairness, the work was designed as blue skies research, but during the study little or no thought had been given to how the data and information might be applied to the practical applications of managing the Humber's processes and decision making. In the end we – the Environment Agency – commissioned a major study costing a significant amount of money to meet our particular needs of managing the future of the Humber, which included as part of its terms of reference the need to incorporate findings from LOIS and other scientific work where possible.

How is marine conservation different from terrestrial conservation?

There are differences, but not as great as one might suspect. For me, the main difference has been the legislative framework and its interpretation, because when I began work for Anglian Water, then subsequently the National Rivers Authority (NRA), there were clear legislative drivers like the Water Act (1974), Control of Pollution Act (1974) and, later, the Wildlife and Countryside Act (1981) which were more or less prescriptive, but framed our work protecting the freshwater environment. The vast majority of marine protection, by contrast, started from nowhere and the legislation has gradually caught up. I think that even now the legal regimes like the Marine Strategy Framework Directive are aspirational and will need many years to be worked through in order to understand them fully and to achieve their aims. Importantly, though, we now have a dedicated Marine Management Organisation (MMO), which is in a much stronger position to regulate the marine environment using a wide range of legal measures.

THEMES

Environmental management and the coast – an agency perspective

The work I was involved in wasn't so much regulatory as using the existing legislation to its maximum to try and secure environmental protection and benefits in order to manage flood defence work at the coast. In the early 1980s there was a good deal of thinking on environmental appraisal for flood defences by both agencies and government, and then the Environmental Impact Assessment (EIA) Directive was introduced in 1985. This helped to formalise the process and language in a more prescriptive approach. It was clear that coastal flood schemes could have various environmental impacts, and although engineers were often initially resistant to these ideas, they gradually saw the importance of understanding things like the dynamics of coastal geomorphology. Putting it simply, the prevailing view was that engineers saw hard defences as the main option, and this was not a view I shared since it seemed to me that many other environmentally sensitive options were beginning to come on stream. The major challenge was to somehow devise methods for valuing the environment, to enable this to be properly included in the cost–benefit analysis that was applied to determining flood defence programmes of work.

Flood management projects were identified by the NRA and the Environment Agency (EA), and consultants produced proposals, which were put to then put to the NRA and EA Flood Project Assessment Boards. If they passed this MAFF (now Defra) decided whether to proceed with grant in aid. The main rule was a cost–benefit calculation, and there was a strict cut-off. The essential problem therefore was how to protect sensitive environments, and the many sites of conservation interest that had a low economic value using conventional measures of cost–benefit analysis. There were two main elements to this: the first was project appraisal, and second the economics of valuation.

Looking at the project appraisal, there was a clear need to see projects put into a more strategic framework of what we were trying to achieve for the coast – a vision for the coast as a whole rather than lots of bits of concrete wall. It was also clear that we needed a stronger and more quantifiable view on human and environmental needs. At the same time, I also wrote the environmental assessment procedures for the NRA Anglian Region's flood protection programme (1989). It took about a year and lots of discussion with engineers, but eventually we came up with clear and agreed procedures. The expectation was that the consultants and engineers producing scheme proposals would adhere to these procedures as the NRA's 'way of working'. I have to say that I became very good at spotting if schemes were trying to avoid following these procedures, and it made conservation and the environment part of their agenda because I and my team would often reject their initial proposals if environmental features and assets were overlooked. Throughout this time we were working closely with civil servants from government who were very receptive to these ideas and promoted them through their own developing approaches to more strategic flood defences.

The second aspect of this was economics, and I became something of an expert in this. One project we worked on was the Norfolk Broads flood alleviation strategy, and of course the essence of the Broads is its environment, but we had great difficulty in ensuring environmental aspects were properly included in the project. Working with government, we had to try and get environmental economics robust enough to comply with the 'Green Book' guidance on cost–benefit procedures. At the time, this was central to decision making, and with no economic assessment of the environmental or societal value it was never going to be possible to win the economic debates on places of high conservation interest.

For example, in 1991 I did presentation to the Institute of Civil Engineers about the freshwater grazing marshes behind the Cley shingle ridge in north Norfolk which was being rapidly eroded by coastal processes. I pointed out that one option was simply to build a flood bank on the coast behind the marshes and shingle ridge to protect people and property, but that would have destroyed the environmental interest of the marshes, which were entirely owned by conservation bodies. We needed economics to support the environmental case and spent considerable effort in trying to do this to support environmental assets. It is easy to put a price on a house, possible to put a price on a wheat field, but the value of a freshwater grazing marsh with similar valuation techniques would be considerably lower and certainly not sufficient to cover the enormous environmental value. What was needed was recognition of what we now call natural capital, and that the environment has a value which can be used in economic determination, if economics is considered essential to funding. This was a fascinating time and there were lots of people involved with this such as John Pethick and Richard Leafe, and collectively we helped to turn round the thinking.

Then in 1992 everything changed when the Habitats Directive (1992) came along and the economic value of habitats became less important than their critical ecological value. This made a huge difference in determining conservation priorities. As a consequence the RSPB and other conservation organisations played a major part in raising the profile of EIA in the coastal context in relation to coastal conservation, including on their own reserves at places like Minsmere. This all was of considerable value in supporting the work we were trying to do in the NRA. Since that time EIA has evolved considerably, with scoping being a valuable addition, to become a completely standard part of the way development is assessed, mitigations devised and licences to operate issued. For the last ten years or so of my work with the EA (1990–2001), I spent a good deal of time working on long-term, strategic views with clear direction for coastal management, including managing all the EA's responsibilities in the Humber estuary – still probably the most satisfying and rewarding work of my career.

Development and deal making

My work for the EA on major flood defence projects, with Associated British Ports (ABP) as a port developer, and subsequently as a consultant, has often involved development work in the coastal and estuarine environment. Port development requires Harbour Revision Orders using a process regulated by the Department for Transport. However, the introduction of the Habitats Directive changed the context of these projects, forcing developers to take the Directive into account for these large, expensive and complex projects. It is fair to say that at the time we were still learning how to apply the Directive and some of the terminology used. For example it was not really until 1999–2000 that we collectively reached an understanding of what mitigation may be able to achieve and when compensation would be needed, or even what comprised either. This coincided with a number of very high-profile port development projects that were fought through the public inquiry process. In these early years, therefore, the reality was that for developers, government and regulatory agencies there wasn't a complete understanding of what the law required, and some very hard-won and expensive learning resulted. Throughout this time and throughout my career I've taken the view that working with people and building constructive relationships is far better than negative and adversarial approaches, an approach which provides the opportunity for negotiation and discussion to work through the detail to produce successful outcomes for all parties and the environment.

At ABP I worked on the Humber estuary with port developments at Immingham Outer Harbour and a medium container port at Hull costing £250m in total, and these are good examples of how we moved on to a more consensual approach. It is also important to recognise that NGOs like RSPB were constantly 'looking over the regulator's shoulder' to ensure that the Habitats Directive was being applied correctly. Channel dredging, habitat loss and land take were the main issues, and both developments were going to have an impact.

Figure 21.1 The Chowder Ness managed realignment site on the Humber, photographed after construction. Source: ABPmer

Early on after our scoping studies we had an internal project group meeting, and I asked the question, 'Will we have an adverse effect?' This was a term critical in understanding and applying the Directive, but only partially understood at the time. Our team was undecided on this, but when the question was asked whether we thought the regulator, English Nature (now Natural England), would think there was an adverse effect, it was clear that they would. At the time this was critical, because it led to the assumption that if you cannot prove you are not having an adverse effect, you had to assume you were; this is now universally understood. When I rang Andy Clements from English Nature to tell him we would go down the adverse effect route, his response was 'Thank goodness now we can work together', and similarly the response was very positive from Gwyn Williams at RSPB. What a difference. This decision provided a constructive setting for a working collaboration to meet the Habitats Directive and get the development approved, which we did by constructing about 70 hectares of mudflat as compensatory habitat for both developments (Figure 21.1).

Openly acknowledging that developments would have an impact on the environment at the EIA stage meant that we, the developers, regulators and NGOs could contribute to discussions about how the compensatory habitats could be designed and set in place. Since we were looking at solutions together, it was possible to work through the issues one by one, step by step. When there was agreement we moved on; when there wasn't we looked for the options to resolve the particular issues. This doesn't mean it was quick or easy, and it involved a lot of hard work, seeking all sorts of clarifications from many parties. It involved lots of discussion and patience, but through negotiation in the end we reached agreement on solutions. After acknowledging the adverse effect, the next step involved preparing the Imperative Reasons of Overriding Public Interest (IROPI), and therefore there would be compensation in terms of habitat creation. I've put this simply, but it took two years.

The next point in the negotiation was about the amount of compensatory habitat, where the issues of scale and type were tested by discussions supported by evidence; but this was a negotiation with both sides taking positions and looking at the options. Later in the process, again because the detail of the legislation is unclear we, ABP, had to prepare a side letter agreeing the commitments we had made to compensatory habitat. This was time-consuming and convoluted, but while this was happening the Dibden

Figure 21.2 Dunlin using the new site on the Humber estuary. Source: ABPmer

public inquiry result came through refusing the development of the port expansion in Southampton and so demonstrating that collaboration looked like a more positive approach. We avoided a public inquiry on one of the Humber developments and held a very local and small public inquiry on other because of residents' concerns, and it proved what I knew all along, namely, that you had to take people with you and provide full environmental inclusion and resolution. The then CEO of English Nature said that we were 'a model of best practice in sustainable development'.

In conclusion, this can seem a long and tedious process, but we have ended up with mudflats which in time will develop into salt marsh, using a managed realignment mechanism, and we have the port development. These are win–wins, and far less costly in terms of finance, reputation and relationships, and more productive in terms of environmental benefits. Fifteen years on, the habitats we created are doing really well and delivering a wide range of multiple benefits for the estuary (Figure 21.2). Increasingly this approach is seen, rightly, as the way to work – but at the time many, both in the regulatory and industry sectors, saw conflict and an adversarial approach as the only way to work.

Environmental management, sustainability and corporate social responsibility

Much of the work I've described so far has been about the environmental assessment of projects, using EIA, and how this can lead to outcomes that protect environmental interests. Another major element of thinking on environmental management arose in the early 1990s around environmental management systems which flowed on from business management systems and the way businesses operate. There has been a steady development to the present day with what is now seen to be mainstream environmental management involving the life cycle of the development process (see Figure 4.2).

In addition to the importance of recognising the value of the environment in port development work, I made ABP also recognise the importance of good environmental management and applying this to the way that ports operated. This can be summarised as sustainability, and my work at ABP and subsequently has involved building sustainability in as a central part of corporate social responsibility (CSR) and company management. This has followed through to SUDG, where there is an explicit recognition of the need to deliver sustainable development. Development and operation are closely linked in reputational terms. I helped persuade the ABP board that they needed to do CSR reports annually, and that these needed to be completely open and honest.

I often had to represent ABP in meetings with external parties, such as investors, with our CEO, Bo Lerenius, to outline the ABP approach to environmental port management and CSR. A staff member from Jupiter, one of the major fund managers, said to me that in relation to investment, once the financial assessments had been made, they looked at probity, legal compliance and CSR to determine whether to invest. The environmental management framework I devised for ABP's 21 ports included lots of the standard environmental management systems (EMS) thinking and, in addition to being able to show that we could manage the ports sustainably, we saved millions in terms of reducing energy usage, litter, waste treatment, contaminated land and many other activities. The CSR logic was applied to every aspect of port activity, and all the staff routinely engaged with their role in this. Importantly, it was part of my job to make sure that everyone in the company had a responsibility for environmental management; it was not something to be left to environment managers. I think that this has been probably the most important concept behind all the work that I have done over the years, namely making the environment and its protection part of everyone's work, not just specialists.

Working with broader partnerships

Having left the EA and ABP, which were 'corporate' in different ways, and working with stakeholders on delivering particular projects, I'm now working with three much broader and bigger partnerships which have evolved in rather different ways, but reflect an approach that is now commonplace across the UK.

The Seabed User and Developers Group (SUDG) comprises the main industry sectors with an interest in the marine environment, including offshore oil and gas, aggregates, ports, renewables and cables. The idea for this arose from Frank Parrish at the Crown Estate, who was then the manager of the marine estate, with help from Bob Earll. It was based on the Wildlife and Countryside Link model and set up in response to the developing work on the Marine and Coastal Access Bill (2008) that was then working its way through parliament. The central idea was that the sector would have an efficient way of liaising with government and a voice for industry during the Bill. I became chairman and we worked on the Bill through extensive negotiation with government. After the Act was passed, our work concentrated on the implementation of the Act and developing a relationship with the newly formed MMO as well as the other regulatory bodies such as Natural England. The strength of SUDG is that it has a wide range of experience to call upon, from policy to day-to-day regulation and development.

All of the sectors represented in SUDG have to meet a wide range of regulations protecting the environment in order to develop and operate. The overall aims of SUDG are to promote sustainable development, and much of our discussion is essentially about increasing clarity and understanding of the legislation – which can be complex and unclear – so that we can agree common targets. Linked to this is the need for investors to have confidence in planning and understanding the way forward for developments. Lots of legislation today is aspirational: it provides a 'framework', it tells you what should be achieved or delivered, but it is not prescriptive in explaining how it might be achieved. This in itself is not a bad thing but it prompts the question whether there is a common understanding of what needs to be done or how it should be done. There is little in the way of guidance from government and very few rules for stakeholders to work with to test and develop measures to deliver suitable outcomes, and extending this beyond case work into policy and wider visions is even more complex and unclear.

There is a wide range of views in industry about the environment, and if this is taken as a normal distribution, industry is broadly happy to take forward environmental measures that are required in the legal frameworks, but it wants to know how to do this so that costs can be determined. In these times of uncertainty, this has become a major role for SUDG. We have also worked closely with the NGOs, through Wildlife and Countryside Link and Joan Edwards in particular, to produce policy statements which reflect the common objectives between the NGOs and industry. This has all been possible because SUDG represents a mature and developed view, based on considerable experience, and whilst governments and regulators come and go we have been able to develop a continuity of views over ten years which has stood the test of time and is used to working with environmental law and the need to protect the environment.

The Welland Rivers Trust (WRT), which also hosts the Welland Valley Partnership (WVP), has taken up the challenge of helping develop the government's idea for catchment management now widely known as the catchment-based approach (CaBA), and I have been its chair for four years. Government saw in the period 2010 onward the need for a broader engagement of stakeholders on a catchment level to help deliver a wide range of measures that were needed to meet the Water Framework Directive. Retaining controls for this within the EA was seen to be inadequate to deal with the range of issues arising. The WRT and WVP took the opportunity to be one of the original eleven pilot catchments and delivered a catchment plan within eighteen months. This had an emphasis on getting on with projects that would make a difference. We have now been working to the CaBA model for five years, and while the EA has been extremely helpful, like other Trusts, we have met with some issues. However, it is increasingly clear that CaBA is the direction of future travel. Along with others in the WVP (which includes organisations such as Anglian Water and the Game and Wildlife Conservation Trust) the reality is that many stakeholders are working towards common goals, with a vision, an action plan and some very positive project outcomes.

WVP is not about one organisation leading; there are at least ten with clear remits in this catchment and the partnership brings them together routinely to discuss their work and progress on the Welland Action Plan. We also communicate routinely with the much wider pool of interested parties, such as parish councils, to create a much more coherent view of the catchment than could ever be derived from a single agency. WVP is not an executive body, but it facilitates what is going on; nobody has to come to us to ask for permission to do x or y, but if they want advice we will happily help. There is a diversity of projects, from working with local authorities on how to deliver the Water Framework Directive to advice on septic tanks and river enhancement projects; we also share a project officer with the River Nene who works on catchment-sensitive farming, and because she is a farmer's daughter she is brilliant at working with farmers.

The Solent Forum works on one of our major estuaries and has been going now for 25 years. This and other coastal partnerships resulted from the Living Coast Initiative in the early 1990s, in part from integrated coastal zone management (ICZM) thinking but also because of wider concerns for estuaries. The Solent Forum has 70+ mainly organisational members, and I became chairman in 2014. I was involved in the Wash Forum and the Humber Estuary Management Partnership in the 1990s, and my sense of these partnerships now is that if are to continue, they must be seen as going beyond their undoubted skills at communicating with hundreds of stakeholders and continuing to carry out work that is relevant to the community they serve.

In addition, we have been able to link people working in relatively small organisations to work with the Solent's wider community. We have the standard planning approaches, a broad vision, a five-year plan and a rolling project programme, but this also importantly says 'If we can help, please ask us.' It also includes being proactive. A few years ago I was involved in conducting a review of the coastal forums, and it was clear that their main value was in facilitating work to produce tangible outcomes. The Solent Forum was one of the leaders in this field and their staff are very collaborative in the way they work, but there are challenges with maintaining momentum, not least in this time of austerity. I've seen my role as helping them to 'sell their success' – to get more partners, especially from industry, and to prepare for the future in relation to funding, project development and essentially adapting to the current changing circumstances.

We've just had our 25-year anniversary meeting, and there was a great deal of support from an enormous range of people all saying that we were working towards a better Solent through what Solent Forum is doing. Funding and reminding stakeholders of the need for their support, even to help meet statutory requirements, however, remains a continuing issue. It is also important to be sensitive to the needs of members – so, for example, we have decided not to reply to formal consultations; our role is merely to facilitate others' contributions to these and act as honest brokers in the process. There is also clearly a role for myself and other chairmen to be knocking on government's door to remind them of the value of the coastal partnerships.

THE FUTURE

Threats and impacts

SUDG has a fairly clear view on Brexit, in particular the Habitats Directive, which has now been in place for 25 years. Whilst these are not perfect – and there are still issues like cumulative impact which continue to pose questions – we take the approach that if it's not broke don't fix it. There are those that would want the Directive replaced, but they don't necessarily understand the huge amount of work that has gone into under-standing and implementing the Directive, or that there will be considerable issues if we start again, working with new laws and new systems. The Marine and Coastal Access Act has now been in place for nine years, and working with the MMO we are all learning and gradually getting the clarity we seek.

Brexit has created the potential for change in environmental management. Are we going to say there is no protection for the marine environment? This is of course highly unlikely, and would be unacceptable, so we must assume that protection will remain; the big question is what it will look like. Funding of regulators is also an issue, because as procedures such as licensing become more complex and interpretations more convoluted, we really need more people and more time, and SUDG has said publicly that we think regulators should be properly funded and resourced to do their job effectively.

Barriers, hidden problems

One area I find problematic wearing an SUDG hat is that there is still occasionally a reticence within the regulators to recognise that industry might want to do the right thing, which on the one hand they acknowledge by saying 'yes we want to work together', but then often maintain a distance, so protecting their role. It may be that we need much clearer commitment from government to environmental protection, not just in words but in meaningful and productive actions. From a personal point of view I think that it is encouraging that we have Marine Conservation Zones coming on stream, and this is very helpful in clarifying how we are to go about protecting the environment. However, we are still unclear about 'what we want' for the marine protected areas (MPAs), in particular in relation to the management measures. If in effect this means that society wants 'no-take' zones then this should be made clear. Linked to this, I also think that we should have scientific reference areas that need to be properly monitored and the lessons learnt from this applied widely through the marine environment.

When you retire, what would you like to have achieved?

I think that I played a significant part in improving the way we managed rivers and coasts and brought environmental management into many different people's work, with real positive outcomes for conservation. I would still like there to be recognition that industry can operate and develop in such a way that it does improve the environment, and that more can be done to develop good practices which can be adopted widely. If we could do this effectively and learn how to do things more cost-effectively, it is possible that more could be invested in positive environmental measures – and that would be fantastic. I'd also like to see much better use of scientific data and a greater understanding about the sea and this converted into useful information which can be applied to improve management of the sea

Optimistic or pessimistic?

Optimistic, or I wouldn't be able to do the job.

CHAPTER 22

Charles 'Bud' Ehler

Charles (Bud) Ehler is a marine spatial planning (MSP) consultant who lives and works in Paris, France. He and his partner Fanny Douvere (now head of UNESCO's World Heritage Marine Programme) were the prime movers who stimulated the development of the new field of MSP around the world. He has been called the 'godfather of marine spatial planning'. The 2009 step-by-step guide to MSP, written by Ehler and Douvere and published by the Intergovernmental Oceanographic Commission of UNESCO, set out a practical process for developing marine spatial plans that has kick-started many MSP initiatives around the world.

This chapter describes a hugely diverse and productive almost fifty-year career on a wide range of topics from marine mapping to oil-spill response and natural resource damage assessment. Bud Ehler has also made considerable contributions to coastal zone management, marine protected area management, and most recently marine spatial planning. This chapter explores the development of many of these ideas and achievements, from the earliest points to their mainstream application around the world.

HOW DO YOU DESCRIBE OR FRAME YOUR BELIEFS AND YOUR APPROACH TO THE COASTAL AND MARINE ENVIRONMENT?

I am not strictly a marine conservationist, although during the middle years of my professional career I was an advocate for marine conservation as the Marine Vice-Chair of IUCN's World Commission on Protected Areas (2000–2005). For most of my career, and certainly now, I have worked toward a pragmatic, balanced approach to marine conservation and development – that's what marine spatial planning (MSP) is all about. I'm more concerned with achieving a balance among the interests of different stake-holders through a broad strategic process, rather than advocating the view of only one sector. Marine protected areas (MPAs) alone cannot conserve or protect the ocean.

LOOKING BACK, HOW HAVE YOUR IDEAS UNFOLDED AND DEVELOPED?

Early years and academia (1968–1973)

As Winston Churchill once wrote, 'History will be kind to me, for I intend to write it.' So here goes. I was born and grew up in the anthracite coal mining area of eastern Pennsylvania in the USA, a great place to find fossils in the mountain of mining waste that burned above my home town of Shamokin. Every weekend most families gathered around mounds of steamed Chesapeake blue crabs. But how did so many blue crabs get to the middle of Pennsylvania coal country? The answer was simple, coal was driven to the port of Baltimore in huge trucks for transport to American and foreign markets through the Chesapeake Bay. Empty coal trucks were then driven to Washington DC and loaded with inexpensive, tax-free alcohol and driven back to Baltimore, and the illegal alcohol covered with hundreds of very aggressive live blue crabs as a deterrent to searches by the Pennsylvania State Police on the return to Shamokin. It was my early introduction to avoiding compliance of rules and regulations.

Like many impressionable youths of the 1950s and early 1960s, I grew up watching TV (when you had to get out of your couch to change the channel), where Jacques Cousteau and the crew of the *Calypso* introduced landlubbers like me to the undersea world, as well as the *Sea Hunt* action adventure series (1958–1961) with Lloyd Bridges as former navy frogman Mike Nelson. At the end of each episode, Bridges would deliver a brief comment, often including a plea to viewers to understand and protect the marine environment – in 1960! My early introduction to political satire, which no doubt helped me through 27 years of working in the American federal government, was watching the cartoon series *Rocky and Bullwinkle and Friends* on TV, about Bullwinkle the moose and Rocky the flying squirrel, outwitting Boris Badenov and Natasha Fatale, Soviet spies operating during the Cold War. Rocky and Bullwinkle always prevailed – unlike today, when the Russians could evidently help influence an American election!

I grew up wanting to be an architect, and went to the Pennsylvania State University (1961–1966) to become one. My undergraduate education was enhanced by a semester's exposure to *avant-garde* architecture (1965) at the Architectural Association in 'Swinging London' and a two-month 'grand tour' of western European architectural monuments, both old and new. It was my first trip to Europe and a real eye-opener for me. Next were two more years (1966–1968) of education in regional planning at the University of Michigan. Graduation led to my first academic jobs as a lecturer at the University of Michigan, and then as an assistant professor of architecture, urban design, and planning at the University of California, Los Angeles (UCLA). UCLA (1968–1971) was especially important because it exposed me to 'systems thinking' and its application to architecture and regional planning. Systems thinking is a holistic approach to analysis that focuses on the way that a system's constituent parts interrelate and how systems work over time and within the context of larger systems. The *systems approach* contrasts with traditional scientific method, which studies systems by breaking them down into their separate elements (Churchman 1968). The American space programme's use of systems thinking inspired me, but at the widest level, the views of the earth from space, especially the NASA image of the 'Blue Marble' in 1972, inspired an entire generation. It got people thinking about the environment of the 'whole earth'.

My professional perspective changed after the Santa Barbara oil spill off California's coast (28 January 1969) from an offshore oil platform blowout and the public response of the spill, 'Earth Day' (22 April 1970), when 20 million Americans took to the streets in coast-to-coast rallies to demand a healthy environment. This awareness also extended to publications like the *Whole Earth Catalog* (Brand 1966–1972). These events led to the creation of the US Environmental Protection Agency (EPA) and the National Oceanic and Atmospheric Administration (NOAA) in late 1970. But they also changed my career path when Dean Harvey Perloff, the Dean of American Urban Planning, chose me to develop the first PhD programme in environmental planning at UCLA. In fact, it was easy to select me – none of the other UCLA Planning Faculty members wanted to work on 'the environment', an issue they considered a fad that would disappear quickly. Since I was a 26-year-old assistant professor with no professional experience, the dean hired three experts to help develop the programme: Blair Bower, an environmental engineer and the deputy director of the Quality of the Environment Program at Resources for the Future (RfF), the pre-eminent Washington-based think tank on natural resources policy; Larry Ruff, a young resource economist at the University of California, San Diego, who in 1970 had already written a classic article on 'The economic common sense of pollution', arguing that pollution was primarily an economic problem; and Ian Burton, a geographer from the University of Toronto specialising in natural disasters and risk assessment.

After three years at UCLA, I was recruited in 1971 by the State University of New York at Stony Brook to be an associate professor and educational director of a new programme of urban and policy sciences focused on applying quantitative analysis to public policy problems and modelled after the approach of Harvard University's Kennedy School of Government. Dan Basta, a systems engineer from the aerospace industry, and someone who would become a close colleague at NOAA, was one of my graduate students. After three years of teaching public policy and planning at Stony Brook, the dean suggested a year in Washington might help me learn how environmental policy really is developed and implemented. I contacted a friend from my UCLA days (economist Larry Ruff) who was setting up a social science research centre within the new EPA and was soon off to Washington. A year in Washington would turn into a 32-year career in the federal government at EPA and NOAA. I never went back to academic life; I was having too much fun getting things done in the real world of environmental policy.

Work at the US Environmental Protection Agency (1973–1978)

My early work at EPA focused on projects assessing the cumulative environmental effects of land use, particularly on the 'costs of sprawl' or unplanned suburban development. This work grew naturally into a programme of integrated (air, water and solid waste) regional environmental planning, greatly assisted by Blair Bower. Blair was a *systems thinker* and a long-time advocate of effluent charges for water quality management. He and his collaborator at RfF Alan Kneese were great assets when I, a non-economist, later initiated a programme on economic incentives for environmental management at EPA, with the Council on Environmental Quality and the Council of Economic Advisors in 1977 (Bower *et al.* 1977).

Work on ocean issues at the National Oceanic and Atmospheric Administration (1978–2005)

I enjoyed my work at the EPA, but loved even more the idea of working on ocean issues. In 1978 I went to NOAA looking for an opportunity to work there – the first time I was job seeking rather than jobs seeking me – and was rejected by NOAA's Office of Coastal Zone Management. Disappointed, but determined, I returned about six months later and interviewed with NOAA's newly created Office of Ocean Management. The office was responsible for providing technical support to NOAA executives, for example, in its reviews of the effects of oil and gas lease sales on the outer continental shelf and coastal areas, and the National Marine Sanctuary programme, particularly the preparation of management plans for marine sanctuaries. The Office of Ocean Management was short-lived, because of questions of its authority from other federal agencies. It was dissolved after less than a year and our small group ended up providing technical assistance to the Office of Coastal Zone Management. My first new hire was Dan Basta, my former graduate student at Stony Brook and a person who would become my creative partner in developing most of the NOAA projects and programmes described below.

Strategic regional data atlases of the EEZ of the USA (1979–1989)

During the 'energy crisis' of 1979 and with the interest of the Carter administration in increasing oil imports and refinery construction, NOAA was asked by the Council on Environmental Quality (at that time Elliott Norse was a young marine biologist at CEQ who became an advocate for our early marine mapping work) where two new refineries should be located on the east coast of the USA to minimise environmental impacts. Our response was that we should not identify where refineries should be located, but where they should *not* be located. One output of our comprehensive assessment of the entire East Coast region was the first region-wide *Eastern United States Coastal and Ocean Zones Data Atlas* published by NOAA (1980). On a common base map we mapped the distribution of marine habitats and species to identify important biological and ecological areas, as well as oceanographic conditions and existing economic activities. With little data available particularly on the life history of marine species in space and time, groups of marine species experts were used to translate what they knew collectively and mapped this information by hand into grid cells on the map, long before the power of today's geographic information systems and data portals (Bower *et al.* 1982). While many of the scientists were reluctant to put lines on maps, eventually they did. Analyses were done by coding data by hand on gridded map overlays and counting 'hits' within grid cell, a task far better and faster done by computers today.

President Reagan declared the first exclusive economic zone (EEZ) for the USA in 1983. Given the poor state of information about the natural resources and human uses of the EEZ and the success of the East Coast *Data Atlas*, we continued the series of regional atlases (Ehler *et al.* 1986). A *Gulf of Mexico Coastal and Ocean Zones Data Atlas* was published in 1985 and a *Bering, Chukchi, and Beaufort Seas Coastal and Ocean Zones Data Atlas* in 1988. While much of the information in these atlases is now dated, in a few cases it still represents the best information available, almost thirty years later.

One memorable event during the production time of the NOAA atlases was a marine atlas colloquium at the Royal Geographical Society (RGS) in London in 1985 organised by John Ramster from the Lowestoft Laboratory of the UK Fisheries Research Directorate and Arthur Lee, former Director of Fisheries for England and Wales (Ramster 1986). Ramster and Lee had produced *An Atlas of the Seas Around the British Isles* in 1981 for the Ministry of Agriculture, Fisheries and Food (Lee & Ramster 1981). Dan Basta and I organised a massive exhibit of examples of our mapping of the US EEZ, which impressed the colloquium attendees (Figure 22.1). Ramster summarised the NOAA work as a 'textbook example to all other countries of how things should be done.' Unfortunately our work did not impress the NOAA administrator, who instructed us to stop production of our final atlas for the west coast because he thought that no

Figure 22.1 Bud Ehler and Dan Basta with resource maps from the Gulf of Mexico on display at the Royal Geographic Society, 1985. Source: Bud Ehler

one would ever need or use information at that scale for marine planning or decision making. He didn't anticipate ecosystem-wide incidents like the Deepwater Horizon in 2010, where our spatial information on the distribution of marine species in the entire Gulf of Mexico was used extensively, let alone the information needs of national-level marine spatial planning today.

Oil-spill response and natural resource damage assessment (1978–2005)

Amoco Cadiz (1978–2003)
In March 1978 the supertanker *Amoco Cadiz* ran aground off the coast of Brittany, France, resulting in the largest oil spill and loss of marine life until that time, and for us work which lasted from 1978 to 1983. NOAA was contracted by the owner of the tanker, Amoco Corporation, to study the environmental and marine resource effects of the spill. However, Amoco was opposed to any assessment of the 'social costs' of the spill, stimulating NOAA to fund an independent assessment of the social costs – the first of its kind – with its own resources, and my organisation was assigned to undertake the assessment. In its 1983 report *Assessing the Social Costs of Oil Spills: the Amoco Cadiz case study* (1983), NOAA estimated the spill caused a total of US$195–284 million in social costs to the world, including Brittany and France. The French government presented claims in US courts totalling US$2 billion; in 1990 France was awarded US$120 million from Amoco.

Exxon Valdez (1989–2000)
Within hours after the supertanker *Exxon Valdez* spilled nearly 11 million gallons of crude oil into Alaska's Prince William Sound on 24 March 1989, a team of NOAA scientists from the Hazardous Materials Response team (a branch of the Office of Oceanography and Marine Assessment that I directed then) arrived on scene. Two days later, I, NOAA Chief Scientist Sylvia Earle, several Coast Guard admirals, and other Washington-based officials arrived in Valdez, Alaska. In its role as scientific advisor to the Federal On-Scene Coordinator (a US Coast Guard commander), NOAA routinely provided spill trajectory, resources at risk, and spill impact information during the initial stages of the

Figure 22.2 Cleaning beaches during the *Exxon Valdez* spill. Source: Bud Ehler and NOAA

spill. During the days, weeks and months after the spill, US Coast Guard and Exxon responders and others trying to control the effects of the huge oil slick used NOAA predictions of the trajectory of the spilled oil. They also employed observations of the oil by NOAA scientists, made during overflights and sampling trips to affected shorelines. Responders used NOAA's environmental sensitivity index (ESI) maps of the region that showed the locations of especially oil-sensitive animal and plant populations to decide which areas were most important to protect and eventually clean up.

When it became clear that NOAA was going to be in Prince William Sound for the long haul, NOAA Administrator Bill Evans assigned me to coordinate all NOAA activities and assets, including spill response, natural resource damage assessment and logistics support (research vessels and helicopters). Many images of the early days of the spill showed workers using disposable diapers to mop up the oil. Exxon had huge resources and it was simply a matter of time before they were mobilised. Exxon was the responsible party, not only for creating the spill, but for clearing it up. Exxon often asked for scientific advice from NOAA, and while we often argued they generally followed our advice. There is no doubt that some mistakes were made, not least of which was allowing the use of high-pressure hot-water hoses to wash the beaches and in the process effectively destroying all living things in their wake (Figure 22.2). Recovery through natural processes might have worked better. But there was also enormous political pressure from the State of Alaska to get rid of the oil. NOAA pressed for reference areas, so we could at least measure natural recovery, but Alaskan government officials disagreed – they wanted all the spilled oil out of Alaska. In the end we did get a few reference areas and documented that the high-pressure washing hadn't made a great deal of difference in the recovery (NOAA website). We learned a lot; 20/20 hindsight is always clear.

The spill affected more than 2,000 kilometres of shoreline, with immense impacts for fish and wildlife and their habitats, as well as for local industries and communities. The oil killed an estimated 250,000 seabirds, 2,800 sea otters, 300 harbour seals, 250 bald eagles, as many as 22 killer whales, and billions of salmon and herring eggs. The NOAA General Counsel Tom Campbell was adamant from the beginning of the spill that a thorough natural resource damage assessment programme should be conducted, and although the Republican administration raised concerns over the cost to business, the idea was implemented. NOAA and three other federal agencies worked together on the natural resources damage assessment. Since NOAA was the only federal agency with marine damage assessment experience, from the *Amoco Cadiz* oil spill, we used our expertise to put together the teams, and the other departments followed our lead. Unlike the clean-up, the damage assessment was confrontational and competitive with Exxon. A compromise was reached among the parties after four years' work. The process proved highly instructive and helped develop methodologies that pioneered future damage assessments in the marine environment.

Exxon eventually paid US$125 million in criminal fines and US$900 million in civil penalties for spill damages to publicly owned natural resources, land, water and wildlife. These numbers pale when compared to the US$40+ billion settlement BP reached with the US government and states of the Gulf of Mexico after Deepwater Horizon. Ultimately, the *Exxon Valdez* spill resulted in a thorough examination of the status of oil-spill prevention, response, clean-up and natural resource damage assessment capabilities in the USA. One lasting output was the passage of the Oil Pollution Act of 1990, which assigned liability for the cost of clean-up and damage, defined responsible parties and financial liability, implemented processes for measuring damages, specified damages for which violators are liable, and established a fund for damages, clean-up and removal costs. The law fundamentally changed oil production, transportation and distribution industries in the USA.

Ocean dumping in the New York Bight (1985–1992)
New York City had been dumping its sewage sludge in the ocean since 1924 and was the only American city still dumping municipal and industrial waste in the ocean by the early 1980s. By 1984 New York City was dumping 6.7 million wet tons of sewage sludge, 40,000 wet tons of acid waste, and 7,650 million cubic metres of dredged spoils at an ocean dump site 12 nautical miles offshore. Under American law the EPA is responsible for issuing permits for ocean dumping of sewage sludge, the US Corps of Engineers for dredged material disposal at sea, and NOAA for monitoring and research on the environmental effects of ocean dumping. I provided expert testimony to Congress on the effects of ocean dumping on the marine environment and its resources on several occasions from 1985 to 1992. Despite significant political pressure from New York City, after a summer during which medical waste washed up on beaches and thousands of dolphins died mysteriously along the Atlantic coast, Congress passed national legislation making in illegal to dump sewage sludge or industrial waste into ocean waters after December 1991. Although New York City insisted there was no connection between the medical debris, dolphin deaths and ocean dumping, the conservation community had a strong election-year issue and few politicians were eager to oppose the ocean dumping ban.

Coastal adaptation to climate change and the IPCC (1992–1996)
I participated in the International Panel on Climate Change (IPCC) Second Assessment because of the track record of the USA in coastal zone management. With Luitzen Bijlsma of the Dutch Rijkswaterstaat, I was co-chair of the working group on coastal zones and small islands that identified and analysed adaptation responses to climate change, especially sea-level change. We had started work on these issues as co-chairs of the World Coast Conference in the Netherlands in 1993 (Bijlsma 1994). After that conference Luitzen and I were asked to lead a team that would write the chapter on coastal adaptation for the IPCC Second Assessment. We put together a great team and produced a well-received chapter (Bijlsma *et al.* 1995). In 2009 I (and other contributors who had made a substantial contribution to the IPCC process) received recognition

from the IPCC for our contributions, which led to its award of the Nobel Peace Prize in 2007 (co-awarded to Al Gore). It was an international experience that I was proud to contribute. But now it's sad to see President Trump ignoring thirty years of excellent science of the IPCC and pulling the USA out of the Paris Agreement.

Coastal zone management in the USA (2001–2003)

I managed NOAA's Office of Ocean and Coastal Resource Management on a collateral basis from 2001 to 2003. Coastal zone management in the USA is a partnership among 34 states, partially funded annually through about US$60 million in grants from NOAA. The original idea for coastal zone management came from Peter Douglas, the main author of the State of California's landmark coastal protection law and for more than a quarter-century executive director of the California Coastal Commission, the powerful regulatory agency he helped create. He later helped write the 1976 California Coastal Act, a landmark law that became a model for other states and countries.

The US Coastal Zone Management Act of 1972, administered by NOAA, provides for the management of the coastal resources of the USA. It left programme implementation to the coastal states under federal guidelines. Under the CZM Act, state plans are regulatory, a requirement not emulated in most other countries, where CZM is advisory. NOAA provides grants to coastal states ranging from US$500,000 to US$2 million. However, the real incentive for states to participate in the federal programme is a legal concept called 'federal consistency'. Federal consistency requires the federal government to comply with a state's approved Coastal Management Programme when taking actions that are likely to affect coastal resources. This provision was unusual in that it gave the states significant influence over federal decisions such as offshore oil and gas permits. Since the jurisdiction of coastal states extends seaward to only 3 nautical miles, marine issues have been rarely considered in American coastal management plans, except in a very few states. In general, land-focused issues prevailed over marine issues. This land-orientation of ICZM was exported from the USA to South America, western Europe and Southeast Asia. Today only a few ICZM programmes anywhere in the world focus on marine issues.

Marine protected areas, the Florida Keys National Marine Sanctuary, and the World Commission on Protected Areas (1990–2005)

Throughout the 1980s, oil drilling proposals, reports of deteriorating water quality, and evidence of declines in the health of coral reef ecosystems in Florida continued to mount. These threats, combined with several large vessel groundings, prompted Congress to act. In November 1990, President George H. W. Bush signed into law the bill establishing the Florida Keys National Marine Sanctuary (FKNMS). It became the ninth and largest site (7,500 square kilometres) to join the national system. Prior to the development of its first management plan from 1990 to 1996, sanctuary management plans were constructed from a simple template and each one was pretty much the same. The FKNMS was a new challenge for NOAA since it was relatively large, multiple-use, and complicated institutionally involving authorities at the federal, state and local levels. In 1991 the NOAA Chief Scientist Sylvia Earle asked me as the director of NOAA's Office of Ocean Resources Conservation and Assessment (ORCA) to provide technical assistance to the director of the FKNMS office in Florida. Dan Basta and his strategic environmental assessment team contributed substantial technical assistance over the next five years. The final plan identified management actions for ten issues and designated five types of marine zones to reduce pressures in heavily used areas, protect critical habitats and species and reduce user conflicts. The first management plan was implemented in 1997. Today the sanctuary is administered by NOAA and is jointly managed with the State of Florida.

From 2000 to 2005, while directing the International Office of NOAA's National Ocean Service, I also served as the Vice-Chair (Marine) of IUCN's World Commission on Protected Areas (WCPA). When Graeme Kelleher, former chairman of the Great Barrier Reef Marine Park Authority, retired as the Vice-Chair (Marine) in 1999, Nancy Foster, then the Assistant Administrator of the National Ocean Service, was selected to succeed him. Sadly, Nancy Foster died unexpectedly in 2000, and I was selected to take on the marine vice-chair responsibility. Based in Switzerland, WCPA is the world's premier network of protected area expertise with over 2,500 members, spanning 140 countries.

In 2003 I co-chaired with Peter Cochrane, Director of National Parks Australia, the marine programme of the World Parks Congress in Durban, South Africa – the largest and most diverse gathering of protected area experts in history. One of the recommendations of the congress was to 'establish by 2012 a global system of effectively managed, representative networks of marine and coastal protected areas ... and that these networks should be extensive and include strictly protected areas that amount to at least 20–30% of each habitat, and contribute to a global target for healthy and productive oceans ... including areas beyond national jurisdiction.' This recommendation was one of the early international actions that contributed to the setting of the 2010 Aichi Biodiversity Targets of the Strategic Plan for Biodiversity of the Convention on Biodiversity. From 2002 to 2004 Simon Cripps, then Director of International Programmes at WWF International in Geneva, and I initiated and co-managed a NOAA/IUCN/WWF project to develop a guidebook of natural and social indicators for evaluating MPA management effectiveness in achieving MPA goals and objectives. The project produced a practical guide, *How Is Your MPA Doing?* that has become the 'bible' for evaluating MPA management effectiveness worldwide (Pomeroy *et al.* 2004).

Marine spatial planning and the Intergovernmental Oceanographic Commission (2005–2018)

In 2005 I voluntarily retired from NOAA and struck out to become an 'American in Paris' 32 years after my one-year work experience in Washington. I had consulted previously to UNESCO's Intergovernmental Oceanographic Commission (IOC) in Paris, had a number of excellent professional and personal contacts in Europe and knew that Europe was also the centre of interest in a new field of interest to me, marine spatial planning (MSP).

In January 2004 I had attended a workshop at the University of Ghent in Belgium, to discuss the work of the Maritime Institute on developing an MSP process. The project was known as 'Gaufre' or 'toward a spatial structure plan for sustainable management of the Belgian part of the North Sea'. Frank Maes, a lawyer and director of the Maritime Institute, managed the project, Jan Schrijvers was the creative marine scientist behind the project and Fanny Douvere was a member of the Gaufre project team. I was impressed with the innovative work of the Belgian team. In October 2005 Fanny and I arrived in Paris and began work as consultants to the IOC. We were invited to attend a meeting of the Working Group on Ocean Zoning, organised by Elliott Norse, at the National Center for Ecological Analysis and Synthesis (NCEAS) at the University of California, Santa Barbara. Our contribution to the working group was to move it away from its

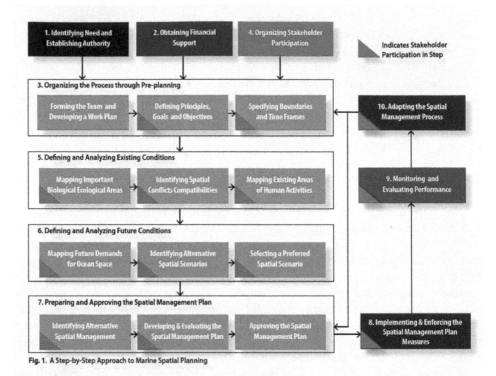

Fig. 1. A Step-by-Step Approach to Marine Spatial Planning

Figure 22.3 Systems diagram of the MSP process. Source: Ehler & Douvere/UNESCO (2009)

focus only on marine zoning and toward the broader approach of MSP. The working group met again in June 2006 and eventually produced several papers, including a short policy article on MSP for *Science*, the first mention of MSP in a prestigious international science journal (Crowder *et al.* 2006). The working group also challenged Fanny and me to publish more about our ideas on MSP.

We returned to Paris excited about starting an IOC initiative on MSP and met with the IOC Executive Secretary to discuss our proposal. He decided that we could move ahead as long as it didn't cost the IOC any resources – not a ringing endorsement, but enough to begin. Thanks to small financial contributions from the Gordon and Betty Moore Foundation, the David and Lucille Packard Foundation, the government of Flanders, WWF and eight other organisations, we raised enough funds to convene the first international workshop on MSP at UNESCO in November 2006 (Ehler & Douvere 2007). This workshop has been characterised as, 'the starting point for the spread of MSP internationally' (Merrie & Olsen 2014). One of the recommendations of the workshop was to develop a step-by-step guide to MSP. A widely read special issue of *Marine Policy* was devoted to 11 papers presented at the workshop (Douvere & Ehler 2008).

When the first International MSP Workshop on Ecosystem-based Marine Spatial Planning was held in 2006, few people beyond a small group of marine experts had heard of marine spatial planning. Only six countries had any MSP activities. The Great Barrier Reef Marine Park was the pioneer in designating zones of allowable human uses within the park, an area of 344,000 square kilometres. From 1982 to 1988 the Great Barrier Reef Marine Park Authority, led by Graeme Kelleher, hired several land-use planners to help zone its large area.

In 2007 the IOC sought and received generous funding from the Gordon and Betty Moore Foundation and the David and Lucille Packard Foundation, both based in California, to develop a step-by-step guide to MSP. In consultation with many practitioners and an advisory panel, we worked over the next 18 months to write and deliver the guide (Ehler & Douvere 2009). We field-tested our approach through workshops in Massachusetts and Vietnam (both at the national level and at Ha Long Bay, a World Heritage site). Published in 2009, the UNESCO MSP guide was an immediate success, not only helping western European countries that were well along in developing their MSP plans, but particularly countries in Southeast Asia, South Asia, Africa and North America, and New Zealand, which were just learning about MSP and starting their own initiatives. An Ecosystem Science and Management Working Group of the NOAA Science Advisory Board concluded that the UNESCO guide 'provides the most comprehensive approach to marine spatial planning' (Collie *et al.* 2013). The guide has now been translated into seven languages and is used worldwide. Fanny and I have promoted the guide both through peer-reviewed publications and hundreds of presentations throughout the world (Figure 22.3).

In 2010 Fanny completed her PhD in marine spatial planning at the University of Ghent (the first MSP doctorate awarded in the world) and left the IOC MSP Initiative to become head of UNESCO's World Heritage Marine Programme. I continued to work on MSP under a new grant from the Moore Foundation to deliver *A Guide to Evaluating Marine Spatial Plans* (Ehler 2014). I also increased the amount of my time consulting with national and regional governments on MSP. I advised or worked on marine plans for Coastal First Nations and the Province of British Columbia, Canada, the USA at national, regional and state level, Costa Rica, Sweden, Israel, Abu Dhabi, China, Vietnam, the Benguela Current Commission (South Africa, Namibia and Angola), the European Commission, the Organisation for Economic Cooperation and Development and the World Bank. I've also enjoyed presenting every year one-day lectures on the basics of MSP to students in the Erasmus Mundus Masters course in maritime spatial planning in Venice and Seville, and the International Ocean Institute (IOI) regional ocean governance course in Malta. Educating young marine planners is one of the largest– and most rewarding – challenges we face to make the MSP profession sustainable.

I've spent the last few years adding content to the IOC MSP website (http://msp. ioc-unesco.org), especially on the global status of MSP by country and substantial revisions to the 2009 UNESCO step-by-step guide. In March 2017 the IOC and the Directorate-General of Maritime Affairs and Fisheries of the European Commission (DG-MARE) organised the second International Conference on Marine/Maritime Spatial Planning at UNESCO headquarters in Paris. Almost 300 people participated in the conference in Paris and another 200 joined online. Participants came from 73 different countries. While over half of the participants were from Europe, the second-largest region participating was Africa. I had the honour of presenting the keynote address on the global status of MSP. One output of the conference was the publication of a Joint DG-MARE and IOC/UNESCO roadmap for international MSP and ocean governance. In October 2017 the EU announced a grant of €1.4 million to IOC-UNESCO to develop international guidelines for MSP. As part of this venture, two MSP pilot projects will be launched in early 2018: one in the Mediterranean and another in the South Pacific.

THE FUTURE

What are the impacts of things happening at present?

A real challenge is the growing 'industrialisation' of the oceans first articulated in 2000 by Hance Smith. We are now clearly seeing this trend. It will be exacerbated by the realities of the rush toward the 'Blue Economy' from which growing sea uses could cause increased harm to the marine environment and increased conflicts among users of the ocean. Building the Blue Economy appeals to politicians, who promise or over-promise substantial increased employment, wages and other indicators of growth in the marine economy – but maritime industries represent only about 5–9% of most national economies. We can watch the inevitable conflicts happen and fix them after they occur or achieve the real benefit of marine planning to avoid the social and environmental costs of conflicting and damaging uses. But it comes down to political will, leadership and public support, because plans will fail when there is none.

What are the hidden barriers?

'We've met the enemy and he is us,' said the famous Pogo poster by Walt Kelly in 1970 celebrating the first Earth Day. Conservationists often take a 'we know best' attitude and are such strong advocates and so hard-headed that they turn off people who would otherwise support them. I think working together is something we have to work on across the board with a willingness to bend a little, to achieve a better collective future – and that in essence is what MSP is about. By meeting most of the interests in a more open, transparent and systematic way, where intentions are not hidden, MSP can deliver better outcomes. People can be brought together to work effectively toward a common future. We know how to do this – and you don't need more natural science to get this done. If we need anything it is more social science in terms of how we make this work more effectively to develop plans that people own and support. You want the people involved in the process to own the outcomes and consider any marine plan as their plan. People want to know how decisions will affect them, how outcomes might look in space and time, and who is going to win and lose. MSP can do this.

What new innovations, technologies do you see making a difference in the future?

Game-changing technologies will continue in the computing, sensing and communications areas. The 'digital ocean' is coming and in some respects is already here. For example, the new low-altitude remote sensing systems – shoebox-sized satellites or 'doves' – being launched by Planet, a small company in San Francisco, California, that will provide unparalleled access to near real-time information about the environment. That capability should provide information for management that will change the way we think about the marine environment from both a planning and an enforcement perspective. We can then see whether the new coal-fired power plant next to the World Heritage marine site in Bangladesh will or will not damage local mangroves as the national government has promised. We will be able to monitor a host of things –illegal fishing, ship discharges and habitat loss – very differently than we can now. Relatively inexpensive remote sensing is going to play an increasingly important part in the way we manage the uses of the marine environment.

Another game changer is Global Fishing Watch, a partnership among Oceana, SkyTruth and Google that analyses spatial data (vessel identification, position, course and speed) from the Automatic Identification System (AIS), a tracking system of ships widely used by national authorities to track and monitor the activities of their national fishing fleets. AIS enables authorities to identify specific vessels and their activity within or near a nation's EEZ. It can be used to identify fishing behaviour, based on the movement of vessels over time. In August 2017 a Chinese fishing vessel was caught by the Ecuadoran navy crossing the protected waters of the Galapagos Marine Reserve (a World Heritage marine site) with a cargo of over 6,000 shark carcasses and fins. SkyTruth and Global Fishing Watch analysed AIS data to trace the vessel's locations in the days and months before it was apprehended. In September 2017 the Ecuadoran government handed down a $5.9 million fine to the vessel owner and a four-year prison sentence to its captain for the illegal transport of sharks and shark fins in the Galapagos Marine Reserve. This new capability to monitor ocean uses is a technology that instead of helping catch fish, can help manage fishing better.

When you retire, what would you like to have achieved?
I've had a pretty productive career. I'm satisfied with what I and my professional colleagues have achieved so far. For almost all of the past forty-five years I have been able to create my own job and often design my own organisation. I have started many initiatives, but have worked with and been helped along by great people. Most of my personal accomplishments have been through teamwork and collaboration. My last twelve years working for the IOC on marine spatial planning have been the most personally rewarding.

Fanny occasionally asks me, 'But what have we really done?' The simple answer is that we have really changed the way many countries plan the use and conservation of their marine areas. We have developed and communicated through publications and hundreds of presentations a planning process for analysing the complex ecological, social, economic and political systems of marine areas, and for identifying management actions that can influence parts of the system in a positive way to deliver desired outcomes. I've been able to work in many of these countries, help guide their initiatives, assist draft their plans and see some early results of implementation. That's a big part of feeling good about my work over the past fifty years.

Are you optimistic or pessimistic about the future?
I'm an optimist. MSP is on a path to be implemented by more than 80 countries by 2050, covering well over 50% of the surface area of the EEZs of the world. In western Europe members of the European Union are bound by a 2014 Directive on Maritime Spatial Planning to have approved maritime spatial plans in place by 2021. Today almost half of the MSP initiatives in the world are located in western European countries; England, Scotland, the Netherlands, Belgium, Germany and Norway already have plans in place. The Netherlands has completed three cycles of MSP since 2005. Even the USA and Canada, both of which have struggled to get MSP off the ground on a national scale, will muddle through in the short run with financial support from the Gordon and Betty Moore Foundation. Australia, a world leader in marine planning in the late 1990s, has devolved its MSP efforts to the implementation of the world's largest national network of MPAs, but has lost its focus on managing multiple uses throughout its EEZ. Not all MSP initiatives are successful.

CHAPTER 23

Elliott Norse

Elliott Norse's career spanned and catalysed the development of marine conservation in the USA, but his focus was always larger: the fate of life on earth, particularly the world's oceans and forests. Over a 38-year career he played a pivotal role in conservation by defining the maintenance of biodiversity as its overarching goal. The books, papers and scientists' statements he wrote and the conferences he organised shaped the world's marine conservation thinking and accelerated action to save species and ecosystems. He helped presidents Carter, Clinton, Bush and Obama create America's largest strongly protected marine areas and built enduring mechanisms for the world's marine conservation community to develop and share ideas. In this powerful essay he looks back on his career, highlighting the thinking behind developments and offering his thoughts on a strategy to save life in the sea.

HOW DID YOU HELP MARINE CONSERVATION HAPPEN?

I am no longer a player. As a retiree I watch my backyard birds, grow salad greens for our household and observe what's happening in the world. Many things I learn are deeply troubling. But I also celebrate seeing swelling numbers of young people who are much more intellectually and emotionally capable of saving marine life. This is deeply heartening, because conserving life in our only home – the earth that shaped and sustains life – is much more important than the immediate concerns of individuals, organisations, professions or countries. Nothing we do is more important. This is the one human endeavour we must not lose.

As luck would have it, circumstances shaped and compelled me to do good things for oceans and forests, which I did as long as my mind and body allowed. It required all of my skills, countless hours and forsaking many things. I succeeded in making many good things happen. The successes I catalysed happened thanks to cooperation and support from really good people, to whom I am eternally grateful.

From early childhood my goal has been conserving life, and my first home on an estuarine canal gave me a special love for life in the sea. Even before entering primary school I read everything I could about living things, and this continued in college, graduate school and for my postdoctoral training. My research on the geographical ecology of swimming crabs (Portunidae) took me to the Caribbean and tropical Pacific coasts of the Americas, where I saw both dazzling diversity of marine life and how people interact with the sea. That proved vital for me, because my academic training did not include conservation. While I was most fortunate to get the education I did, not all learning happens in schools, nor do schools have authority to govern key human activities beyond the cloisters.

From the vantage point of the twenty-first century's second decade, it seems strange that marine conservation did not exist as a coherent field of thought or action in the 1960s through to the 1980s. Conservation (including the science of conservation) focused almost completely on lands and freshwaters then. It is ironic that the US environmental movement is largely traceable to a land-focused book, *Silent Spring*, by a marine biologist, Rachel Carson (1962), but apart from the events that followed the Santa Barbara Oil Spill of 1969, examples of marine conservation were exceedingly scarce. Only the thinnest scattering of people were working to save life in the oceans. The institutional framework was woefully inadequate. Marine conservation was a very lonely field.

To progress, it was essential to understand which canonical principles worked much the same in all the terrestrial biogeographic realms and which were distinct to the sea. One crucial difference was that almost no coastal waters or oceans were under private ownership, leaving nearly all authority to governments, yet governments provided very little effective governance of human activities in the marine environment. The institutional framework was not up to the task. So I realised the need to learn how governments function (or fail to function) and how to get their attention, because they make the decisions that determine the fate of our oceans.

From what I saw, the status and trends for marine life were far more perilous than government and UN agencies thought. I could choose to wait for needed change or devote myself to catalysing whatever change was needed. Time did not favour waiting. There was urgent need for changing the institutions that are the proximate drivers affecting our oceans. True, underlying causes (e.g., population, consumption of resources, why so many other issues take precedence over the earth's habitability) are no less urgent. But facing the finiteness of this one human's effort, I consciously avoided devoting myself to efforts that drive the need for conservation. Rather, my single-minded focus was to catalyse more effective marine conservation institutions. Building a movement strong enough to save marine life seemed the far better choice.

To succeed, I had to confront the obvious: victories feel good, and small victories are far more tangible. But given the oceans' vastness and diversity, it was also clear to me that local efforts to protect small areas, however essential, were not sufficient; conserving the earth's largest and increasingly imperilled ecosystems required marine conservation to accelerate much faster and to function at much larger scales. So while I sometimes worked on regional or national issues, I always saw these as precedent-setting for conservation on a global scale. Clearly coalescing a successful marine conservation movement (Norse 1996) required success soon enough at a scale large enough to make the needed difference.

To learn what universities couldn't teach and to start on the path I envisioned, in 1978 I left academia and became the marine biologist in the Ocean Programs Branch of the US Environmental Protection Agency (EPA), analysing impacts of proposed oil and gas drilling in continental shelf waters. It quickly became obvious that being an ecologist allowed me to envision impacts that my EPA colleagues (mostly engineers) did not. My marine biology classes had really helped, but in hindsight I wish I had taken appropriate training in psychology, sociology, economics and political science, because catalysing change requires understanding people and institutions even more than fishes and seaweeds.

I quickly discovered that marine conservation was not playing on a level field. We have always been vastly outnumbered and outspent. Marine industries were huge economic enterprises single-mindedly devoted to their own growth, while advocates for the marine environment were far fewer, usually isolated and always woefully underfunded. As a result, governments' decisions nearly always reflected the loudest voices, not ones dedicated to maintaining living oceans. So for the oceans to win, marine conservationists had to have far more effective voices than those from the marine industries.

At EPA, after 18 months of learning how a regulatory agency works (and doesn't), I was fortunate enough to be chosen as the Staff Ecologist of the White House Council on Environmental Quality (CEQ) under President Jimmy Carter. The once-in-a-lifetime privilege of working as part of a small, elite team of experts in environmental law, economics and ecology who advised the President allowed me to do two consequential things for the oceans. The first was an idea: defining a new goal for conservation, one that went beyond conserving only species that humans used as consumable resources or only those in imminent danger of disappearing entirely (Norse & McManus 1980). Rather, inspired by heroes including Rachel Carson and Aldo Leopold, the goal I envisioned was maintaining the integrity of nature, which provides the environment humankind needs to survive and prosper. My co-author Roger McManus and I called it conserving biological diversity. It became the theme of all I did in my career. Not accidentally, the first paragraph of the biological diversity chapter in CEQ's 1980 annual report begins with the tragic story of the scientific discovery and rapid extinction of Steller's sea cow. I was determined to show that conservation was not only about land.

New ways of thinking have real power only if they lead to real, enduring changes in the world. The second thing being in the right place at the right time positioned me to accomplish was to recommend that President Carter designate four proposed National Marine Sanctuaries in 1980–1981: Gray's Reef off Georgia, Looe Key off Florida, and Channel Islands and Gulf of the Farallones off California. Happily the President did so, thereby tripling the number of US National Marine Sanctuaries and keeping oil drilling out of these special marine places. It was my special honour to write the first draft of President Carter's sanctuary designation statements, explicitly comparing these extraordinary ocean areas to America's great national parks.

In 1981, when the newly elected President Ronald Reagan fired the CEQ's entire professional staff, thanks to Michael Weber I landed at a non-profit advocacy organisation called the Center for Environmental Education (later called the Center for Marine Conservation, now the Ocean Conservancy). My job was to integrate CEE's science and policy efforts in a climate made far more difficult because so many of the Reagan administration's political appointees came from extractive industries and lobbying organisations opposed to conservation. Given the unfavourable environment for progress, I felt compelled to learn from the far more advanced field of terrestrial conservation. So in

1983 I became the Public Policy Director of the Ecological Society of America (ESA), North America's professional society of ecologists. The great majority of my colleagues worked in terrestrial and freshwater ecosystems and few of them worked in conservation, but from those I learned more about conservation principles, laws and people that could be applied to the sea.

Fortunately my position at ESA allowed me to work with other conservation scientists including Michael Soulé, Ted LaRoe, Tom Lovejoy and Bill Conway to strategise about the political environment and the need for a new scientific professional society to help conservation. The Society for Conservation Biology (SCB) became a global organisation dedicated to advancing the science and practice of conserving the earth's biological diversity, with more than 12,000 members in over 140 countries. Building effective institutions takes time; SCB only took its first breath in 1985. But effective institutions take more than time: they need a driving idea.

To me the idea was biological diversity. At the request of the Wilderness Society, with some colleagues in ecology, wildlife biology and environmental law, I assembled and edited *Conserving Biological Diversity in Our National Forests* (Norse *et al.* 1986), the first book about conservation having biological diversity in the title, which was the origin of the three-level definition (genes, species and ecosystems) of biological diversity widely used ever since. That same year Ed Wilson held a widely attended conference that brought together large numbers of biologists focused on what then from then on became known as biodiversity, a more euphonious and memorable term for this idea (Wilson 1988).

Fortunately for me, in 1987 the Wilderness Society asked me to write a second book focusing on a region struggling to determine the fate of its last unlogged forests (Norse 1990). I called them 'ancient forests' and learned about them from brilliant teachers such as Jerry Franklin and Dave Perry. *Ancient Forests* became a primary source used by President Bill Clinton's administration in formulating the Northwest Forest Plan to save spotted owls and their old-growth forest ecosystems on public lands. The strange story of a marine biologist writing his first two books on forest conservation in effect gave me a PhD in a second conservation arena, one whose beliefs, rules and scientific basis proved invaluable for my later work in marine conservation.

In 1989, Roger McManus (whom I had brought to the Ocean Conservancy's ancestor) hired me to make marine biodiversity a national and global public policy issue. I saw my mission as creating the science and policy foundation for a global marine conservation movement. To start, I organised a workshop of leading scientific experts on marine biodiversity at the Smithsonian Institution's National Museum of Natural History. It generated ideas about what was needed to put marine conservation on the world map, including the need to raise its profile at the coming 1992 Earth Summit in Rio de Janeiro. I then worked with Kenton Miller, Walt Reed, Kristina Gjerde, Sue Gubbay and many others to assemble a book called *Global Marine Biological Diversity*, a compendium of relevant policies for conserving living oceans (Norse 1993). From the United Nations Conference on Environment and Development (UNCED, more often called the Earth Summit) in Rio to Washington to Geneva, we put drafts or published copies in the hands of people such as the UN delegates who were working on the new Convention on Biological Diversity (UNCBD, 1992–1993).

One of the key recommendations of *Global Marine Biological Diversity* was the need for a new science of marine conservation biology. Inspired by what I saw from the idea (Soulé & Wilcox 1980) and organisation that Michael Soulé founded, in 1996 I founded the Marine Conservation Biology Institute (MCBI, later renamed Marine Conservation Institute). Its two main aims were advancing the science of marine conservation biology and securing protection for the oceans' ecosystems. It became the most influential organisation shaping the mindscape of marine conservation and proved astoundingly cost-effective in winning protection for marine life and its habitats, key next steps for the marine conservation movement. Its success attracted many other non-governmental organisations to marine conservation, some of which have been very helpful.

The marine conservation movement did not take shape by happenstance. Like other movements I had witnessed, its clear, timely purpose attracted other people and organisations. But succeeding requires synergies that come from cooperation, not competition. The goal is larger than the players.

Given the steep odds against marine conservation – resulting from the power of opposing industries, the inertia of government and intergovernmental organisations, advocacy organisations' competition for press attention and money and myopic, short-term thinking of many funders – my efforts have failed more often than not. Saving the earth is neither a simple nor an easy task. Yet it feels good to have done things that have advanced marine conservation, notably through things that the Marine Conservation Institute achieved while I headed it or helped Lance Morgan, including:

1. To evaluate a seriously under-researched marine ecosystem disturbance, in 1996 Les Watling and I organised the first scientific meeting examining impacts of bottom trawling on the world's marine ecosystems. The participating scientists from five countries provided data and insights that led to our 1998 cover article in *Conservation Biology* that compared bottom trawling with forest clear-cutting (Watling & Norse 1998). Bottom trawling is far more widespread, probably making it the world's most extensive severe human-caused disturbance. Nearly two decades later this is still the most cited scientific paper on the impacts of mobile fishing gear and bottom trawling, and has led to many efforts to reduce bottom trawling around the world.

2. To launch the movement into a higher orbit, I organised the first Symposium on Marine Conservation Biology in Victoria, British Columbia, Canada in 1997. The symposium's 1,051 scientists from thirty nations examined ways to protect and recover marine biodiversity and established ties with one another. One eminent marine conservationist who witnessed marine science luminaries conferring between sessions said 'This is Woodstock.' Indeed, it was so successful that Lance Morgan and I organised the Second Symposium on Marine Conservation Biology in San Francisco in 2001. These were models and forerunners of the SCB's ongoing biannual International Marine Conservation Congresses.

3. To call attention to scientists' concern for our oceans, I wrote *Troubled waters: a call for action*, a statement signed by 1,605 conservation biologists and marine scientists from seventy countries (Norse *et al.* 1998). It said that the sea is vital to humankind but is in serious trouble, and it urges governments and individuals to take five key steps to reverse these threats. This unprecedented indication of scientific concern got major news coverage in the USA, Canada, the UK, France, Russia, India, China and Australia.

4. To advance ecosystem protection in US waters, in 2000 I organised and led a workshop of natural and social scientists on marine protected areas (especially

no-take reserves) as a way to maintain the sea's biodiversity. I then wrote to President Bill Clinton a letter that was signed by participating scientists, calling upon him to establish a national system of marine protected areas (MPAs). With drafting help from MCBI, President Clinton then issued Executive Order 13158, calling for a national system of MPAs, establishing the national MPA Center and the MPA Federal Advisory Committee. Nine years later the Obama Administration established the USA's national system of 225 MPAs.

5.   To establish the precedent that marine invertebrates can be protected under the US Endangered Species Act, in 2001 I led the team gathering existing scientific information on the status of white abalone, which became the first marine invertebrate listed as endangered.

6.   To curtail devastation from bottom trawling by creating 'poster children' for deep-sea conservation, in 2005 I wrote the *Scientists' statement on protecting the world's deep-sea coral and sponge ecosystems*, which was signed by 1,452 scientists from 69 countries (Norse *et al.* 2005). It signifies unprecedented concern by experts. IUCN then submitted the statement to the UN General Assembly in New York to show scientists' support of protection for structurally complex ecosystems in international waters.

7.   To further the growth of marine conservation biology, Larry Crowder and I assembled and edited the first textbook in this new science (Norse & Crowder 2005). Thousands of marine conservationists around the world who used it are now working on better ways to save marine biodiversity.

8.   To establish the precedent that marine areas can be both strongly protected and very large, I persuaded President George W. Bush's top environmental advisor James Connaughton to urge the President to give strong protection to a colossal area surrounding the northwest Hawaiian Islands. In 2006 President Bush did so by designating Papahānaumokuākea Marine National Monument, home to an incredible diversity of coral-reef species, millions of seabirds and endangered Hawaiian monk seals. At nearly 362,000 square kilometres (larger than Germany or New Mexico), it became what was then the world's largest no-take MPA (Figure 23.1). Having liked the win–win dynamic and results, I helped the Marine Conservation Institute and White House staff to work together until President Bush designated Pacific Remote Islands Marine National Monument in 2009, just weeks before he left office. Later Lance Morgan's Marine Conservation Institute team provided the scientific basis for President Barack Obama's further enlargement of Papahānaumokuākea and Pacific Remote Islands Marine National Monuments, leading other world leaders to protect huge ocean areas.

9.   To facilitate a quantum increase in place-based marine conservation, in 2011 I envisioned and began drafting the outline for the Global Ocean Refuge system (GLORES), which Lance Morgan and the Marine Conservation Institute are now advocating to catalyse strong protection for 30% of the ecosystems in each marine biogeographic region of the world's oceans by 2030. GLORES will be the world's network of safe places to maintain our seed bank of marine biodiversity. GLORES can succeed if it receives sufficient sustainable funding from private donors and public institutions.

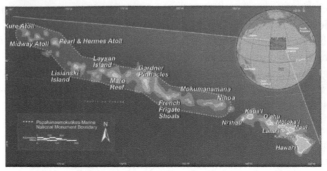

Figure 23.1 Papahānaumokuākea, the Hawaiian Islands Marine National Monument: one of the world's largest MPAs. Source: NOAA

10. To facilitate more protection for vulnerable marine species and ecosystems, I organised and led a team of eminent ecologists, fisheries biologists, economists, mathematicians and international lawyers to write the first comprehensive inter-disciplinary synthesis on the potential for fisheries sustainability in the deep sea (Norse *et al.* 2012). It showed that economics drives humankind to mine life in the earth's largest and least protected ecosystems, precluding sustainability. With marine shallows now severely depleted, these findings are being used to reduce fishing pressure in the deep sea.

LOOKING BACK, WHAT ARE THE MAIN LESSONS YOU TAKE FROM YOUR LONG INVOLVEMENT IN MARINE CONSERVATION?

'To be or not to be?' Humankind's existential challenge is sustaining the earth's life-support capacity. Despite many heartening successes in temperate zones, the tropics and polar regions, on land, in nearshore waters and open oceans, the preponderance of evidence shows that conservationists are not winning. Billions of human consumers armed with rapidly growing technologies have already severely reduced marine biodiversity. Much worse is yet to come unless the small but uniquely capable subset of humankind – the marine conservation movement – finds the way to make maintaining life on our planet a top global priority. I do not think it is too late, but the clock is ticking. That means marine conservationists must relinquish the every-scientist/political operative/organi-sation-for-itself mode that critically impedes this movement. Only recognising that our current course is failing can push us to transcend ourselves as a movement, to do more than we have done thus far. That will not be easy, because working together to gain influence is not a situation many marine conservationists find familiar.

Do we have what this requires? Do we love our grandchildren – many of whom are yet unborn – enough to devote ourselves to winning together? Can our movement coalesce to save marine mammals, seahorses and ourselves?

Do we have the vision, strength and will to go for the GLORES, the scientifically designed worldwide system of Global Ocean Refuges that will be the seed bank for species that drive the sea's ecosystem processes?

I retired because it was time for me. The movement to which I devoted my professional life was not yet working effectively as one. But if the most knowledgeable people work together to save the oceans to work together, we still have a chance. Can we do it?

To be, or not to be, that is the question.

CHAPTER 24

Action

This chapter looks to the future, setting out some of the key themes to emerge from the interviews. The first part looks at the main threats to the marine environment, and this is followed by an exploration of the barriers to effective action. The final section describes the range of approaches that have been used to achieve change.

THREATS

It is now widely recognised that we have entered into a period which has been termed the Anthropocene mass extinction. This is the sixth mass extinction in geological history, and the rate of species loss recorded is on a par with those of the previous five. It is occurring at every spatial scale, from local to global, and the causes are entirely man-made. The marine environment has not been spared in this process, and indeed, as Dan Laffoley points out, 'We already know that the combination of ocean acidification, ocean warming, and deoxygenation accompanied the last five major extinction events in earth's history.'

The damage caused by some activities and events is very obvious and immediate: for example, the effects of an oil pollution incident, or the devastation caused by seabed trawling or mining. Others take a long time to show us their true danger, the so-called 'slow train issues' (Olson 2016), and these have been elaborated by the European Environment Agency in the report *Late Lessons from Early Warnings: Science, Precaution and Innovation* (2013). These issues include climate change, ocean acidification, the effects of plastics and the 'homogenising effect' of invasive species (Callum Roberts). Unfortunately many of these impacts are working in a synergistic way to produce cumulative effects that are becoming increasingly difficult to counter.

Extended lists of threats and their impacts have been compiled (Figure 2.2), and Jon Day refers to a matrix of threats compiled from a management perspective for the Great Barrier Reef. Reviews such as that by Jackson *et al.* (2001) and Rogers (2013) take a step back and review the state of play. In asking about threats, I was keen to get the interviewees' sense of the future priorities and how we might address these.

Economic imperatives driving development with little care for the marine environment

An all-pervasive and insidious theme that commonly emerged in the narratives is the impact of economic development and resource exploitation. This, combined with greed, with little or reckless regard for the environment, is a huge challenge. The continuing growth of the human population and the priorities of economic development and of consumerism, not least in developing countries, are placing huge pressures on the space and natural resources of the coastal and marine environment. Hance Smith (2000) wrote about the industrial revolution in the ocean, and Bud Ehler and many others have identified growing industrialisation of the oceans – so-called 'blue growth'.

Bud Ehler argues that using marine spatial planning (MSP) should help to resolve conflicts between users and 'achieve the real benefit of planning, to avoid the social and environmental costs of conflicting and damaging uses'. Joan Edwards expressed her concerns about the UK government's disregard for many of the processes that usually regulate development, and how politicians all too often run roughshod over environmental concerns. When the Cameron coalition government took power there was a fundamental change in outlook towards the environment, but it takes time realise this. What to do about it when tried legal measures like judicial review fail is quite another matter. The long-standing and still prevailing attitude of large corporations, many of which operate internationally, is described by Paul Horsman. These corporations' approach is frequently to impose their world view on developing countries using outdated technologies that have been rejected on environmental grounds by more developed countries, and to pursue development with scant regard for recognised environmental management practices. In the USA, Donald Trump personifies this thoughtless disregard for the environment, and it is clear that this attitude is a huge issue that will occupy the marine community for many years to come.

The methodologies around economic valuation and assessment of the environment have been around for thirty years and will continue to develop and be better understood in the context of supporting decision making at an operational level. However, as was illustrated with the introduction of the Habitats Directive, legal backing to protect important wildlife sites proved a far stronger lever than previous attempts to value nature in achieving environmental gains (Peter Barham). In reality, the immediacy of economic gain still often overrides more considered economic assessments that include the environment.

Climate change

Climate change was the most often cited threat that interviewees raised, noting the sheer scale of the related issues and their daunting and depressing nature. There is little doubt that the impacts of climate change are real, occurring now and increasing in impact. Bud Ehler, recognised for his contributions to climate change science, put it in perspective: 'There is no doubt in my mind that the biggest issue is climate change, bigger than the ocean, and it is happening faster than we thought. Climate change is going to be a difficult to turn around for at least twenty to forty years.' It is manifest in terms of increased water temperature, ocean acidification and the impact of more powerful storms, especially in the shallow coastal waters of the tropics. The marine conservation community is fully aware of the issues, not least from impacts like coral bleaching (Figure 24.1). Euan Dunn highlighted the regime shift in northern waters which has resulted in ecosystem-scale changes in plankton, population shifts in marine fish stocks and changes to seabird populations, with the North Sea warming faster than adjacent sea areas and being cited as 'a cauldron of climate change' (Rutterford *et al.* 2015).

The role of the marine scientists, especially oceanographers, in monitoring global changes is vital, as is improving our understanding of changes such as ocean acidification. For the marine conservation NGOs, the range of options for action on a global issue such as climate change is relatively limited. Encouraging public awareness, 'bearing witnesses' and lobbying with their terrestrial colleagues is perhaps all that can be done, since many of the drivers of change come from land-based policies on energy. Apart

Figure 24.1 Coral reef: the sequence from living through bleached to dead coral illustrates the immediate threat to coral reefs across the world. Source: the Ocean Agency, XL Catlin Seaview Survey

from the very poor quality of emissions from shipping, and issues of offshore energy, there is relatively little that NGOs and governments working on the marine environment can do about emissions. Interestingly, whilst the NGOs have been generally supportive of renewable energy in the UK, the sheer scale of offshore wind development is of concern, and Joan Edwards and Euan Dunn highlight this. Given the scale of the threats posed by climate change to the environment and conservation in general, it is perhaps surprising that the marine NGOs, which have a vast membership to draw on, have not made more of this resource to try and encourage more collaborative action by civil society as well as by government.

Pollution

From the 1960s to the 1990s in European waters, marine pollution was the highest issue on the agenda, and although it has slipped down today it has not gone away. Some pollutants, such as plastics and oil, are highly visual, but others, including toxic chemicals, are out of sight – though their effects may be far more insidious. A clear example is the situation with persistent organic pesticides that are simply accumulating in the food chain, and in particular in marine mammals (Jepson *et al.* 2016). A stark illustration of the effects of pollution was the death of Lulu, a female killer whale, off the UK in 2017 (Figure 24.2). This individual had never given birth and had the highest levels of PCBs ever recorded in a marine mammal, a level so high that it was believed to be sufficient to sterilise the animal. In the developing countries all the usual pollution suspects are very much in evidence, and are covered in the interviewee's narratives: eutrophication in China (Callum Roberts) and Australia (Jon Day), and dumping waste at sea in South America (Paul Horsman).

More recently the identification by oceanographers of the amount of plastics in the oceans has alerted the world to the scale of plastics pollution. As plastics break down to ever smaller pieces, and as we use 'microplastics' in many products, these contaminate *every* marine and coastal environment from the poles to the deep sea (Figure 24.3). There is a massive sting in this plastics tail, not only because a wide range of species, including plankton, consume them, but also because plastics attract other toxic chemicals such as pesticides. Both plastics and their toxic companions are building in the food chain. Pollution in certain circumstances can be just as problematic as in the past, such as when storm-drain overflows flush raw sewage into rivers, or when eutrophication, extensive seabed organic contamination causing 'dead zone' and medicinal chemical pollution come from caged salmonid fish farms. The wide-ranging effects of underwater noise

Figure 24.2 Lulu the killer whale, stranded in 2017 after probable net entanglement. She had the highest levels of toxic PCBs recorded for the species. Source: Nick Davison, Stranding Coordinator, Scottish Marine Animal Stranding Scheme, SRUC Veterinary Services

pollution are gradually being understood. Vigilance – not least by the NGOs – is going to be needed if more pollutants are not to surprise us and cause serious problems in the future.

Fishing

From the 1980s, concern, and then gradually understanding, grew as to the scale of impacts of fishing practices on the seabed and mobile species. In summary, fishing is a major long-term threat to the marine environment, a threat that takes many forms and is not being addressed with sufficient urgency. The fishing sector is unlike any of the big industrial sectors operating in the marine environment. In Europe, for example, every large sector from aggregates to ports and from renewables to oil and gas operates in a highly professional manner, which is heavily prescribed by regulation, licensing requirements and comprehensive understanding of the life cycle of environmental management

Figure: 24.3 Razorbill entangled in the plastics from a balloon release. Source: Christine McGuinness.

requirements (see Box 3, page 38). The same cannot be said for the fishing sector, where there is often scant regard for environmental consequences.

The progress that has been made has been painfully slow, gradually breaking down the barriers between government agencies and then with the fishing industry. Garcia, Rice and Charles, in their book on the governance of marine fisheries and biodiversity conservation (2014), explore the tension that arises between two interconnected but historically separate streams of governance. Unregulated fisheries, especially on the high seas, still prosper because the economic drivers are so huge. Deep-sea fishing using highly destructive technologies is clearly unsustainable and akin to mining the resource (Sarah Fowler, Heather Koldewey). The massive fisheries for sharks, in particular for their fins, but also for other large fish species, are fundamentally changing the marine ecosystem by fishing down the food web. This is having a wide range of consequences for many species of fish as well as other species and habitats. In addition to its effects on the environment, this has a worrying effect on public perception, as it shifts the baselines which people take as normal. There is also the new challenge of addressing the impact of climate change on dynamics of fish stocks and their geographic distribution.

There is a growing trend, not least among the scientific communities, and particularly across intergovernmental agencies such as ICES, to understand the shared issues of fishing and conservation. In the Western fisheries model there seems to be a growing acceptance of a number of ideas, including that science needs to be taken seriously, that work on the supply chain to help reporting and accountability to consumers is essential, and that the greater involvement of stakeholders including the NGOs in the debates about fishery management is important. There also need to be very deliberate measures that recognise the long-standing nature of the problem and bring these two interest groups together, for example by means of the European Marine Fisheries Fund (EMFF).

Many of the trends and ideas developed recently have flowed from the application of the precautionary principle and the ecosystem approach. Conservationists and fisheries managers have recognised ecosystem-based management around the world and the potential for joint working, but there is scope for more guidance on this.

Lack of clarity of objectives and political weakness often led to catastrophic over-exploitations of fish stocks in the past, although some management systems – in New Zealand, for example – are draconian in terms of their punishment for non-compliance. However, politicians, with fisheries in particular, tend not to change their spots and in many settings a lack political will, combined with fishermen gaming the system, causes stocks to continue decline. The EU's Common Fisheries Policy is beginning to show signs of reform, but the UK, post Brexit, will have to address a host of fisheries issues from scratch.

Cumulative effects and the need to take an ecosystems based approach

The basic idea behind cumulative effects, many impacts working together to cause damage, effectively 'death by a thousand cuts', is easy to understand. Quantifying and manging it, however, is a major challenge. We can now see cumulative effects at different geographic scales and in a range of habitats. On the Great Barrier Reef, Jon Day describes fifteen or more major threats, including many from land-based sources, damaging the reef environments. On a smaller geographic scale Sue Sayer describes threats to the Cornish seal population that include the damage caused by particularly strong sequences of winter storms, disturbance by tourism, underwater noise, as well as shooting and injury to seals in harbours from boats and where people feed seals. Taken together, these can have a significant effect on the population. There is a growing awareness of cumulative effects, and more integrated and holistic ideas like the ecosystem approach and marine spatial planning should give us the principles, frameworks and tools to try and get to grips with this, to reverse the historic damage which has been done.

BARRIERS TO ACTION

At its core the challenge of marine conservation is to achieve change. There are many barriers to change, and sometimes working together to recognise and successfully address those barriers is at the heart of conservation work. It is also the case that progress on marine conservation can be tortuously slow, so the interviewees were asked to describe their experience of the real and hidden barriers they had encountered. Since these barriers to action are unlikely to go away, a better understanding of them and how they might be addressed is key to future progress.

Understanding resistance to change

Understanding why there is resistance to change is not clear-cut. Conservationists are often challenging long-held attitudes that run through the psyche of the governments or large corporations. 'Vested interests run deep and reach far into the past; the relationships between government and business are more baroque than you might ever suspect, and unravelling these interests is very difficult' (Callum Roberts). Similar sentiments were expressed by Chris Rose, who put it like this: 'Economic and technical solutions are often perfectly possible in practice, but they are not perceived to be politically possible. If you start digging into why this is, you often find it is based on outdated perceptions and assumptions that are not founded on very much and certainly no rational answer.' Recognising this complexity is important even if it doesn't lead to easy resolution of issues.

The scale of the problem

The sheer scale of the issues we face is huge, in particular at the larger geographic scales, so much so that this can be a significant barrier to change. Dan Laffoley cites the ideas of oil-industry strategist Adam Kahane (2008) on 'tough problems' as a way helping to resolve these issues. Marshalling the evidence and then organising the stakeholders at appropriate geographic scales is a big job which takes high levels of commitment, organisation and funding.

Time and timing

Common to all the narratives is the question of time, and the timing of interventions. Why change takes so long is a recurring question. In the UK it took eleven years to protect the basking shark (Speedie 2017) and 25 years to get adequate protection from destructive fishing practices for marine protected areas (MPAs) (Clark et al. 2017). The Beachwatch anti-litter campaign began in the USA and was taken up in the UK in the early 1990s. It has taken thirty years for the plastics issue to reach the top of the environmental agenda, largely because the plastics industry, in framing the issue in terms of 'litter', shifted the responsibilities from the producers to the users. The impacts of some threats, like climate change, by their very nature take a long time to show themselves.

Politicians can be the ultimate procrastinators, and in the face of almost non-stop noise from multiple issues they can suffer from 'attention deficit disorder' where issues simply drop off their radar (Sarah Fowler). Continuity and persistence from NGOs are particularly important. The political process, with its five-year election cycle – 'not in my term' – and in the short term the delays caused by elections, can have a significant effect on projects.

By contrast, when governments wish to, they can act astonishingly quickly. Joan Edwards describes the electric pace at which the UK government stopped poly-isobutanes (PIBs) being released from shipping. NGOs spend lots of time trying to get issues onto the political agenda, but it is also the case that governments can suddenly take the lead – and this is the time to seize the opportunity and engage with the political process. Understanding and recognising that this can occur, as with other types of unplanned events, is important – as shown, for example, by the UK government's recent response to the plastics issue highlighted by the *Blue Planet* series.

Understanding what is happening on a timeline can be difficult. When do you accept delay or press for more action? 'Conservation is a long game' (Roger Mitchell) and 'it takes time to bring people along with you' (Keith Hiscock) both have more than a grain of truth. If you look at society's change in attitude to smoking and seatbelt use, these have both taken several generations. But these views prompt us to ask how long one needs to wait before other tactics or approaches are used to try and achieve change. Finally, whilst a battle might be won, a new Act passed, a campaign won or a new species added to a listing, the delays in implementation can be enormous. These can often be deliberately delayed, as a function of the lack of political will.

Politicians, political will and understanding the political dynamics

It is a given, described by the majority of interviewees, that to achieve change in democracies you end up working with politicians or coming into conflict with them. There is no better introduction to the realities of this than books in the *Yes Minister* series, which have the merit of being both funny and too true to life, capturing the labyrinthine way that politics and the civil service work (Lynn & Jay 1986). Politicians are subject to a huge array of influences, and the process of reconciling these competing voices is problematic (Figure 24.5).

Figure 24.5 Science is only one of the inputs that influences political decision making. Source: Larcombe (2006)

Coming to terms with such different political agendas under different government regimes is fraught with problems. In the UK the contrast between working with the almost utopian Blair/Brown governments, who were pro-environment and achieved the major harmonising reform of marine legislation, contrasts starkly with the recent Cameron 'cut the green crap' and May administrations. Austerity is the excuse they have used for cutting Defra's funding by 50%, but can be no excuse for gagging its agencies and tightening the rules of charities including conservation NGOs on speaking out, and doing away with their commitments to sustainable development. The deregulation agenda has 'hollowed out' (Euan Dunn) and weakened environmental regulation under an attack on 'red tape'. Although Trump has gained headlines for acts to reduce environmental protection, the current UK government, in its more low-key, very 'British' way, has degraded it just as much.

Not surprisingly, therefore, there is often a failure of political will to take forward measures. Ignoring scientific advice and not adopting strong enough environmental measures to protect fish stocks under the CFP process is a stark example of a continuous failure of political will. Chris Rose is less flattering in putting some of this down to a lack of scientific literacy among MPs, 'which continuously leads to a systematic failure to act on evidence, simply because they do not understand it'.

There is a host of experiences with politicians described by the interviewees, including:

- Wild-card decision making where government axes a project for no rational reason (Dan Laffoley)

- Politicians saying one thing to your face and another in public (Sue Gubbay)

- Politicians signing up to conventions and agreements to fanfares and then doing nothing to implement them (Heather Koldewey)

Conversely, on the very positive side, Bud Ehler and Elliott Norse describe the process of helping a number of US presidents create very large marine sanctuaries as a part of their legacy programmes.

Using the lack of information as a way to block progress

NGOs often produce reports to prompt action, and the interviewees all supported the evidence-based approach to decision making and management. Governments take a host of positions to suit their political ends, not least in collecting *policy*-based evidence: decide on your policy, then collect the evidence to support your decision. Calling for more evidence is one of the classic ways that governments delay action and take the sting out of issues. They set up a review, calling for more evidence, and then if they don't like the outcomes they deploy various strategies to stifle its recommendations. The choice to ignore the advice of a host of chief scientists in instigating the politically expedient badger cull is a perfect example of the contrarian way governments act. 'Lack of evidence' is a common excuse to delay action, and the UK government for example has sought ever greater levels of information about where to site MPAs, as a way of delaying action. In different contexts the need to use 'existing evidence' is a mantra frequently used by civil servants when government is proposing a new development, with a view to reducing the cost to the developers and government agencies (Roger Mitchell, Keith Hiscock). Essentially, governments can do what they like, and take entirely contrary positions on evidence and information when it suits them.

Funding

Anyone who gets involved with conservation for any length of time will experience a host of funding issues, ranging from seeking funds for projects through to developing sustainable funding models in capacity building. The lack of funding can be a significant barrier, and for NGOs in particular this can be a constant challenge, since fund raising is a time-consuming and demanding task. Seed funding to lever change to get government action is a tactic deployed by NGOs to break down barriers (Sue Gubbay), and independent foundations have recognised this and the need to provide an adequate time frame to achieve the desired outcomes (Sarah Fowler).

On the downside, funders can switch off funding almost on a whim, arguing that 'we've done our bit'. The austerity approach (death by a thousand cuts), by means of which government departments fail to fulfil some of their basic commitments, is 'symptomatic of decay in a government infrastructure that has taken years to set up' (Euan Dunn). A major problem in the current financial climate in the UK and USA is what happens when government cuts funding to its environment agencies, leading to areas of work that have been a mainstream activity by those agencies being dropped. This is a political choice, but it is difficult to reconcile in many ways and can leave the NGOs once again having to step into the breach.

The failings of marine conservationists and their organisations

One of the surprising themes which was developed in a candid manner by many of the interviewees was the failure of conservationists themselves, not least the NGOs. This is also the one area which is within their gift to solve.

1. **Competition between organisations**. Organisations are very competitive: the problem of 'egos and logos' (Heather Koldewey). NGOs by their very nature claim issues as their own, and this does not lead to effective collaboration. This is a very large issue for a host of reasons, not least the need for NGOs to explain themselves to their members and the public and for fundraising. The issue of competition between organisations is not confined to the NGOs, as government departments also have turf wars over particular issues, when collaboration would lead to a much more successful outcome.

2. **A failure to agree on objectives**. A given of successful collaborative working and negotiation is developing common objectives. This is key to enabling the variety of styles and tactics of different organisations to come into play with effective collaboration. For NGOs this can be critical, because a standard government tactic is to divide and conquer. The interviewees highlighted this critical point in a host of settings, from MPA negotiations to agreeing habitat compensation (Peter Barham) or bringing issues into the open in order to foster joint working (Bud Ehler). Not having shared objectives is a recipe for very slow progress.

3. **Being perfect**. A 'we know best' attitude and being such a strong advocate of conservation that you turn off people who would support you is highlighted by Bud Ehler. Similarly Dan Laffoley highlights the 'perfect being the enemy of the good' scenario which achieves nothing compared with a compromise which may offer a much better return. Recognising the value of people with the right skill sets, especially with negotiation, can be incredibly important to breaking down barriers.

4. **Keeping on doing the same things, even though they don't work.** After a while the impact of using the same approach will start to level off. Chris Rose put it thus: 'In NGOs there is often resistance to trying new things, approaches that work, because they like doing the same old things, which they would like to work but don't.' There is an ongoing need to carefully assess the progress of campaigns, and to change tactics if necessary. It is also important to try new things, and have the freedom to do this, because even if they fail you learn a great deal.

5. **Not communicating effectively.** Dan Laffoley elaborated on the failure to recognise the gap between what conservationists know ('we know too much') and the lack of understanding in the people whose views we are trying to change. Recognising this and taking steps to address it are described by many of the interviewees. In the early years it was the terrestrial conservationists who simply didn't get the marine environment. Breaking down barriers between conservationists and fishermen can be done, but it often requires very deliberate processes that are aimed to achieve this (Joan Edwards, Euan Dunn). As Chris Rose points out, conservation organisations frequently pay much too little attention to how and what they communicate.

MAKING A DIFFERENCE – ACHIEVING EFFECTIVE CHANGE

Acting to change and if necessary reverse the worst human effects on the marine environment is at the heart of marine conservation. All the interviewees found the status quo unsatisfactory and challenged it, and this has translated into change and tangible protection. This section describes the different ways people have gone about their work, but trying to describe and rationalise this is difficult. It goes beyond a simple classification of the disciplines applied to marine conservation such as that shown in Figure 2.2, since they have used many different strategies and tactics as individuals, within organisations, or by working with coalitions or partnerships of like-minded organisations. Achieving change is demanding, and the examples and the lessons they describe reveal the difficulties, the nuances and the stark realities of working in conservation.

The five categories described below represent a pragmatic attempt to cover the recurring themes of the narratives:

- Change management
- Models of action
- Conservation styles of different organisations
- Common work themes
- Innovation and new opportunities

There is of course some overlap between these categories, and other actions, including capacity building and working with people in different ways such as collaborations, are major and recurring themes throughout the narratives (see Chapter 4).

THE PROCESS AND STAGES OF CHANGE MANAGEMENT

In Chapter 4 there is a description of the life-cycle approach to project management. This logic is being applied to the management of conservation projects too, and this section elaborates on some examples given by the interviewees.

Starting – bearing witness

On an early survey of fish farming in Scotland we were able to photograph a seal trapped in an anti-predator net, which was hugely helpful in raising the issue of predator control on fish farms (Figure 24.4). More recently I was introduced to eXXpeditions (http://exxpedition.com), an organisation which attracts crews of women to sail the world's oceans to record and raise the profile of what they see, from wildlife to plastics pollution. This approach has a long tradition. Paul Horsman describes how Greenpeace drew on its Quaker traditions of 'bearing witness' as a key point in its mission to show the public what is going on in remote parts of the oceans. Before the 1960s the marine environment was often taken to be 'out of sight, out of mind', but since then, and coincident with the growth of diving and many other new technologies, it has become much easier to show people what is happening beneath the surface of the sea. The influence of television and 'A-list' celebrities was emphasised by almost all of the interviewees, as outlined in Chapter 2.

Collecting the information and building the evidence

There is a strongly shared belief among all the interviewees about the need for an evidence-led approach. Science has played a key part in the development of marine conservation, not least from the ideas and observations it has generated, but the relationship between science and marine conservation is complicated. The starting point for many campaigns and conservation programmes has been to collate or collect information on the issue.

Conservationists often undertake science and produce reports in order to enable change, rather than simply advance science. For example, Sue Gubbay explains how the production of the Coastal Directory in 1985 eventually levered change by shaming the NCC into starting a long-delayed review of the UK's shallow marine habitats and species. Greenpeace used the evidence of the need to protect deep coral reefs in its high-profile and successful court case against the UK government to get clarification that the Habitats Directive extended to the limits of the UKs exclusive economic zone.

Figure 24.4 Seal trapped and killed by a fish-farm anti-predator net in Scotland. Source: Gordon James

Two rather interesting conversions of scientists into conservationists illustrate the grim realities of why just doing science and hoping other people will act won't suffice. Sarah Fowler told me how Samuel H. Gruber, a highly respected shark biologist, worked for many years at the Bimini Biological Field Station. When the population of sharks he was working on crashed to virtually nothing he realised that being a scientist was not enough, and he became one of the world's leading advocates for shark conservation. Patricia Majluf, the founder of Peru's leading marine conservation organisation, had a similar experience. She had spent a career studying fur seals when a major collapse of their population, caused by the decline of the Peruvian anchoveta, prompted her to fundamentally re-assess what she was doing. She went on to play a leading part in the rejuvenation of the Peruvian anchovy fishery and many of its dependent species.

This leads to another question, about whether there will ever be enough science to allow for rational exploitation of species, and this takes us into arguments about relative risks and the precautionary approach. The question of whether to wait for sufficient knowledge to accumulate whilst uncontrolled exploitation continues is difficult, especially because, for many species, we will probably never have adequate funding to understand their ecology.

Conservation is not just about doing science. It involves the application of science to further the protection of biodiversity and the marine environment. Roger Mitchell makes this point very strongly, because he always saw the science as a means to an end, not an end in its own right. Callum Roberts is a scientist, but he sees conservation as an applied science which takes him into strong debates with others who think that championing particular ideas undermines his scientific credibility. He takes the contrary view, seeing this approach as something that makes his science important. Measuring things, knowledge, or the collective will of the scientific community are often not enough to effect change, and this is acknowledged by many of the interviewees. Once a certain amount of information has been gathered there is the requirement to act.

Recognising the problem and deciding to act

Recognising that there is a problem is a critical step. There is a range of scenarios that determine how people make this decision. In some cases the overwhelming evidence prompts a clear strategic decision to initiate a conservation programme. In others, the decision to act arises from a chance meeting, a report of unusual activity, or an unplanned event. Whatever the impetus, the action moves on to activities that are geared to achieve change (see *Models of action*, below). The interviewees have described a range of examples in these two scenarios.

The need for strategic planning and decision making is described throughout the book, not least by Paul Horsman and Dan Laffoley. Good examples of strategic programmes are provided by Joan Edwards on the long haul and focus on trying to achieve the Marine and Coastal Access Act, and by Euan Dunn and Sarah Fowler on the work done by the Pew Trusts to get European NGOs to collaborate on getting the EU to act.

Examples of events that have prompted action include the leak of information that prompted Greenpeace to campaign against the dumping of the Brent Spar (Chris Rose, Paul Horsman); the successful work on the PIB spills that led to tighter controls on shipping (Joan Edwards); the response to the phocine distemper virus in the North Sea, which led to the establishment of British Divers Marine Life Rescue and what is now the routine response to marine mammal incidents (Alan Knight).

Mainstreaming, implementation and maintenance

The objective of many NGO campaigns is to get government – society – to take responsibility for issues. A common theme of the narratives is work on legislative process so that issues become 'mainstream'. Once legislation is passed, then implementation becomes a challenge – and this can be protracted and time-consuming. Jon Day describes how it took thirteen years to complete the first zoning of the Great Barrier Reef after the GBR Act was passed. Marine conservation also requires a great deal of ongoing vigilance to ensure that management measures are being maintained.

MODELS OF ACTION

The interviewees describe a combination of methods and ideas to achieve particular ends. Those described below by no means constitute an exhaustive list, and there is considerable scope to develop this idea.

The science–policy–management model

A traditional view of marine conservation uses the science–policy–management paradigm, which assumes that if one does the science, and collects the evidence, this will translate into a logical and objective policy eventually leading to effective management. Joan Edwards and Dan Laffoley describe ten years' work leading to the Marine and Coastal Access Act (2009), involving numerous committees, reviews and reports. In the end, following the collaborative work of a host of sectoral stakeholders, there was an almost unanimous and overwhelming desire to introduce this legislation in support of marine management fit for the twenty-first century, uncluttered by forty years of legislative baggage.

Nevertheless, whilst it might be ideal, this model does not necessarily bring results. In practice, an array of barriers and vested interests, whose objectives are often economically driven, oppose this. Piers Larcombe (2006) highlights the range of influences that affect political decision making, showing that science and evidence are only one of 22 other interests competing for politicians' attention (Figure 24.5).

The problem–solutions model and consumer-led change

This approach involves promoting the problem and the solution simultaneously, without waiting for government to support it. One of Buckminster-Fuller's quotes captures the idea perfectly: 'You never change anything by fighting the existing reality. To change something, build a new model that makes the existing model obsolete.' Chris Rose refers to the Greenfreeze fridge campaign, in which Greenpeace encouraged the manufacture of fridges that did not use the highly polluting HCFCs at the same time as campaigning to ban their use, and Paul Horsman describes how Greenpeace demonstrated the solar-panel solution while simultaneously pressing for the greater use of renewables. Working directly with fishermen to change their gear and innovate new designs to reduce or eliminate bycatch are other examples of this approach, described by Euan Dunn in relation to seabirds and Alan Knight in relation to whales.

Simon Brockington describes another version of this model: 'Conservation is about change management, and it strikes me that a lot of conservation issues gain traction at the point where people engage with them. MCS's work on fisheries was a consumer-led programme using the Good Fish Guides, and when a certain number of consumers engaged, the supermarkets followed suit. When the supermarket buying policies changed, that led to a change in the way the fisheries were managed in order for the products to be sold.'

Community-based models
Heather Koldewey's description of the development of local networks in the Philippines to take forward seahorse conservation, MPAs, mangroves and then disaster relief is a brilliant example of a community-based model. Sue Sayer's work on setting up networks of volunteers to observe and record seals and other marine wildlife, and more recently to help understand entanglement in fishing debris, is also an example of a community-based scale of working.

Direct action
A number of environmental organisations make use of non-violent direct action in the marine environment. Greenpeace and Sea Shepherd are among these. Paul Horsman describes the rationale behind this in some detail in his interview.

Making best use of unplanned events
Interestingly, unplanned events that are bad in themselves may at the same time provide a massive opportunity for raising awareness and subsequently stronger legislation to prevent them recurring, or being able to respond more effectively if they do happen again. Some notorious examples are oil spills (see Box 4, page 43), each of which can act as a major spur towards greater public awareness and more effective environmental management. As discussed above, British Divers Marine Life Rescue also arose from an unplanned event, the mass mortality of seals in the North Sea in 1988.

Joan Edwards has been involved in a range of activities arising from unplanned events, and she points out that disasters 'can provide a major opportunity. They have the ability to raise the public's awareness of the marine environment … The events provide a hook to put the marine environment in the news, and remind people that it should not be taken for granted.'

CONSERVATION STYLES OF DIFFERENT ORGANISATIONS
Marine conservation organisations are enormously diverse, including marine departments in much larger organisations and single-issue groups. The development of these organisations over the last fifty years is one of the key themes of the book, and it is clear that they have very different styles of working. The first big distinction is based on whether the organisation is governmental or non-governmental, then the diversity of style within the different non-governmental organisations (NGOs) is enormous. The interviewees have worked in a very wide range of these organisations, and their observations on the styles in which they operate is a key feature of their narratives.

Non-governmental organisations
Marine conservation NGOs have very different styles from governments and their agencies, and they have greater freedoms and are far less constrained in the ways they can conduct their work. Even though their underlying ethos often varies considerably, they often have common objectives. Their different styles when applied to the same problem can be complementary and enable more pressure to be brought to bear. The 'space' that direct-action groups create for other NGOs to operate, for example, is described by Paul Horsman, Euan Dunn and Sarah Fowler. Conservation organisations need lots of different tools in the toolbox, and need to know how to apply these. Some examples:

- **Innovation.** Often there is no apparent precedent, so 'innovation' is required, although it is often not called this. Virtually every interviewee has pioneered new ideas, projects and ways of doing things.

- **Communication – providing a voice.** It has always been the case that NGOs can communicate what cannot be easily or comfortably communicated by government departments. This is more important than ever in today's world, when government conservation agencies in the UK are effectively gagged by the government, and agencies in the USA are subject to Orwellian language control by the Trump administration.

- **Leverage by embarrassment.** Publishing the first Coastal Directory demonstrated that it could be done, shaming the NCC into starting the long-postponed Marine Nature Conservation Review.

- **Persistence.** Governments play a long game, often deferring action until it becomes politically very difficult not to act. NGOs can be very persistent in their pressure to deal with issues. If it does not work the first time, they keep on trying.

- **NGOs working with government.** Governments periodically change their spots, inviting NGOs to work closely *with* them. Whilst on the one hand it is flattering to be invited 'into the tent', there can be a serious downside – and Callum Roberts tells us how the NGOs' position can be fundamentally compromised. Euan Dunn describes this as making the NGOs 'partially pregnant'. This is why NGOs like Greenpeace and Friends of the Earth often deliberately choose to stay out of joint programmes with government.

Government organisations – mainstreaming

About half the interviewees have worked for government agencies, both full-time and as consultants. As the marine conservation agenda has developed and work areas have been recognised, many topics have moved from being largely the preserve of the NGOs into becoming part of the mainstream work of government and its agencies. Simon Brockington describes the differences between working for an NGO, where the members and volunteers were the 'customers', and Natural England, where the government was the customer. Dan Laffoley points to how the ethos of government nature conservation departments can change depending on the prevailing government view. Government agencies can, however, get access to business in a way the NGOs cannot. Roger Mitchell describes how the major oil companies were persuaded by NCC that the continuing bad PR from oil spills was unsatisfactory and agreed to a number of initiatives which substantially reduced the amount of oil washing up on shores, as well as funding oil-spill sensitivity maps and seabird assessments.

COMMON WORK THEMES

The marine conservation structure diagram (Figure 2.2) shows how the subject spans a wide range of disciplines and approaches. The three that are described here in more detail were strong recurring themes in the interviews. They cover work on legislation, marine protected areas and the wider marine environment.

Legislation and international conventions

Working on international conventions, legislative process and policy making is a very common theme for the vast majority of the interviewees, at a UK, European and international level. Six months into my job at MCS I had to go to meeting with MPs in the House of Commons. This was my introduction to working on legislation, and to say that it was a culture shock, after my previous work as a lecturer, would be an understatement. A degree in natural sciences does not really prepare you well for working with legislators and policy makers.

The interviewees gave many examples of engaging in policy making at these levels, including working on the UK's Marine and Coastal Access Act and the EU Marine Strategy Framework Directive; working with US presidents to put in place the big legacy of marine sanctuaries; contributing to CITES, CMS, ICCAT and North Sea ministerial meetings; attending IWC scientific meetings; and working to promote the ecosystems approach in European fisheries.

Key points to emerge concerning this legislative work were as follows:

1. Creating primary legislation can take an enormous amount of time. For example it took nearly eight years of preparatory work before the Marine and Coastal Access Act was passed.

2. Even after conservation legislation and conventions are in place it can be very difficult to get marine species included – for example basking sharks in the Wildlife and Countryside Act (Sarah Fowler), and fish in the CITES listings (Heather Koldewey, Sarah Fowler)

3. Implementation can also take time, especially if an incoming government does not like its predecessor's legislation. It took 22 years for the UK government to address the damaging effects of scallop dredging in its European marine sites, and then only after major legal challenges (Clark *et al.* 2017).

4. High-level principles provide an important framework. Sustainable development was at the core of the Marine and Coastal Access Act, for example, and is also key to the way marine spatial planning (MSP) and large multi-sectoral stakeholder groups work together in partnership. Principles like precaution have application to pollution control, fisheries and habitat protection. Packages of principles, like the ecosystem approach, integrated coastal zone management (ICZM) or MSP, have been key to a host of conservation actions.

5. Building trust and working relationships with people in other sectors is important, as outlined by Joan Edwards and Euan Dunn in the context of fisheries, and by Peter Barham in describing the way port developers worked with the conservation agencies and government to get deals that worked for all parts and secured significant mitigation of major developments.

Marine protected areas as a tool for furthering marine habitat and ecosystem conservation

Marine protected areas (MPAs) have a long tradition, and they are one of the tools most widely used by marine conservationists. Key points in their development are shown in Timeline 4 (page 83). Not surprisingly, many of the interviewees describe MPAs, and a number of points stand out:

1. There are key differences between marine and terrestrial conservation, as exemplified by the concepts of 'wildlife gardening' and 'restoration' – described by Callum Roberts and discussed in Chapter 2.

2. The UK lagged behind much of the world, and it is clear that taking the wrong steps at the start can distort the process (see Box 5, page 117). New Zealand had its Marine Reserves Act (1971) four years before the UK even had a government marine conservation official. The New Zealand reserves were not special places but set up as no-take zones (NTZs) to look at the science underpinning what happened when the influence of human activities was removed. The results confounded marine biologists, upsetting many of their preconceptions about what would happen: mobile species stayed put and grew and invertebrate communities changed (Shears & Babcock 2003). As a result of this wrong step, the UK still only has a network of partially protected sites which are increasingly expensive and cumbersome to manage and monitor.

3. There is a growing number of meta-studies on MPAs (Callum Roberts) revealing their performance in relation to a variety of factors, and in all sorts of settings from coral reefs to northern seas, which is providing a sound and reliable body of evidence upon which to base decisions. The evidence is also building on the positive role of stakeholder engagement in MPAs (Jones 2014).

4. The relationship between marine conservationists and fisheries scientists on the role of MPAs is becoming more positive and productive as the results of studies reveal benefits to fisheries. Even in the UK, the recovery of lobsters and scallops in NTZs is clear for all to see.

5. Capacity building: up until 2005 the international MPA interests met under the auspices of the World Parks Congress. Now they meet independently every two years, and the most recent meeting in Chile in 2017 had over 800 delegates.

6. But – and it is a big but – MPAs are not the only solution. Other approaches are necessary to address a host of other issues that arise. For example, MPAs are horribly exposed to climate change and pollution, as plastic pollution is currently demonstrating.

Marine conservation in the context of wider marine management

The interviewees support the idea of rational and sustainable management of the seas, management that provides the context for specific conservation measures. In the UK and other parts of Europe there has been considerable progress in recognising and understanding the work of the major industrial sea-use sectors and the need for effective marine conservation, while also accepting the need for sustainable development. There are clearly debates about the scale and impact of development, but there is now a very clear and internationally recognised framework for environmental management across the life cycle of projects (see Box 3, page 38). The marine conservation NGOs have led the way in helping to promote a rational framework for marine conservation to be set in the context of wider coastal and marine spatial management.

America led the way on ICZM, looking at the integration of land and sea interests on a wide range of issues. This was taken up in the UK and other parts of Europe in the late 1980s and early 1990s, with an initial emphasis on getting better protection for estuaries. Jon Day describes the current challenges of working with multiple sectors across the land–

sea boundary, not least in relation to issues such as agricultural pollution and land-use management. The ability of stakeholders from different sectors to meet and discuss a wide range of issues is now considered completely normal and is a product, in the UK at least, of the early moves on ICZM and the development of partnerships at local and national scales. This was the basis for the development of marine spatial planning, which embodies many of the same high-level principles as ICZM, and provides a framework for marine management including marine conservation and protection. Stakeholder engagement and sustainability are at the core of this idea, and Peter Barham discusses this in describing his work with professionals from all the sea-use sectors, including conservationists, and how they interact constructively. Both he and Chris Rose describe the upsurge in the work on corporate social responsibility (CSR) after the first Rio Earth Summit, providing practical ways for corporations to do more for the environment.

INNOVATION – NEW TOOLS AND OPPORTUNITIES

A major theme running through the development of marine conservation to date has been innovation. I sought the interviewees' views on this, and the strength of their responses to this question gave a very clear sense of how important they felt innovation to be. Whilst the challenges of protecting the marine environment remain broadly the same, the new tools and technologies – 'more tools in the tool box' – enable greater productivity and different ways of achieving goals. Three common themes, the developing technologies, from digital to DNA, social media, and the continued expansion of capacity building, emerged from the interviews and will play an increased role in the future; these are discussed below.

Developing technologies transform the way we do research and make the public aware of marine conservation

The development of technologies over the last fifty years has been astounding and has revolutionised many difficult and time-consuming tasks – and what is more important is that it looks set to continue to transform what can be achieved. Alan Knight highlighted the value of technology in triaging rescue options for stranded mammals by providing photographic records and accurate locations. Digital photography – in mobile phones, with drones and also in more specialised underwater settings – is enabling huge changes in marine science as well as enhancing our ability to communicate images to a much wide audience. We are just beginning to realise the possibilities of using autonomous underwater vehicles, and their use is clearly set to include a wide array of tasks including habitat mapping and monitoring on ever larger geographic scales. The applications of satellite tracking of marine species continue to develop and become more effective at enabling much clearer and often astounding insights into their spatial distributions (Euan Dunn). Bud Ehler and Dan Laffoley describe the new generation of small satellites, which will give whole-earth coverage and allow remote monitoring of everything from coastal developments to ship pollution and illegal high-seas fishing. Developments in DNA analysis are also enabling scientists to work out what organisms are feeding on, from their feathers or faeces. The environmental DNA sciences will no doubt be more informative as time goes on.

Social media and its impact on routine communication, public awareness, citizen science and campaigns

It is easy to forget that much of what is described in this book took place before email and the World Wide Web, and that what we now know as the social media only started to develop in the late 1990s. Making sense of social media can be confusing, but Chris Rose summarised it perfectly as follows: 'Social media is like old media, only organised differently, with the producers and consumers all mixing up different roles and functions in a new media ecosystem – but it is settling into patterns of use that people are coming to terms with.' For the conservation sector, whose funds were always limited, the increase in productivity that these computing technologies have allowed in reducing tedious administration has been transformative, not least in allowing audiences on previously unimagined scales to be reached. At the same time the power and limitations of social media (fake news) are increasingly being recognised.

There is no doubt that social media will have a growing part in many roles from public awareness to campaigning. Its potential is increasingly being realised, as shown by the power of short online videos by groups like Greenpeace or online petitions from organisations like Avaaz and 38 Degrees with campaigns that galvanise huge new audiences.

Capacity building

The simple recognition that you cannot do it all on your own, and that you need to engage more people from a wide range of disciplines as well as create new organisations, is a theme that has marked the development of marine conservation (see Chapter 3); but it was also a key response to the question on innovation. Whilst one might think that the number of marine conservation organisations has reached some upper limit, it seems likely that an ever-increasing stream of new organisations and networks will develop to meet the challenges of particular locations and issues. The new social media campaigning organisations seem set to expand, as does the development of collaborations and networks between organisations. So capacity building in its many guises, from simply getting enough people involved through to the creation of new organisations, will continue to be a major part of the way protecting the marine environment is achieved.

CONCLUSIONS

We have long-standing and detailed evidence on many of the *threats*, and we now appreciate the full scale of their impact. But there are many worrying aspects of what we currently face, not least of which is that many of the issues are not directly to do with the marine environment – issues like population growth or emissions leading to climate change, or growing land-based pollution reaching the sea. There is also clear evidence of denial, backtracking and disregard for the legal norms and environmental management by governments and corporations which might previously have kept in check the worst excesses of human activity. It points to the need for people concerned with the marine environment to make effective connections with the broader environmental movement and requires major collaborations at every geographic scale.

The *barriers*, like the threats, are not going away. Recognising these is one thing, but the question then arises of how they are factored into different ways of achieving change. Being more systematic about the ways conservation projects are timed and how their progress is assessed may well save considerable amounts of time and resources in the long run, avoiding the problems of insuperable barriers.

How people have taken *action* to change the status quo has been one of the major themes of the book, and many different approaches to effecting this change are evident in the interviews. Before attempting to synthesise these in this chapter, I sought advice – but I found that the answers you get depend on the person you ask and their speciality. There are different taxonomies of action. For example, Valerie Kapos *et al.* (2013) come at this from a predominantly UK terrestrial nature conservation perspective. Aric McBay and colleagues, authors of the *Deep Green Resistance: Strategy to Save the Planet* (2011), take a very much more societal view of the range of actions they describe. These classifications overlap but cover different approaches, and neither entirely fits the range of actions described by the interviewees – suggesting that there is scope for a merged marine protection and conservation version.

One of the difficulties in compiling a synthesis of what makes thing work in practice is that the 'weight' of the examples included in this chapter also varies considerably. There are many examples of the sheer weight of work – custom and practice – on marine protected areas, for example, but fewer examples in other areas such as the individual tactics and tips. The 'opt in device' used by Wildlife and Countryside Link (Box 2, page 32) is as simple as they get – and yet it has enabled almost forty years of collaborative working by very different organisations on the most political of issues. Factors such as 'trust' – which was often highlighted by the interviewees – are essential but in many ways intangible. There is a miscellany of valuable lessons, tips and tactics scattered throughout the narratives that have made a difference, and yet these are difficult to categorise.

I was surprised by the strength of views expressed by the interviewees on the need for innovation, and how deeply embedded it was in many people's thinking. The importance, not least for the NGOs, of having the freedom to try new things is critical. Some ideas will obviously fail, but others will take hold and prosper.

How *individuals* make things happen is a key theme of this book but it is not easy to describe. Chris Rose highlighted the importance of individuals as innovators, people like Jack O'Neill the inventor of the wetsuit and David McTaggart's business-based approach that he applied to Greenpeace because he wanted, above all else, to win. Achieving change is about their personalities as much as the methods they use. It is perhaps rather like cookery – reading the recipe and having the skill set isn't necessarily going to deliver the goods. Keith Hiscock's description of Bill Ballantine's personality encapsulates this rather well: 'Those of us who had the privilege to know him, will remember an individual passionate about conservation who did not suffer bureaucrats gladly, who "managed" opponents, who could be cantankerous and seemingly impossible, but who made things happen.'

BOX 7. RETIREMENT, ACHIEVEMENTS AND OPTIMISM

To conclude the interviews I asked some questions about retirement, achievements and optimism. My presumption about retirement was hopelessly wrong. A lot of the interviewees are in their late sixties and still going strong, with no sense of retirement in sight. Only one had actually retired. Earlier I reflected that conservation was probably best regarded as a vocation, and this was borne out by their responses. The interviewees continue to find their work interesting and rewarding, and so even if 'work' is not paid for by conventional employment within an organisation many are still engaged on projects that maintain their interest in the subject, which they do voluntarily or on a consultancy basis.

I was interested to explore how the interviewees regarded their achievements. There was the satisfaction of a job well done, or, as Roger Mitchell put it, 'a warm feeling about the projects I helped set up'. Sue Sayer spoke of getting a 'great buzz' when people refer to her work. One significant achievement mentioned by a number of interviewees was that many of the *ideas* which in their own time were new had been developed and embedded into routine thinking. There was also the recognition of having brought lots of people into working on marine conservation. The succession management issue of ensuring that there would be people skilled enough to carry on and enabling a great diversity of people to play their part was also recognised as a success in its own right (Sarah Fowler, Heather Koldewey). This sentiment also applied to the organisations that had been created and the desire to leave them with a sustainable model (Alan Knight).

However, there was also the recognition of simple goals unfulfilled, like highly protected marine sites around the UK (Sue Gubbay, Joan Edwards) or for species to be protected more effectively and their populations to recover. Views about achievements were also moderated by an understanding of the bigger challenges ahead – or, as Dan Laffoley put it, 'There is still everything – quite literally "our world" – to play for.'

At the end of each interview I asked whether the interviewee was optimistic or pessimistic about the future. The overwhelming response was positive even when it was qualified. From a personal perspective being positive and optimistic was seen as being essential to do the job – as Perter Barham said, 'I couldn't do the job if I wasn't optimistic.' Once again the qualifications revolved around the scale of the challenges ahead. What was also clear was that the sense of successes achieved feeds into an ongoing determination to continue to push for change and succeed. The idea of *ocean optimism*, described by Heather Koldewey, has been a highly successful reminder of the need to convey positive messages.

So, if you're involved in marine conservation, don't plan on retiring. You'll have a life's work covering many different projects – and despite the bad news you're still likely to be optimistic.

CHAPTER 25

Conclusions

THE BOOK AND THE PROCESS

At a personal level, preparing this book, and in particular conducting the interviews, has been both a great privilege and hugely enjoyable, and it has helped me put many things into a wider context. Importantly, it shows how action to change the status quo can take many forms, and there are many useful pointers to future action contained in the narratives. The initial idea for the book was based around a chapter on each interviewee together with a systematic 'read across', forming a synthesis of common themes from the responses. This is broadly how it has worked out, and the process of asking my questions and documenting the interviewees' answers has allowed me to explore particular questions from very different perspectives. There is much more that could be written about a whole range of issues like trust, strategy, diplomacy and communication that make conservation initiatives work. The format of the questions developed as the interviews progressed and as I learnt more about what worked and what didn't. I'd like to thank the interviewees for their patience and support in seeing this process through.

I did not know what this process would reveal, and the outcomes have been both interesting and in some cases unexpected. For example, I was surprised by the number of different elements that interviewees cited when asked about how they frame and explain their work in marine conservation. This extended way beyond formal definitions of marine conservation. Another was the candour of people's views on the work of the NGOs. For most of the sections the process of synthesising the answers to the questions was straightforward, but the exception to this was Chapter 24 describing the different approaches people had used. The structure and the categories that emerged, whilst they may not be perfect, at least start to reflect the way such an analysis could be undertaken. There is much more that could be written about subjects like trust, strategy, diplomacy and communication that make conservation projects work.

Putting this book together has given me lots of ideas, many shared here, and has made me re-evaluate what I do. One idea has already borne fruit, in that it has helped with the first steps in setting up a marine social sciences network in the UK. If I did the project again I'd do it more quickly and much more collaboratively from the outset. Refining the questions is key. As these pages demonstrate, there are lots more interesting people to listen to and learn from, whose stories would be revealing.

WHAT IS MARINE CONSERVATION? AN OVERVIEW

Most people active in marine conservation focus on particular topics in order to achieve change. I knew from the outset of the project that deliberately selecting interviewees with a wide range of viewpoints and disciplines on how we protect, manage and conserve the ocean would lead to a broad overview of the field. The interviewees' narratives are compelling, and I found their coherent views of their areas of experience fascinating and inspiring. They share common beliefs, and the aim of the book has been to explore these. Every interviewee added significantly to the chapters on common themes. I hope that readers will find the outcomes and insights gained from the overview of a wide diversity of interviewees helpful.

Throughout the book I have used the term *marine conservation* in a very broad way, often linking it to the words *protection* and *management*. Whether one calls the subject *biodiversity conservation* or *marine protection*, it is probably pointless trying to disentangle actions in terms of strict definitions or types of specialty. As the timelines and many narratives reveal, this blurring of the edges of wider environmentalism and biodiversity conservation has been a common feature of the development of marine conservation. An interesting twist to this was the number of interviewees who, despite their achievements in protecting the marine environment, don't think of themselves as 'marine conservationists'. This confirmed to me that we should be encouraging a very wide diversity of *committed* people from many disciplines to contribute to taking action to address the threats we face.

In 2014, I attended my first International Marine Conservation Congress (IMCC). The sheer diversity of the contributions prompted me to question what a structure of marine conservation might look like. I developed Figure 2.2 to express this diversity in a systematic way, addressing the topic of marine protection, management and conservation as a whole. Over time, structure diagrams such as this will also enable a more holistic way of collating and accessing information in this field.

When interviewees were asked about the differences between terrestrial and marine conservation, the strong support for common and shared beliefs emerged spontaneously. There is, however, an important caveat when applying terrestrial conservation thinking to the marine environment, since land-based ideas do not necessarily work at sea. This is an area that has been fraught with problems, which continue to this day – not least with protected areas, species conservation and the requirements for information (data deficiency). Given the rates of destruction of marine species and habitats, we simply cannot wait until issues of data deficiency have been resolved before acting.

I wasn't surprised by the diversity of different sets of ideas that we have drawn on to protect the marine environment (Chapter 4). The fact that there are at least six major sets to this should give people who wear a purely biodiversity conservation hat food for thought. The concept of systems thinking, stretching back at least a century, underpins many of the ideas we have today. This includes understanding the dynamics of ecology and many of the systems approaches to environmental management, conservation, integrated coastal zone management and marine spatial planning. The role of sustainability and the principles that emerge from it, especially the precautionary approach, and packages of principles like the ecosystem approach and corporate social responsibility, have also been hugely influential in providing the backdrop to change. The changes achieved by groups like Greenpeace, who take a different philosophical position in relation to action, with organisational and individual commitment to societal change going beyond the legal norms of the day, has also produced key outcomes.

The contradiction of holding apparently opposing views is interesting. Marine conservationists may be subjectively inspired and passionate, but then objective and scientific in pursuing solutions. A set of contradictory views and beliefs has been recognised by the International Whaling Commission and its members as a basis for its ongoing work, enabling it to cover a range of issues from resource use to ecotourism and welfare (Simon Brockington). Given the plight of many of our larger marine vertebrate groups such as sharks and rays, recognising this reality could be a pointer to how to develop more effective programmes in future. Might new international organisations that recognise these contradictions provide a better way of stopping the decline of these species?

Working with people is at the core of marine conservation, and the idea behind this book. The richness of the interviewees' contributions can be seen both in their individual chapters and when brought together in the framing chapters. When asked to explain their approach to marine conservation, many of the interviewees developed the dual idea of the importance of working *with* people and conservation *for* people. Protecting, managing and conserving the marine environment is achieved by a very wide range of disciplines, styles, approaches and personal skills. Understanding and *working with* people – not least with those who are *not* conservationists – is crucial in protecting the ocean. It was interesting how interviewees saw more collaboration as key to achieving greater change more quickly.

While the natural sciences have dominated marine conservation in its first fifty years, social sciences and their applied disciplines such as business management, communication, campaigning, behaviour change, involvement and participation are of growing importance. Quite independently, many of the interviewees mentioned key inputs from people using insights from the social sciences, and it seems clear that this will play an increasing role over the years to come.

Whilst the conservation of species, habitats and sites is one of the main objectives of marine conservation, this sits in the wider context of the management of the seas as a whole, and the need for effective mitigation of damaging human activities in every sector of operations. Recognising that every major use sector can contribute to the protection of the oceans, and indeed that those sectors often have the resources to make a much greater contribution once they apply themselves, is important.

COMING TO TERMS WITH THE CHALLENGES – SCALE, AMBITION, ACTION

What do we do about the scale of the challenges we face and the speed and scale of response required? How should we set out our actions in this context? A simple version of this dilemma is embodied in the expression 'think globally, act locally'. The reality in practice is much more complex and involves many elements, and this is reflected in the interviewees' narratives. This section describes the need to meet the global challenges we face and to encourage ambitious programmes of action, and discusses ways of visualising this.

The global scale of challenges to the marine environment

As I wrote the section on the development of marine conservation in Chapter 2, it became clear that marine conservation had tackled some huge challenges over the last fifty years (see Box 1, page 18), and that many of these remain. Some of the biggest achievements have been in combating the deep-rooted societal attitudes that allowed the worst forms of marine pollution and exploitation of marine natural resources. There have also been many successes in species and habitat conservation which are described by the interviewees. Whilst much of the effort to date has focused on the coastal and nearshore marine environment, it is clear that the future challenges are moving thinking on biodiversity and ecosystem protection out to the high seas and the deep oceans.

The challenges of marine conservation are never-ending. We now know the scale

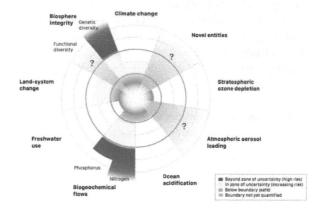

Figure 25.1 Planetary boundaries: Stockholm Resilience Centre diagram. Source: F. Pharand-Deschênes/Globaïa Stockholm Resilience Centre

of the task, and that the pressures of economic drivers are relentless, and we also know that hard-won gains can be wiped out quickly. As if we need to be reminded of this, the events in 2017, when much of this book was prepared, provided a backdrop highlighting the need for action. The vaquita, a small porpoise, was on the verge of extinction – fewer than thirty animals left – through entanglement in fishing nets. Our awareness of the scale of plastics pollution went exponential as more evidence from every part of the marine environment showed contamination and damage. At the same time Donald Trump and his administration dismembered many parts of environmental protection in the USA and signalled their withdrawal from the Paris climate change agreement, while the UK's decision to leave the European Union highlighted how environmental protection might be undermined. These events remind us to take nothing for granted.

Many of the interviewees literally have a world view. Their work over the years has taken them to many different locations around the world. They have seen how particular issues play out in many different settings, and this has led to their involvement in producing global reports, plans and international agreements. With a growing understanding of issues on a world scale comes an understanding of the true scale of the problems we face. This is a daunting reality. Many of the issues, such as economic development, climate change and plastics, are 'environmental' in the sense that they cover both land and sea. Others, such as the damaging effects of fishing or the growing issue of underwater noise, are more strictly marine.

How can we represent the scale of the risks to the global environment? The Stockholm Resilience Centre planetary boundaries diagram provides a valuable visual representation (Rockström *et al.* 2009, Figure 25.1). The issues are scaled in relation to our understanding of how human activities threaten the earth's ability to deal with the problem, and the colour coding is also helpful in prioritising these issues. This diagrammatic approach could also be used to demonstrate the scale of risk posed to a wide range of marine natural resources from various human threats. Kate Raworth in her book *Doughnut Economics* (2017) also draws on the inspiration of the planetary boundaries diagram in providing different economic models to tackle the scale of international problems. An interesting and important aspect of this is that she addresses the threats to the environment whilst invoking both economic and social arguments in a way that builds on the ideas of sustainability.

The need to encourage ambitious action

How do we organise to get the scale of change necessary? This question struck me at the Glasgow IMCC meeting in 2014, where many of the talks gave no sense of fitting into a context for action. This issue recurred in pulling the concluding chapters together and considering the bigger picture, especially when it came to the question of how actions could be organised and visualised. Figure 25.2 merges the natural resources and threats shown in the structural diagram (Figure 2.2) with the geographic scale elements of Amanda Vincent's onion model (Figure 13.2). It provides a way of visualising how we need to cover the range of topics, and the geographic scales that need to be included. The diagram recognises a number of elements:

- The sheer diversity of topics that people and organisations work on to protect the marine environment is huge.

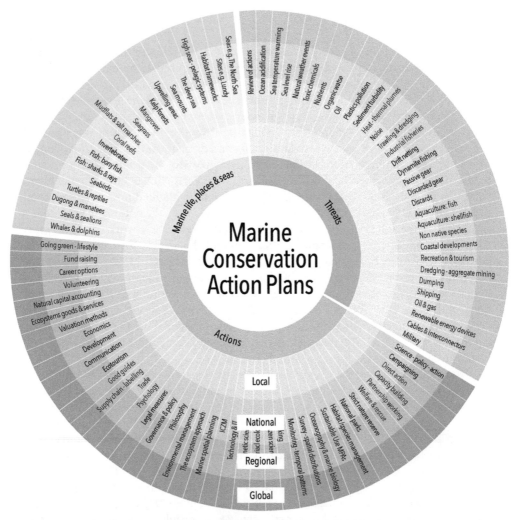

Figure 25.2 This diagram merges the wide range of topics covered by the marine conservation structure diagram (Figure 2.2) with the onion world model developed by Amanda Vincent (Figure 13.2). It highlights the need for action across marine natural resources, threats and approaches.

- For all of these topics we need ambitious plans that provide targets, direction and priorities.

- Geographic scale, from local to global, is important, not least because people and organisations working at local and national scales can draw inspiration from plans at higher scales.

- A key aim must be to encourage a greater range of people from a much wider range of disciplines to be actively involved and to collaborate in these plans and the programmes to deliver them.

The sheer diversity of topics

The range of work performed by people and organisations protecting, conserving and managing the marine environment is enormous. The IMCC meeting in Glasgow was revealing in many ways, because whilst roughly a third of the 500+ presentations were organised in themes, most of the rest were not arranged thematically. Sitting in the non-thematic sessions, the sheer diversity of topics – from fiordic habitat conservation to wardening the Galapagos Islands, from shark ecotourism to Indian dolphins – was stimulating but raised questions of focus, direction and priorities. Simply getting people to meet on ever larger scales is not the way forward. To achieve things, people tend to focus on specific topics. Figure 25.2 and filling in the details of the plans and priorities in all of the wide range of topics covered by marine conservation could provide a useful way forward for those topics that tend to get overlooked.

The need for ambitious plans

We understand the global threats, and it is simply not good enough to ignore the bigger picture. Marine conservation is about acting on information to change the status quo for the better. But how ambitious should we be? This also raises the question of whose ambition (or strategy) is being followed. You could apply this question to all the big issues facing the marine environment. It is important to think about what you want to achieve. For individuals starting new organisations aided by the digital media, this is a particularly important question. It is important that the NGOs have plans, policies and strategies that are more ambitious and demanding than those likely to be adopted by governments. International collaborations have a massive part to play in this, not least for NGOs with limited geographic reach. The critical thing is that these larger plans encourage collaborative actions at a range of scales right down to the local level. Resolving questions that arise over slower and incremental change versus much more ambitious step changes is difficult, but this really shouldn't impede the setting of ambitious goals.

Elliott Norse, in his chapter, poses questions about our level of ambition and whether we are winning in our efforts to protect the marine environment. These words – *winning* and *ambition* – are perhaps not raised often enough. Posing the questions of success and whether efforts are achieving their objectives could well be applied much more routinely to marine conservation issues. The concept of adaptive management – plan, act, review – provides us with the tools to do this routinely. Whilst governments often don't particularly relish clear objectives, because it means being held to account, there is little to stop the environmental NGOs posing these questions.

Geographic scale – local to global

Marine conservationists inevitably operate at a wide range of geographic scales, and this is to be encouraged. It is clear from numerous examples in the text that action is needed at every geographic scale from the global to the local. At a global scale the Paris Agreement on climate change (Paul Horsman), MPAs (Dan Laffoley, Elliot Norse), the Aichi targets (Jon Day, Bud Ehler) and the shark action plan (Sarah Fowler) are good examples of a truly 'world view'. Interviewees also describe their work at a regional or whole-sea scale – the Great Barrier Reef (Jon Day), the North Sea (Euan Dunn), the Southern Ocean (Euan Dunn) and marine spatial planning for ocean states. Throughout the narratives there are also many examples of work at the national and the local scale. 'Local' for most marine conservationists, however, is still huge in comparison with terrestrial conservation. For example, Sue Sayer's work on seals in Cornwall spans their ecology and distribution around England, France, Ireland and Wales and in the English Channel and Celtic Sea, on the basis of routine movements within a population. The key thing is to look for the links across the geographic scales that help frame the actions needed.

Encouraging a great range of people from different disciplines

The number of people involved in protecting the marine environment has increased enormously over the years, but just getting more people together, on however large a scale, isn't the answer. Focus is what most successful people bring to the resolution of issues, and action needs to be more focused on strategic targets. Given the range of threats, addressing the priorities and direction of travel to achieve significant change is really important. It would be interesting to review across the totality of topics who is doing what, both in terms of intergovernmental strategies and also, more importantly, what agendas the environmental NGOs are setting at larger geographic scales.

FINAL THOUGHTS

I recently came across a short biographical summary and tribute to Rachel Carson and her book *Silent Spring* (1962) by Linda Lear (1998). Rachel Carson's books in the 1950s described the wonder of the marine environment, but *Silent Spring* was her greatest work – and what she wrote then still has resonance today, as pointed out by Linda Lear:

> *Rachel Carson was a prophetic voice and her 'witness for nature' is even more relevant and needed if our planet is to survive into a twenty-second century. Silent Spring inspired the modern environmental movement, which began in earnest a decade later. It is recognised as the environmental text that 'changed the world'. She aimed at igniting a democratic activist movement that would not only question the direction of science and technology but would also demand answers and accountability, not least from government. Carson's passionate concern in Silent Spring is with the future of the planet and all life on earth. She calls for humans to act responsibly, carefully, and as stewards of the living earth. Carson identified human hubris and financial self-interest as the crux of the problem and asked if we could master ourselves and our appetites to live as though we humans are an equal part of the earth's systems and not the master of them.*

Fifty-six years after the publication of *Silent Spring*, Elliott Norse revisits some of the same themes, putting it like this:

Humankind's existential challenge is sustaining the earth's life-support capacity. Despite many heartening successes ... the preponderance of evidence shows that conservationists are not winning. Billions of human consumers armed with rapidly growing technologies have already severely reduced marine biodiversity. Much worse is yet to come unless the small but uniquely capable subset of humankind – the marine conservation movement – finds the way to make maintaining life on our planet a top global priority. I do not think it is too late, but the clock is ticking.

Rachel Carson's identification of 'human hubris and financial self-interest' as the crux of the problem is particularly apposite. It was interesting how this theme of economic development emerged almost insidiously from the interviewees' narratives.

With marine protection, conservation and management, as Winston Churchill might have said, we are perhaps at 'the end of the beginning'. We now realise the scale of the challenges we face. We clearly need to galvanise a much wider response from many people and from different disciplines to help resolve these challenges, and this will be no small task for those who follow.

The threats are real and can be depressing, but the antidote to this is of course the sea itself. Being by, on or under the sea provides exhilarating and amazing experiences. That is where the inspiration comes from, and why, for the interviewees, 'retirement' is proving to be an illusory concept. We need to be confident and to quote Dan Laffoley once again, 'individuals still count, and can make a difference even in this highly populated world.'

Figure 25.3 Part of a large group of fin whales at the surface off Greenland, 2017. Source: Alan Davis

The interviews in this book demonstrates the truth of that assertion. We need to make our own luck. The group of over one hundred fin whales shown in Figure 25.3 would simply not have existed in 2017 had it not been for the success of the Save the Whales campaigns of the 1970s and 1980s. Optimism is essential, and we need to build on and share our successes.

References

Advocates for Animals (2009) *A Seal's Fate: the Animal Welfare Implications of Shooting Seals in Scotland.* Edinburgh: Advocates for Animals. https://www.onekind.scot/wp-content/uploads/seals-fate.pdf (accessed 31 May 2018).

Agardy T., Claudet J. and Day J.C. (2016) 'Dangerous targets' revisited: old dangers in new contexts plague marine protected areas. *Aquatic Conservation: Marine and Freshwater Ecosystems* 26 (Suppl. 2): 7–23. http://onlinelibrary.wiley.com/doi/10.1002/aqc.2675/epdf (accessed 31 May 2018).

Allen, R., Jarvis, D., Sayer, S. and Mills, C. (2012) Entanglement of grey seals *Halichoerus grypus* at a haul out site in Cornwall, UK. *Marine Pollution Bulletin* 64: 2815–19.

ANZECC Task Force on Marine Protected Areas (1998) *Guidelines for Establishing the National Representative System of Marine Protected Areas.* Canberra: Environment Australia. http://nepc.gov.au/system/files/resources/378b7018-8f2a-8174-3928-2056b44bf9b0/files/anzecc-gl-guidelines-establishing-national-representative-system-marine-protected-areas-199812.pdf (accessed 31 May 2018).

Ashworth, J., Aish, A. and Stoker, B. (2010) *Marine Conservation Zone Project Ecological Network Guidance.* JNCC and Natural England document.

Australian Government (1975) Great Barrier Reef Marine Park Act No. 85 (1975) www.legislation.gov.au/Details/C2017C00279 (accessed 31 May 2018).

Australian Government (2017) Great Barrier Reef Marine Park Zoning 2003 (GBRMPA). https://data.gov.au/dataset/great-barrier-reef-marine-park-zoning-2003-gbrmpa (accessed 31 May 2018).

Ballantine, W. (2014) Fifty years on: lessons from marine reserves in New Zealand and principles for a worldwide network. *Biological Conservation* 176: 297–307.

Beaugrand, G., Reid, P.C., Ibanez, F., Lindley, J.A. and Edwards, M. (2002) Reorganization of North Atlantic marine copepod biodiversity and climate. *Science* 296: 1692–4.

Beaugrand, G. (2004) The North Sea regime shift: evidence, causes, mechanisms and consequences. *Progress in Oceanography* 60: 245–62.

Bijlsma, L., Ehler C.N. and Mimura N. (eds.) (1994) *Proceedings of the World Coast Conference: Preparing to Meet the Coastal Challenges of the 21st Century: Conference Report, World Coast Conference 1993, Noordwijk, The Netherlands.* The Hague: Rijkswaterstaat.

Bijlsma, L., Ehler C.N., Klein, R. *et al.* (1995) Coastal zones and small islands. In *Climate Change 1995: Impacts, Adaptations, and Mitigation of Climate Change: Scientific-Technical Analyses. Contribution of Working Group II of the Second Assessment Report of the Intergovernmental Panel on Climate Change.* Cambridge: Cambridge University Press: pp. 289–324.

Blue Marine Foundation (2018) *Best Practice in World Fisheries: Lessons for Brexit.* London: Blue Marine Foundation and the Fishmongers' Company. www.bluemarinefoundation.com/wp-content/uploads/2018/01/Best-Practice-in-World-Fisheries-Lessons-for-Brexit-FINAL.pdf (accessed 31 May 2018).

Bowen, J. and Bowen, M. (2002) *The Great Barrier Reef: History, Science, Heritage.* Cambridge, UK: Cambridge University Press

Bower, B.T., Ehler C.N. and Kneese, A.V. (1977) Incentives for managing the environment. *Environmental Science & Technology*, 11: 250–4.

Bower, B.T., Ehler, C.N. and Basta, D.J. (1982) *Coastal and Ocean Resource Management: a Framework for Analysis.* Solna, Sweden: International Federation of Institutes for Advanced Study (IFIAS). (Also published by Delft Hydraulics Laboratory, Delft, The Netherlands).

Boyes, J. and Elliott, M. (2016) Brexit: the marine governance horrendogram just got more horrendous! *Marine Pollution Bulletin* 111: 41–4.

Brand, S. (1968–1972) *The Whole Earth Catalog.*

Bräutigam, A., Callow, M., Campbell, I.R. *et al.* (2015). Global Priorities for Conserving Sharks and Rays: a 2015–2025 Strategy. Global Sharks and Rays Initiative. https://portals.iucn.org/library/sites/library/files/documents/2016-007.pdf (accessed 31 May 2018).

Carson, R. (1962) *Silent Spring.* Boston: Houghton Mifflin.

Ceballos, G., Ehrlich, P.R, Barnosky, A.D. *et al.* (2015) Accelerated modern human–induced species losses: Entering the sixth mass extinction. *Science Advances* 1(5): e1400253. DOI: 10.1126/sciadv.1400253.

Churchman, C.W. (1968) *The Systems Approach.* New York: Dell Publishing.

Clark, J.R., Salm, R.V. and Siirila, E. (1982) *Marine and Coastal Protected Areas: a Guide for Planners and Managers*. Gland: IUCN. Revised 2000.

Clark, R.B. (1973) *Marine Wildlife Conservation: Report of the NERC Working Party*. Natural Environment Research Council Publications Series 'B' No. 5.

Clark, R.C., Humphreys, J., Solandt, J.L. and Weller, C. (2017) Dialectics of nature: the emergency of policy on the management of commercial fisheries in English European Marine Sites. *Marine Policy* 78: 11–17.

Clover, C. (2004) *The End of the Line: How Overfishing Is Changing the World and What We Eat*. London: Ebury Press.

Collie, J.S., Adamowicz, W.L., Beck, M.W. et al. (2013) Marine spatial planning in practice. *Estuarine, Coastal and Shelf Science*. 117: 1–11.

Commonwealth of Australia (1998). *Australia's Oceans Policy. Volume 1: Caring, Understanding, Using Wisely*. Canberra: Commonwealth of Australia.

Cornwall Council (2017) *Cornwall's Environmental Growth Strategy: 2015 to 2065*. Truro: Cornwall Council. www.cornwall.gov.uk/environmentalgrowth (accessed 31 May 2018).

Crowder, L.B., Osherenko, G., Young, O.R. *et al*. (2006) Resolving mismatches in U.S. ocean governance. *Science* 313: 617–18.

Day, J.C. (2008) The need and practice of monitoring, evaluating and adapting marine planning and management: lessons from the Great Barrier Reef. *Marine Policy* 32: 823– 31.

Day, J.C. (2015) Marine spatial planning (MSP): one of the fundamental tools to help achieve effective marine conservation in the Great Barrier Reef. In Hassan, D., Kuokkanen, T. and Soininen, N. (eds.), *Marine Spatial Planning and International Law: a Transboundary Perspective*. Abingdon: Routledge, pp. 103–31.

Day, J.C. (2016) The Great Barrier Reef Marine Park: the grandfather of modern MPAs. In Fitzsimmons, J. and Wescott, G. (eds.), *Big, Bold and Blue: Lessons from Australia's Marine Protected Areas*. Clayton South: CSIRO Press, pp. 65–98.

Day, J.C. (2017) Effective public participation is fundamental for marine conservation: lessons from a large-scale MPA. *Coastal Management* 45: 470–86.

Day, J.C., Roff, J.C. and Laughren, J. (2000) *Planning for Representative Marine Protected Areas: a Framework for Canada's Oceans*. Toronto: WWF Canada. http://assets.wwf.ca/downloads/planning_for_representative_mpas.pdf (accessed 31 May 2018).

Douvere, F. and Ehler, C.N. (eds.) (2008) Special issue on ecosystem-based, marine spatial management. *Marine Policy* 32 (August).

Dudley, R. *et al*. (2017) Cornwall's bottlenose dolphins photo ID project. www.cornwallwildlifetrust.org.uk/sites/default/files/fact_sheet_bottlenose_dolphin_nov_22.pdf (accessed 31 May 2017).

Dunn, E. (1998a) The impact of fisheries on sea birds in the North Sea. In Symes, D. (ed.), *Northern Waters: Management Issues and Practice*. Oxford: Fishing News Books, pp. 208–15.

Dunn, E. (1998b) The Shetland sandeel fishery: the ecosystem approach in action. *El Anzuelo* (Institute for European Environmental Policy Newsletter on Fisheries and the Environment). Vol 1: 4–5.

Dunn, E.K. (2005) The role of NGOs in fisheries governance. In Gray, T.S. (ed.), *Participation in Fisheries Governance*. Dordrecht: Springer, pp. 209–18.

Dunn, E. (2011) Reducing seabird bycatch: from identifying problems to implementing policy. In Vidas, D. and Schei, P.J. (eds.), *The World Ocean in Globalisation: Climate Change, Sustainable Fisheries, Biodiversity, Shipping, Regional Issues*. Leiden: Martinus Nijhoff, pp. 247–61.

Dunn, E. and Steel, C. (2001) The impact of longline fishing on seabirds in the north-east Atlantic: recommendations for reducing mortality. Sandy: RSPB/NOF Report no 5-2001.

Earll, R.C (1992) Commonsense and the precautionary principle: an environmentalist's perspective. *Marine Pollution Bulletin* 24: 182–6.

Earll, R.C. (2016) *What is Marine Conservation? A Framework for Action*. Kempley: CMS.

Ehler, C.N. (2014) *A Guide to Evaluating Marine Spatial Plans*. IOC Manuals & Guides 70; ICAM Dossier 8. Paris: UNESCO.

Ehler, C.N. and Douvere, F. (2007) *Visions for a Sea Change. Technical Report of the International Workshop on Marine Spatial Planning, 8–10 November 2006*. IOC Manual & Guides 46; ICAM Dossier 3. Paris: UNESCO.

Ehler, C.N. and Douvere, F. (2009) *UNESCO Intergovernmental Oceanographic Commission and Man and the Biosphere Programme Marine Spatial Planning: a Step-by-Step Approach toward Ecosystem-Based Management*. IOC Manual and Guides No. 53, ICAM Dossier No. 6. Paris: UNESCO.

Ehler, C.N., Basta, D.J., LaPointe, T.F. and Warren, M.A. (1986) New oceanic and coastal atlases focus on potential EEZ conflicts. *Oceanus* 29 (3): 42–51.

Eno, C. (1991) *Marine Conservation Handbook*, 2nd edition. Peterborough: English Nature/Nature Conservancy Council.

European Environment Agency (2013) *Late Lessons from Early Warnings: Science, Precaution and Innovation*. EEA Report.

FAO (1999) *International Plan of Action for Reducing Incidental Catch of Seabirds in Longline Fisheries*. Rome: Food and Agriculture Organization of the United Nations.

Fernandes, L., Day, J., Kerrigan, B. *et al*. (2009) A process to design a network of marine no-take areas: lessons from the Great Barrier Reef. *Ocean and Coastal Management* 52: 439–47.

Frederiksen, M., Moe, B., Daunt, F. *et al.* (2012) Multicolony tracking reveals the winter distribution of a pelagic seabird on an ocean basin scale. *Diversity and Distributions* 18: 530–42.

Galil, B.S. and Goren, M. (2014) Metamorphoses: bioinvasions in the Mediterranean Sea. In Goffredo, S., Dubinsky,Z. (eds.), *The Mediterranean Sea*. Dordrecht: Springer.

Garcia, S.M., Rice, J. and Charles, A. (eds.) (2014) *Governance of Marine Fisheries and Biodiversity Conservation: Interaction and Co-Evolution*. Chichester: Wiley.

GBRMPA (2004). *Great Barrier Reef Marine Park Zoning Plan 2003*. Townsville: Great Barrier Reef Marine Park Authority. www.gbrmpa.gov.au/__data/assets/pdf_file/0015/3390/GBRMPA-zoning-plan-2003.pdf (accessed 31 May 2018).

GBRMPA (2014). *Great Barrier Reef Outlook Report 2014*. Townsville: Great Barrier Reef Marine Park Authority. http://elibrary.gbrmpa.gov.au/jspui/handle/11017/2855 (accessed 31 May 2018).

Gilding, P. (2011) *The Great Disruption: How the Climate Crisis Will Transform the Global Economy*. London: Bloomsbury.

Gislason, H. and Kirkegaard, E. (1997) The industrial fishery and the North Sea sandeel stock. Summary of a presentation at the Seminar on the precautionary approach to North Sea Fisheries Management, Oslo, 9–10 September 1996. *Fisken og Havet* Nr 1. Bergen: Institute of Marine Research / Norwegian Ministry of Fisheries.

Gislason, H. and Kirkegaard, E. (1998) Is the industrial fishery in the North Sea sustainable? In Symes, D. (ed.) *Northern Waters: Management Issues and Practice*. Oxford: Fishing News Books, pp. 195–207.

Gordon, D.P. (ed.) (2009–2012) *New Zealand Inventory of Biodiversity*, 3 vols. Christchurch: Canterbury University Press.

Gordon, D.P., Beaumont, J., MacDiarmid, A., Robertson, D.A. and Ahyong, S.T. (2010) Marine biodiversity of Aotearoa, New Zealand. *PLoS ONE* 5(8): e10905.

Gormley, A.M., Slooten, E., Dawson, S. et al. (2012). First evidence that marine protected areas can work for marine mammals. Journal of Applied Ecology 49: 474–80.

Gubbay, S. (1986a) *A Coastal Directory for Marine Nature Conservation*. Ross-on-Wye: Marine Conservation Society.

Gubbay, S. (1986b) *Conservation of Marine Sites: a Voluntary Approach*. Ross-on-Wye: Marine Conservation Society.

Gubbay, S. (1989) *Coastal and Sea Use Management: a Review of Approaches and Techniques*. Ross-on-Wye: Marine Conservation Society.

Gubbay, S. *et al.* (2017) *European Red List of Habitats. Part 1. Marine Habitats*. European Commission.

Helliwell, J., Layard, R. and Sachs, J. (2017). *World Happiness Report 2017*. New York: Sustainable Development Solutions Network.

Hiscock, K. (1973) *A Seminar on Marine Conservation*. University of North Wales, Department of Marine Biology, Marine Science Laboratories, Menai Bridge.

Hiscock, K. (ed.) (1996) *Marine Nature Conservation Review: Rational and Methods*. MNCR Series.

Hiscock, K. (2014) *Marine Biodiversity Conservation: a Practical Approach*. London: Routledge.

Hiscock, K. (2018) *Exploring Britain's Hidden World: A Natural History of Seabed Habitats*. Plymouth: Wild Nature Press.

Hiscock, K. and Irving, R. (2012) *Protecting Lundy's Marine Life: 40 Years of Science and Conservation*. Lundy Field Society.

House of Commons (1992) *Coastal Zone Protection and Planning*. House of Commons Environment Select Committee Report (HC 17). London: HMSO.

Horsman, P.V. (1982) The amount of garbage pollution from merchant ships. *Marine Pollution Bulletin* 13: 167–9.

Horsman, P.V. (1985) *The Seafarer's Guide to Marine Life*. London and Sydney: Croom Helm.

Hoyt, E. (2009). Whale watching. In Perrin, W.F., Würsig, B. and Thewissen, J.G.M. (eds.), *Encyclopaedia of Marine Mammals*, 2nd edition. London: Academic Press.

IMCRA Technical Group (1998) *Interim Marine and Coastal Regionalisation for Australia: an Ecosystem-Based Classification for Marine and Coastal Environments*. Version 3.3. Canberra: Environment Australia. www.environment.gov.au/resource/interim-marine-and-coastal-regionalisation-australia-version-33 (accessed 31 May 2018).

IUCN (1980) *World Conservation Strategy: Living Resource Conservation for Sustainable Development*. Gland: IUCN.

Jackson, J. (1997) Reefs since Columbus. *Coral Reefs* 16: S23–32.

Jackson, J.B.C., Kirby, M.X., Berger, W.H. *et al.* (2001) Historical overfishing and the recent collapse of coastal ecosystems. *Science* 293: 629–38.

Jackson, W. (1997) Designing projects and project evaluations using the logical framework approach. IUCN Monitoring and Evaluation Initiative. Gland: IUCN.

Jepson, P.D. Deaville, R., Barber, J.L. *et al.* (2016) PCB pollution continues to impact populations of orcas and other dolphins in European waters. *Scientific Reports* 6: 18573.

Jessopp, M.J., Cronin, M., Doyle, T.K. *et al.* (2013) Transatlantic migration by post-breeding puffins: a strategy to exploit a temporarily abundant food resource? *Marine Biology* 160: 2755–62.

Jones, P.J.S. (2014) *Governing Marine Protected Areas: Resilience through Diversity*. Abingdon: Routledge.

Kahane, A. (2008) *Solving Tough Problems: an Open Way of Talking, Listening, and Creating New Realities*. California: Berrett-Koehler.

Kapos, V., Balmford, A., Aveling, R. *et al.* (2008) Calibrating conservation: new tools for measuring success. *Conservation Letters* 1: 155–64.

Laffoley, D.d'A. and Baxter, J.M. (eds.) (2016). *Explaining Ocean Warming: Causes, Scale, Effects and Consequences*. Gland: IUCN. https://portals.iucn.org/library/sites/library/files/documents/2016-046_0.pdf (accessed 31 May 2018).

Laffoley, D.d'A. and Freestone, D. (2016) Part 9. World heritage in the high seas: an idea whose time has come. In Casier, R. and Douvere, F. (eds.), *The Future of the World Heritage Convention for Marine Conservation. Celebrating 10 years of the World Heritage Marine Programme*. UNESCO, Paris. *World Heritage Papers*, 45, pp. 123–35.

Laffoley, D.d'A. and Grimsditch, G. (eds.) (2009) *The Management of Natural Coastal Carbon Sinks*. Gland, Switzerland: IUCN.

Laffoley, D.d'A., Maltby, E., Vincent, M.A. *et al.* (2004) *The Ecosystem Approach. Coherent actions for marine and coastal environments. A Report to the UK Government*. Peterborough: English Nature.

Lakoff, G. (2004) *Don't Think of an Elephant!* Vermont: Chelsea Green Publishing Company. Revised 2014.

Larcombe, P. (2006) Continental shelves. In Perry, C. and Taylor, K. (eds.) *Environmental Sedimentology*. Oxford: Blackwell.

Larsson, N. (2015) How to write a logframe: a beginner's guide. *Guardian Online* 17 August 2015. https://www.theguardian.com/global-development-professionals-network/2015/aug/17/how-to-write-a-logframe-a-beginners-guide (accessed 31 May 2018).

Lear, L. (ed.) (1998) *Lost Woods: the Discovered Writing of Rachel Carson*. Boston, MA: Beacon Press.

Lee, A. and Ramster, J.W. (eds.) (1981) *Atlas of the Seas around the British Isles*. London: Ministry of Agriculture, Fisheries and Food.

Lester, S.E., Halpern, B.S., Grorud-Colvert, K. *et al.* (2009) Biological effects within no-take marine reserves: a global synthesis. *Marine Ecology Progress Series* 384, 33–46.

Lourie, S.A., Vincent, A.C.J., Hall, H.J. (1999) *Seahorses: An Identification Guide to the World's Species and Their Conservation*. London: Project Seahorse.

Louv, R. (2011) *The Nature Principle: Reconnecting with Life in a Virtual Age*. Chapel Hill, NC: Algonquin Books.

Lundy Management Forum (2017) *Lundy Marine Management Plan 2017*. Written by R. MacDonald, revised by R. Irving. Produced for Natural England by the Landmark Trust, Lundy Island.

Lynn, J. and Jay, A. (1986) *Yes Prime Minister*. London: Guild Publishing.

Macintosh, A., Bonyhady, T. and Wilkinson, D. (2010). Dealing with interests displaced by marine protected areas: a case study on the Great Barrier Reef Marine Park Structural Adjustment Package. *Ocean and Coastal Management* 53: 581–8.

Macleod, K., Fresne, S. du, Mackey, B., Boyd, I. and Mundie, D. (2010) *Approaches to Marine Mammal Monitoring at Marine Renewable Energy Developments. Final Report*. Sea Mammal Research Unit.

McBay, A., Lierre, K. and Jensen, D. (2011) *Deep Green Resistance: Strategy to Save the Planet*. New York: Seven Stories Press.

McCarthy, M. (2009) *Say Goodbye to the Cuckoo*. London: John Murray.

McKibben, W. (1989) *The End of Nature*. New York: Random House. www.billmckibben.com/end-of-nature.html.

Merrie, A. and Olssen, P. (2014) An innovation and agency perspective on the emergence and spread of marine spatial planning. *Marine Policy* 44: 366–74.

Michael, D.T. (1968) *The Unprepared Society: Planning for a Precarious Future*. New York: Basic Books.

Mitchell, R. and Pritchard, T. (1979) *Nature Conservation in the Marine Environment*. Nature Conservancy Council Report of the NCC/NERC joint Working Party on Marine Wildlife Conservation.

Mitchell, R. (1987) *Conservation of Marine Benthic Biocenoses in the North Sea and Baltic. A Framework for the Establishment of a European Network of Marine Protected Areas in the North Sea and Baltic*. Nature and Environment Series No. 37. Strasbourg: Council of Europe.

Monbiot, G. (2013) *Feral*. London: Penguin.

National Oceanic and Atmospheric Administration (1980) *Eastern United States Coastal and Ocean Zones Data Atlas*. Washington, DC: NOAA.

National Oceanic and Atmospheric Administration (1983) *Assessing the Social Costs of Oil Spills: the Amoco Cadiz Case Study*. Washington, DC: NOAA.

National Oceanic and Atmospheric Administration (1985) *The Gulf of Mexico Coastal and Ocean Zones Data Atlas*. Washington, DC: NOAA.

National Oceanic and Atmospheric Administration (1988) *Bering, Chukchi, and Beaufort Seas Coastal and Ocean Zones Data Atlas*. Washington, DC: NOAA.

National Oceanic and Atmospheric Administration. Office of Response and Restoration (n.d.) Lessons learned from the Exxon Valdez spill. https://response.restoration.noaa.gov/oil-and-chemical-spills/significant-incidents/exxon-valdez-oil-spill/lessons-learned-exxon-valdez.html (accessed 31 May 2018).

Nature Conservancy Council and JNCC (1987–1998) *The Marine Nature Conservation Review*. Peterborough: NCC/JNCC.

Nature Conservancy Council (1984) *Nature Conservation in Great Britain*. Peterborough: NCC.

Nicholson, M. (1967) *The System: the Misgovernment of Modern Britain*. London: Hodder & Stoughton.

Norse, E.A. (1990). *Ancient Forests of the Pacific Northwest*. Washington, DC: Island Press.

Norse, E.A. (ed.) (1993) *Global Marine Biological Diversity: a Strategy for Building Conservation into Decision Making*. Washington, DC: Island Press.

Norse, E.A. (1996). A river that flows to the sea: the marine biological diversity movement. *Oceanography* 9: 5–9.

Norse, E.A. and Crowder, L.B. (eds.) (2005). *Marine Conservation Biology: the Science of Maintaining the Sea's Biodiversity*. Washington, DC: Island Press.

Norse, E.A. and McManus, R.E. (1980). Ecology and living resources: biological diversity. In *Environmental Quality 1980: the Eleventh Annual Report of the Council on Environmental Quality*. Washington, DC: CEQ, pp. 31–80.

Norse, E.A., Rosenbaum, K.L., Wilcove, D.S. *et al.* (1986) Conserving Biological Diversity in Our National Forests. Washington, DC: Wilderness Society

Norse, E.A. (chief author) and 1,604 other scientists (1998). Troubled waters: a call for action. Released at the US Capitol, January 6, 1998. https://mcbi.marine-conservation.org/publications/pub_pdfs/Troubled-Waters.pdf (accessed 31 May 2018).

Norse, E.A. (chief author) and 1,451 other scientists (2005). Scientists' statement on protecting the world's deep-sea coral and sponge ecosystems. Released at the AAAS Annual Meeting, Seattle, February 15, 2005.

Norse, E.A., Brooke, S., Cheung, W.W.L. *et al.* (2012). Sustainability of deep-sea fisheries. *Marine Policy* 36: 307–20.

Northridge, S., Kingston, A. and Thomas, L. (2016) Annual Report on the Implementation of Council Regulations (EC) No 812/2004 during 2015.

Olson, R. (2016) *Missing the Slow Train: How Gradual Change Undermines Public Policy and Collective Action*. Wilson Centre, Forgotten Problems project. www.wilsoncenter.org/sites/default/files/slow_threats_report_0.pdf (accessed 31 May 2018).

Packard, V. (1957) *The Hidden Persuaders*. New York: Ig Publishing. Republished 2007.

Pauly, D. (1995) Anecdotes and the shifting baseline syndrome of fisheries. *Trends in Ecology and Evolution* 10 (10), 430.

Pauly, D., Christensen, V., Dalsgaard, J., Froese, R. and Torres, F. (1998) Fishing down marine food webs. *Science* 279: 860–3.

Pew Charitable Trusts (2015) The Virtual Watch Room: pioneering technology to monitor and protect marine reserves. www.pewtrusts.org/en/research-and-analysis/fact-sheets/2015/01/virtual-watch-room (accessed 31 May 2018).

PISCO (n.d.) *The Science of Marine Reserves* booklets. Partnership for Interdisciplinary Studies of Coastal Oceans. www.piscoweb.org/science-marine-reserves (accessed 31 May 2018)

Pomeroy, R.S., Parks, J.E. and Watson, L.M. (1994) *How Is Your MPA Doing?* Cambridge: IUCN.

Porritt, J. (2005) *Capitalism as if the World Matters*. London: Routledge/Earthscan.

Probert, P.K. (2017) *Marine Conservation*. Cambridge: Cambridge University Press.

Ramster, J.W. (1986) The management of an exclusive economic zone (EEZ): the role of the marine resource atlas. Exclusive economic zones: resources, opportunities, and the legal regime. London: Society for Underwater Technology.

Ratcliffe, D.A. (ed.) (1977) *A Nature Conservation Review: The Selection of Biological Sites of National Importance to Nature Conservation in Britain*. 2 vols. Cambridge: Cambridge University Press.

Raworth, K. (2017) *Doughnut Economics: Seven Ways to Think Like a 21st-Century Economist*. London: Random House.

Roberts, C. (2007) *The Unnatural History of the Sea*. London: Gaia.

Roberts, C.M., Andelman, S., Branch, G. *et al.* (2003a) Ecological criteria for evaluating candidate sites for marine reserves. *Ecological Applications* 13 (Supplement): S199–214.

Roberts, C.M., Branch, G., Bustamente, R. *et al.* (2003b) Application of ecological criteria in selecting marine reserves and developing reserve networks. *Ecological Applications* 13 (Supplement): S215–28.

Rockström, J., Steffen, W., Noone, K. *et al.* (2009) Planetary boundaries: exploring the safe operating space for humanity. *Ecology and Society* 14 (2), 32. www.ecologyandsociety.org/vol14/iss2/art32.

Rogers, A.D. (ed.) (2013) The global state of the ocean: interactions between stresses, impacts and some potential solutions. Synthesis papers from the International Programme on the State of the Ocean 2011 and 2012 workshops. *Marine Pollution Bulletin* 74: 491–552.

Rondinini, C. (2010) *Meeting the MPA Network Design Principles of Representation and Adequacy: Developing Species–Area Curves for Habitats*. JNCC Report No. 430. Peterborough: JNCC.

Rose, C. (1990) *The Dirty Man of Europe: Great British Pollution Scandal*. London: Simon & Schuster.

Rose, C. (1998) *The Turning of the Brent Spar*. London: Greenpeace.

Rose, C. (2010) *How to Win Campaigns: Communications for Change*. London: Earthscan

Rose, C. (2011) *What Makes People Tick*. Leicester: Troubador Publishing.

Rose, C. (2012) *How to Win Campaigns: 100 Steps to Success*. London: Earthscan.Rose, C. and Markham, A. (1985) An EEC Directive on wildlife conservation – letter. *ECOS* 6(4).

Rose, C., Dade, P. and Scott, P. (2008) *Qualitative and Quantitative Research into Public Engagement with the Undersea Landscape in England*. Natural England Research Report NERR019. Sheffield: Natural England.

RSPB (1990) *Turning the Tide: a Future for Estuaries*. Sandy: RSPB.

Ruff, L.E. (1970) The economic common sense of pollution. *The Public Interest* 19: 69–85.

Rutterford, L.A., Simpson, S.D., Jennings, S. *et al.* (2015) Future fish distributions constrained by depth in warming seas. *Nature Climate Change* 5: 569–73; doi:10.1038/nclimate2607.

Sayer, S. (2012) *Seal Secrets*. Cornwall: Alison Hodge Publishers.

Sayer, S. and Millward, S. (2016) Polzeath Seal Photo Identification Project and Marine Life Survey Harbour Porpoise Report. Unpublished.

Shears, N.T. and Babcock, R.C. (2003) Continuing trophic cascade effects after 25 years of no-take marine reserve protection. *Marine Ecology Progress Series* 246: 1–16.

Shiffman, D.S. and Hueter, R.E. (2017) A United States shark fin ban would undermine sustainable shark fisheries. *Marine Policy* 85, 138–40.

Smith, H.D. (2000) The industrialisation of the world ocean. *Ocean and Coastal Management* 43: 11–28.

Smith, H.D. and Vallega, A. (eds.) (1991) *The Development of Integrated Sea-Use Management*. London: Routledge.

Soulé, M.E. and Wilcox, B.A. (eds.) (1980). *Conservation Biology: an Evolutionary-Ecological Perspective*. Sunderland, MA: Sinauer Associates.

Speedie, C. (2017) *A Sea Monster's Tale: in Search of the Basking Shark*. Plymouth: Wild Nature Press.

Thompson, R.C., Swan, S.H., Moore, C.M. and vom Saal, F.S. (2009) Our plastic age. *Philosophical Transactions of the Royal Society B* 364, 1973–6.

Turner, E.S. (1953) *The Shocking History of Advertising*. New York: Dutton.

Vincent, A.C.J. (1996) The international trade in seahorses. Traffic International. www.projectseahorse.org/conservation-tools/2015/4/1/the-international-trade-in-seahorses (accessed 31 May 2018).

Vincent, A.C.J. (2008) Keynote: reconciling fisheries with conservation on coral reefs: the world as an onion. Reconciling fisheries with conservation. Fourth World Fisheries Congress. *American Fisheries Society Symposium* 49: 1435–67.

Waldock, R., Rees, H.L., Matthiessen, P. and Pendle, M.A. (1999) Surveys of the benthic infauna of the Crouch Estuary (UK) in relation to TBT contamination. *Journal of the Marine Biological Association of the United Kingdom* 79, 225–32.

Warren, L. and Gubbay, S. (1991) *Marine Protected Areas. A Discussion Document compiled for the Marine Protected Areas Working Group*. English Nature Research Report 688.

Watling, L. and Norse, E.A. (1998) Disturbance of the seabed by mobile fishing gear: a comparison with forest clearcutting. *Conservation Biology* 12: 1180–97.

Wells, S., Ray, G.C., Gjerde, K.M. *et al.* (2016) Building the future of MPAs – lessons from history. *Aquatic Conservation: Marine & Freshwater Ecosystems* 26 (Suppl. 2), 101–25.

Willoya, W. and Brown, V. (1962). *Warriors of the Rainbow: Strange and Prophetic Indian Dreams*. Healdsburg, CA: Naturegraph.

Wilson, E.O. (1988) *Biodiversity*. Washington DC: National Academy Press

Wootton, E.C., Woolmer A.P., Vogan C.L. *et al.* (2012) Increased disease calls for a cost-benefits review of marine reserves. *PLoS ONE* 7(12): e51615. doi:10.1371/journal.pone.0051615.

World Commission on Environment and Development (1987) *Our Common Future*. The Brundtland Report. Oxford: Oxford University Press.

World Wildlife Fund (1961) *Morges Manifesto*. Switzerland: WWF. Referenced at http://wwf.panda.org/who_we_are/history/sixties.

Index

CPSIA information can be obtained
at www.ICGtesting.com
Printed in the USA
BVHW09s1618220918
528133BV00001B/1/P